Biology of Archaea 1

SCIENCES

Biology, Field Director – Marie-Christine Maurel

Microbiology, Subject Head – Marie-Christine Maurel

Biology of Archaea 1

Discovery, Evolution and Diversity of Archaea

Coordinated by
Béatrice Clouet-d'Orval
Bruno Franzetti
Philippe Oger

iSTE

WILEY

First published 2025 in Great Britain and the United States by ISTE Ltd and John Wiley & Sons, Inc.

ISTE Ltd
27-37 St George's Road
London SW19 4EU
UK

www.iste.co.uk

John Wiley & Sons, Inc.
111 River Street
Hoboken, NJ 07030
USA

www.wiley.com

Library of Congress Control Number: 2024945709

British Library Cataloguing-in-Publication Data
A CIP record for this book is available from the British Library
ISBN 978-1-78945-168-9

ERC code:
LS8 Ecology, Evolution and Environmental Biology
LS8_10 Microbial ecology and evolution

Printed in Singapore
M WEP342733 040225

Contents

Preface

Béatrice CLOUET-D'ORVAL[1], Bruno FRANZETTI[2] and Philippe OGER[3]

[1] *MCD-CBI, CNRS, Université de Toulouse, France*

[2] *IBS, CNRS, Université Grenoble Alpes, France*

[3] *MAP, CNRS, INSA Lyon, Villeurbanne, France*

More than 40 years ago, *archaea* were proposed as a new domain of life alongside bacteria and eukaryotes. It is now accepted that these organisms constitute a distinct group, unique in many cellular and molecular aspects. The discovery of these universally important organisms represents one of the greatest advances in modern biology. Archaea have evolutionary links with early eukaryotic cells and are now used to elucidate fundamental biological questions. As champions of extremophilia, archaea have also shed light on the limits of life on Earth. Over the years, our understanding of archaeal microorganisms has evolved considerably. Far from being exotic forms of microbial life, archaea are found to be omnipresent in all terrestrial sites (> 20% of marine microbial biomass) and are notably present in the human microbiota. As new genomes are sequenced, it becomes evident that archaea represent a complex, extremely diverse world and also possess a largely unexplored virosphere. In soils and oceans, archaea play a key role in the planet's major geochemical cycles. In particular, methanogens, exclusively represented by archaea, are central to issues of climate change and energy challenges. To understand their impact on the environment and human health, it is imperative to decode their unique features at a molecular level. The molecular machinery deciphered in archaea will provide key paradigms for understanding fundamental biological processes conserved across life forms. Notably, the molecular bases of genetic information processing in archaea are often very similar to their counterparts in eukaryotes (translation, transcription, replication, recombination and DNA repair), even though

Biology of Archaea 1,
coordinated by Béatrice CLOUET-D'ORVAL, Bruno FRANZETTI and Philippe OGER.

archaea have a cellular structure similar to bacteria. Moreover, the regulatory systems for gene expression through post-transcriptional mechanisms remain poorly understood.

Our knowledge of archaea is rapidly evolving thanks to the advent of high-throughput sequencing of entire genomes. It is time to introduce a wide audience to the advances that illuminate our understanding of biology and reveal the originality of this domain of life.

The 14 chapters, divided into two volumes, of this book review the discovery and evolution of the domain of archaea and summarize our current knowledge of cellular and molecular processes to better integrate this domain into the broader understanding of life.

November 2024

1

The Discovery of Archaea

Patrick FORTERRE[1,2]
[1] *Institut Pasteur, Université Paris Cité, France*
[2] *Université Paris-Saclay, Gif-sur-Yvette, France*

1.1. Introduction

The discovery of archaea by Carl Woese and his collaborators was one of the greatest breakthroughs in 20th-century biology. This finding went largely unnoticed by many biologists, particularly those focused on eukaryotes, and remains unknown to nearly the entire general public. By unveiling a previously unknown segment of the living world, the discovery of archaea has, for the first time, allowed us to consider the history of life on Earth in its entirety. This discovery has spectacularly revived debates on the nature of the last common ancestor of current living beings and the origin of eukaryotes (our origin). Challenging the prokaryote–eukaryote dichotomy was one of Carl Woese's major endeavors. This struggle appears far from over, especially as the emphasis is once again on phenotypic similarities between archaea and bacteria at the genomic level. The scenario favored by most biologists continues to reflect the classical view that eukaryotes emerged from the association of different prokaryotes.

These reflections recently led me to write a review titled "Carl Woese: Still ahead of our time" on the 10th anniversary of his death (Forterre 2022a). In this context, it seems particularly important to revisit the history of the discovery of

For a color version of all figures in this chapter, see www.iste.co.uk/clouet/archaea1.zip.

Biology of Archaea 1,
coordinated by Béatrice CLOUET-D'ORVAL, Bruno FRANZETTI and Philippe OGER.

archaea to understand the intricacies of the major debates which continue to captivate 21st-century evolutionists.

1.2. Prokaryotic–eukaryotic dichotomy

In the late 1960s, molecular biologists imposed the division of the living world into two major groups: eukaryotes and prokaryotes (Sapp 2005). This classification was long considered a reflection of the history of life. The simpler prokaryotes were thought to have appeared first, followed much later by eukaryotes. The term "prokaryote" ("before the nucleus") reflects this idea. Quickly, the terms "prokaryotes" and "bacteria" became synonymous for most biologists. The term "prokaryote", which originally referred to a type of cellular structure (absence of a true nucleus), eventually took on an evolutionary connotation: all bacteria were believed to be close relatives, descending from the same ancestral "prokaryotic" bacterium. Prokaryotes were considered "primitive" compared to eukaryotes (Stanier and Van Niel 1962; Whittaker 1969). Most biologists accepted that eukaryotes descended from prokaryotes similar to current bacteria, possibly through a symbiotic merger of several different bacteria (Margulis 1975) (Figure 1.1(a)).

For many biologists, the molecular mechanisms of prokaryotes were assumed to be similar across the group, and studying one bacterium at the molecular level was thought sufficient to understand them all! Consequently, discoveries made on the model bacterium *Escherichia coli* were believed to be applicable to all bacteria. This view, still tacitly accepted by many biologists and taught in many universities, proved to be false. The work initiated in the late 1970s by the American molecular biologist Carl Woese revealed this (for reviews and books summarizing the history of this work, see Woese (1981), Forterre (2007, 2016, 2022b), Sapp and Fox (2013) and Quammen (2018)).

In the early 1970s, Carl Woese set himself the goal of constructing an evolutionary tree of life that would include microorganisms, bacteria and protists. Until then, microorganisms had never been integrated into a comprehensive universal phylogeny, as these were primarily based on the fossil record and the comparison of phenotypic characteristics. However, this approach was only feasible for "macro-organisms", mainly animals and plants. Consequently, the evolutionary models of microorganisms could only be based on theoretical considerations, primarily focused on the presumed evolution of metabolic types (from anaerobiosis to aerobiosis) and certain morphological characteristics considered more or less "primitive". For instance, mycoplasmas were sometimes regarded as the most primitive living beings because they lacked a cell wall (Figure 1.1(a)).

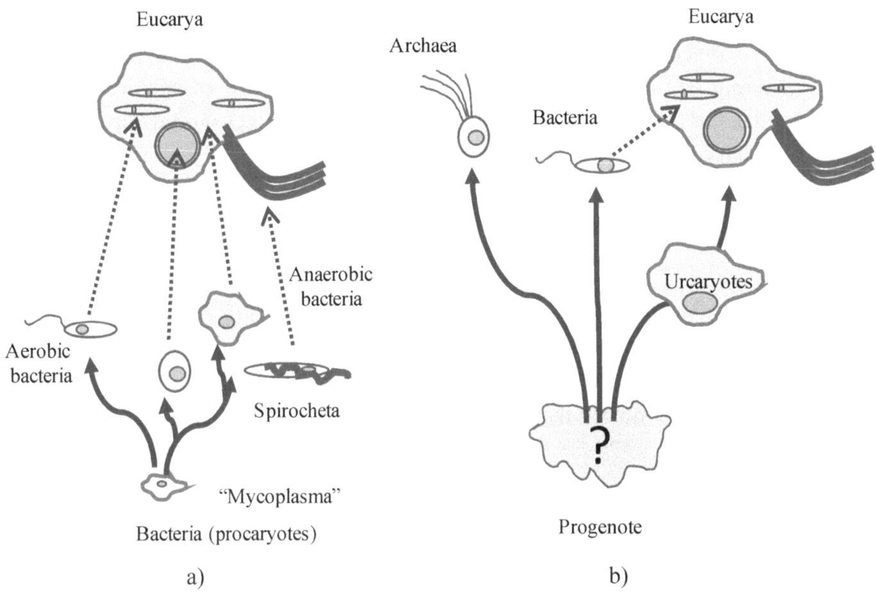

Figure 1.1. *Two visions of evolution in the late 1970s*

COMMENT ON FIGURE 1.1.– *(a) According to Lynn Margulis, eukaryotes result from the fusion of various types of bacteria, assimilated to prokaryotes (Margulis 1975). (b) According to Carl Woese, three kingdoms of prokaryotes emerge from an ancestor more "primitive" than current prokaryotes, the progenote. Eukaryotes originate from an extinct prokaryote lineage, the urcaryotes (Woese and Fox 1977).*

1.3. Two domains for prokaryotes: archaebacteria and eubacteria

1.3.1. *Ribosomal RNA as a molecular marker: a historical choice*

To establish a classification of microorganisms on an objective basis that reflects their actual, rather than presumed, kinship relationships, Carl Woese and his collaborators at the University of Urbana in Illinois began in the early 1970s to determine the kinship relationships between different bacteria by comparing the RNAs present in the small subunit of their ribosomes, specifically the 16S ribosomal RNA (16S rRNA) (Figure 1.2(a)). Woese and his team chose ribosomal RNAs because these informational macromolecules are present in all organisms (with the exception of viruses, which will be discussed in Chapter 2). Additionally, the various functions they perform have been conserved throughout evolution, suggesting a "neutral" mode of evolution unaffected by changes in function. In other words, these macromolecules would have consistently evolved at the same rate across different cell lineages. This concept allows for the use of a molecular clock, meaning that the differences observed at the sequence level would directly reflect the evolutionary distances between organisms.

Why 16S rRNA rather than 5S or 23S rRNA? When Carl Woese began his research program in the early 1970s, studying all three ribosomal RNA molecules simultaneously was not feasible. He chose 16S rRNA because it contains more sequence information than the smaller 5S rRNA, while being easier to analyze than the (larger) 23S rRNA. Since the development of PCR and massive genome sequencing, the advantage of 16S rRNA over 23S rRNA has disappeared. Today, it is possible to quickly obtain the sequences of both molecules. However, this was not the case in the early 1970s. At that time, DNA sequencing was not yet developed, and RNA sequencing was a very lengthy and complex technique. Carl Woese and his collaborators overcame this difficulty by using a technique developed by the English biochemist Frederick Sanger (Sanger et al. 1965). The idea was to determine the evolutionary distance between two organisms by comparing the catalogs of oligonucleotides derived from the digestion of their RNA by a specific ribonuclease, ribonuclease T1. The genetic fingerprint of an organism corresponded to all the oligonucleotides identified in it. These oligonucleotides, previously made radioactive, were separated from each other by chromatography according to their size and composition, and then visualized by autoradiography (Figure 1.3(a)). It was then necessary to compare the catalog of the studied species with those of all the other species already analyzed.

This procedure required extensive development work, which explains Carl Woese's choice of an RNA molecule of intermediate size, neither too large nor too small (Figure 1.3(c)). First, the organisms had to be grown in large quantities to isolate their 16S (or 18S in eukaryotes) rRNAs. These molecules were then digested with ribonuclease T1, which systematically cuts the RNA sequence to the right of a G. For example, the portion of an RNA molecule corresponding to the sequence AUUUCGUUGAAU would have yielded the three oligonucleotides: AUUUCG, UUG and AAU. The different oligonucleotides present in the mixture obtained (a few dozen) were then separated by two-dimensional chromatography on very large sheets of paper. The cells from which the 16S RNA molecules were derived had been cultured in a medium containing radioactive phosphorus (P-32). The radioactive oligonucleotides then appeared as black spots on the photographic films placed for some time above the sheet of paper (Figure 1.3(a)). The arrangement of these spots was characteristic of the species studied. The oligonucleotides corresponding to each spot could then be eluted and their sequence identified by a combination of chemical and enzymatic methods. Fortunately, this laborious last step quickly became unnecessary, as the positions of the spots were reproducible from one analysis to the next, allowing for the visual identification of the corresponding oligonucleotide.

Figure 1.2. *Carl Woese and the Triumph of Reductionism*

COMMENT ON FIGURE 1.2.– *(a) The secondary structure of 16S rRNA, highlighting variable regions useful as an evolutionary marker in blue. The inset depicts Carl Woese's finger pointing to a specific spot of the archaea, corresponding to an oligonucleotide containing a modified base. (b) Carl Woese receives the Crafoord Prize in 2003 from the hands of the King of Sweden, Gustav XVI. (c) Lecture by Carl Woese at the symposium organized for the 30th anniversary of the 1977 publication; the two-domain, prokaryotic/eukaryotic tree was replaced by the three-domain tree, with the archaea sister group of eukaryotes.*

Many small oligonucleotides (AG, CCG, UAAG, etc.) were found by chance in the catalog of all species and were therefore not useful for analysis. To mitigate their influence, Carl Woese chose to consider only oligonucleotides that were at least six letters long (AUUUCG in the previous example). Under these conditions, if the same oligonucleotide appeared in two different species, it was likely inherited from a common ancestor, rather than occurring by chance. Consequently, the closer two species were, the greater the number of large oligonucleotides they shared.

To quantify the similarity between two organisms A and B, Carl Woese and his colleagues devised a similarity coefficient, S_{AB}, calculated as the number of large oligonucleotides common to both species (N_{AB}) multiplied by two and divided by the total sum of the large oligonucleotides present in the catalog of each species, N_A and N_B, that is, $S_{AB} = 2N_{AB}/(N_A + N_B)$ (Figure 1.3(c)). When two species were identical, $N_A = N_B = N_{AB}$, S_{AB} equaled 1, while the coefficient approached 0 as the species grew more distant. Below 0.1, the value of S_{AB} was no longer significant, as it was statistically expected to always obtain, even by chance, a few common oligonucleotides of at least six nucleotides.

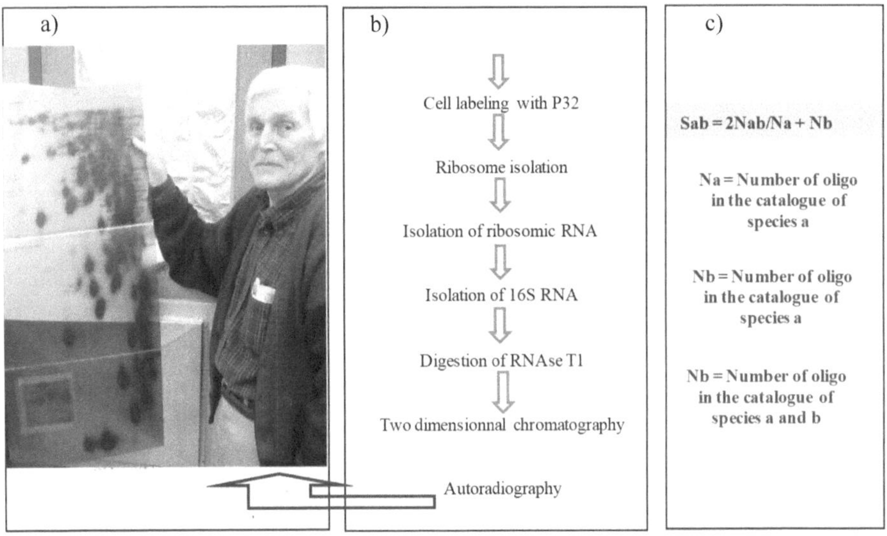

Figure 1.3. *Analysis of 16S rRNA by the Oligonucleotide Method. (a) Carl Woese showing the film of a two-dimensional chromatogram and diagram. The black spots correspond to the positions that the radioactive oligonucleotides occupied on the chromatography sheet. (b) Protocol of the experiment. (c) Method of calculating the SAB between two species, a and b*

Initially, by comparing the catalogs obtained with bacterial 16S rRNAs and those obtained with 18S rRNAs from three eukaryotes, this new approach allowed Carl Woese and his team to rediscover the separation between "prokaryotes" and eukaryotes. Pairs of bacteria or eukaryotes always had S_{AB} scores above 0.1 (e.g. 0.25 between *Escherichia coli*, a Gram-negative bacterium, and *Bacillus firmus*, a

Gram-positive bacterium, or 0.33 between humans and the yeast *Saccharomyces cerevisiae*). However, when comparing a bacterium and a eukaryote, S_{AB} scores were equal to or less than 0.1, which was below the threshold where the values obtained could be considered significant (Table 1.1).

The work of Carl Woese's team challenged several misconceptions. For instance, microbiologists had become accustomed to classifying bacteria according to Gram staining. This classification proved to be irrelevant from an evolutionary standpoint; indeed, the Gram-negative cyanobacterium *Aphanocaspa* was as distant from *Escherichia coli* as it was from the Gram-positive *Bacillus firmus*, with an S_{AB} of 0.26 in both cases. Gram-negative bacteria were even grouped with Gram-positive bacteria. This was the case with mycoplasmas, the smallest known bacteria at the time, characterized by the absence of a cell wall. Some considered mycoplasmas to be possible descendants of the first bacteria that appeared on the planet (Figure 1.1(a)). The analysis of their 16S rRNAs showed that mycoplasmas actually originated from Gram-positive bacteria by losing their cell wall (Woese et al. 1980). This evolution by simplification was not surprising, as mycoplasmas are intracellular parasites living inside eukaryotic cells.

Furthermore, using 16S rRNA, Carl Woese and his team were able to demonstrate that chloroplasts in green plants originated from ancient bacteria, as predicted by Lynn Margulis a few years earlier (Zablen et al. 1975). The comparison between the 16S rRNA of chloroplasts and that of cyanobacteria, their ancestors, yielded an S_{AB} value of 0.31, while the comparison between the 16S rRNA of chloroplasts and that of plants resulted in values below 0.1. The same result was independently obtained by Ford Doolittle's team in Canada (Bonen and Doolittle 1975), indicating that the approach initiated by Carl Woese had already gained widespread adoption.

1.3.2. *Atypical ribosomal RNA of methanogenic "bacteria"*

The big surprise came when Carl Woese became interested in methanogenic "bacteria", that is, methane producers. Methanogens are strict anaerobic organisms that were very difficult to culture. In 1976, for the first time, four strains of methanogens had just been successfully cultured in the laboratory of Ralph Wolf, who worked at the same institute as Carl Woese. When George Fox, a postdoctoral researcher in Carl Woese's team, analyzed the 16S rRNA of the first Gram-positive methanogenic "bacterium", *Methanobacterium thermoautotrophicum* ΔH, in May 1976, he noticed that the distribution of spots corresponding to radioactive oligonucleotides on the photographic film was very different from those of all the bacteria they had studied so far, including Gram-positive bacteria. Many spots typically observed in bacteria were absent, while new, previously unseen spots were present (inset in Figure 1.2(a)). For a time, George Fox even wondered if he had sequenced the correct rRNA or if he had properly separated the 16S and 23S rRNAs before digesting the 16S rRNA with ribonuclease T1. However, when he repeated the experiment on the other three species of methanogens (one of which was Gram-negative) during 1976, he obtained the same result. The spots absent in the first methanogen studied were also absent in the other three; the new spots that appeared were common to all four methanogens. Today, we know that some of the specific spots observed by George Fox corresponded to oligonucleotides with modified bases specific to archaea (inset in Figure 1.2(a)).

This initial visual impression revealing the peculiarities of methanogenic "bacteria" was confirmed by the calculation of S_{AB} values. In pairs formed by one of the four methanogenic "bacteria" and any other bacteria or eukaryotes, the S_{AB} values ranged from 0.05 to 0.13, which was not significant (Table 1.1). According to these analyses, methanogens, presumed to be bacteria, did not show a more detectable kinship with other bacteria than with eukaryotes. In contrast, the S_{AB} values for methanogen pairs were significant, ranging from 0.24 to 0.51 (Table 1.1). Therefore, the four methanogens formed a group of prokaryotes as evolutionarily distinct as bacteria, on the one hand, and eukaryotes, on the other hand.

1.3.3. *The concept of archaeobacteria*

To explain this unexpected observation, Carl Woese proposed abandoning the old binary division of the living world between eukaryotes and prokaryotes, replacing it with a partition into three major "primordial kingdoms" (*urkingdoms*): eukaryotes, classical bacteria and a third group corresponding to "methanogenic bacteria" (Woese and Fox 1977) (Figure 1.1(b)). A name had to be found for this third group. Carl Woese chose "archaeobacteria[1]" – literally, "ancient bacteria" – based on the anaerobic nature of methanogenic bacteria and their requirement for an atmosphere rich in carbon dioxide and hydrogen. It is known that the atmosphere of early Earth, about 4 billion years ago, did not contain oxygen. Most geologists believe that this primitive atmosphere was rich in carbon dioxide. It was therefore tempting to imagine that methanogens could have already populated such an environment. Additionally, the low S_{AB} value observed between certain pairs of methanogens (0.24) was of the same order as those observed between the most distantly related bacteria (Table 1.1). According to this result, methanogens had already diversified very early in the history of life on Earth, indicating that their appearance dates back to the very early stages of evolution. While proposing the

1 This name was suggested to Carl Woese and George Fox by their collaborators Linda Magrum and David Nanney (Woese and Fox 1977).

new name "archaeobacteria" for the primordial kingdom represented by methanogens, Carl Woese also suggested the name "eubacteria" for "classical" bacteria and "urcaryote" for the lineage of prokaryotes that led to eukaryotes with the emergence of the nucleus (Figure 1.1(b)). The discovery by Carl Woese and George Fox thus challenged the prokaryote–eukaryote dichotomy and proposed a new vision of the universal tree of life (Figure 1.2(c)).

1.3.4. *The grouping of halophilic and thermo-acidophilic "bacteria" with methanogens within archaeobacteria*

Were all archaeobacteria methanogens? This seemed unlikely if archaeobacteria truly represented a primordial kingdom on par with bacteria and eukaryotes. Carl Woese quickly devised a strategy to identify other types of archaeobacteria based on the available knowledge about methanogenic "bacteria". One notable characteristic of methanogens caught his attention: their natural resistance to most antibiotics known to inhibit bacterial growth. Carl Woese identified a small number of other "bacteria" in the literature, which were also known to be insensitive to most antibiotics and prioritized analyzing their 16S rRNAs. Initially, in 1978, Woese and his collaborators demonstrated that the 16S rRNA of an aerobic "bacterium" called *Halobacterium halobium* was very similar to that of methanogens (Magrum et al. 1978). This halophilic (salt-loving) microorganism had long been known for its ability to grow in environments with very high salt concentrations, up to 4 M NaCl, a value close to saturation (Figure 1.4(c)). To compensate for the osmotic pressure of such an environment, *H. halobium* accumulates equal concentrations of KCl in its cytoplasm. All its proteins are thus adapted to function in these extreme halophilic conditions, which had greatly intrigued biochemists (Lanyi 1974).

Furthermore, *H. halobium* could perform a unique type of non-chlorophyll photosynthesis based on a single protein called "bacteriorhodopsin". This protein formed a purple-colored membrane on the cell surface (see Figure 1.4(b)).

Other "halobacteria" without bacteriorhodopsin exhibited red pigmentation due to bacterioruberins, proteins that protect them from light exposure at the surface of salt lakes. At first glance, *H. halobium* did not appear to share any similarity with the four methanogenic archaeobacteria that led Carl Woese to propose the concept of archaeobacteria.

However, the analysis of their 16S rRNAs revealed that *H. halobium* was particularly close to one of these four methanogens, *Methanosarcina barkeri*, with an S_{AB} of 0.34. We now know that halophilic archaea have indeed evolved from a subgroup of methanogenic archaea, which includes *M. barkeri* (see Chapter 5).

In 1979, still searching the literature for microorganisms known to be naturally resistant to antibiotics, Carl Woese's team identified a third type of archaeobacteria: thermo-acidophiles. These were two atypical microorganisms discovered a few years earlier by the American microbiologist Thomas Brock, namely, *Thermoplasma acidophilum* and *Sulfolobus acidocaldarius*. *T. acidophilum* had been isolated from coal mine heaps where heat-generating combustion reactions occurred (Darland et al. 1970). It thrived in the laboratory in a very acidic medium (pH 1) and at 60°C. On the other hand, *S. acidocaldarius* was found in Yellowstone Park in the United States (Brock et al. 1972). It was slightly less acidophilic than *T. acidophilum* (with an optimal pH around 3), but it set records for thermophily at the time, with a maximum growth temperature of 85°C! *T. acidophilum* and *S. acidocaldarius* were aerobic, like *H. halobium*, and acidified their environment by producing sulfuric acid through the oxidation of elemental sulfur. The S_{AB} values between *T. acidophilum* and the four methanogens ranged from 0.19 to 0.23, and those between *S. acidocaldarius* and the four methanogens ranged from 0.13 to 0.19 (Table 1.1). These sometimes very low values, especially in the case of *S. acidocaldarius*, reinforced the idea that archaeobacteria were indeed very ancient organisms.

Compared species	Range of S_{AB} values
Eukaryotes among themselves (3)	0.29–0.36
Eukaryotes (3) and bacteria (6)	0.05–0.11
Bacteria among themselves (6)	0.20–0.34
Methanogens (4) and bacteria (6)	0.05–0.13
Methanogens (4) and eukaryotes (3)	0.06–0.11
Methanogens among themselves (4)	0.24–0.51
Methanogens (4) *and Halobacterium*	0.28–0.34
Halobacterium and *Thermoplasma*	0.23
Methanogens (4) and *Thermoplasma*	0.19–0.23
Methanogens (4) and *Sulfolobus*	0.13–0.19
Sulfolobus (S), *Thermoplasma* (T) and *Halobacterium* (H)	0.13–0.19
STH and eukaryotes (3)	0.05–0.13
Methanogens (4) or STH and eukaryotes (3)	0.05–0.13

Table 1.1. *Ranges of S_{AB} values obtained by comparing different species with each other. The numbers in brackets indicate the number of species tested for a given group. These values were taken from the review published by Woese in 1981*

The grouping of methanogens with *H. halobium*, *S. acidocaldarius*, and *T. acidophilum* in the new kingdom of archaeobacteria validated the prediction of Carl Woese and George Fox that the third group of living beings on Earth were also characterized by a great variety of organisms and lifestyles (for the first review summarizing all this work, see Fox et al. (1980)).

1.4. The living trilogy

1.4.1. *The concept of archaeobacteria confirmed by the uniqueness of their membrane phospholipids*

If archaeobacteria truly represented a primordial kingdom, equivalent to bacteria and eukaryotes, their biochemistry and/or molecular biology should exhibit unique characteristics distinguishing them from the other two "primordial kingdoms". This prediction was soon confirmed. In the 1960s, Canadian researchers discovered a new type of membrane phospholipid in *H. salinarium* (Kates et al. 1965). Unlike those in bacteria and eukaryotes, which are composed of a glycerol molecule linked to two long linear fatty acid chains by ester bonds, *H. salinarium* phospholipids consist of a glycerol molecule linked by ether bonds to two complex long fatty alcohol chains called isoprenols. These chains are characterized by the presence of a methyl group (CH_3) replacing a hydrogen atom every five carbon atoms in the chain (see Chapter 2). Additionally, and most surprisingly, the lipids of *H. salinarium* exhibited a reverse stereochemistry compared to those known in lipids in bacteria and eukaryotes[2].

How might these atypical lipids benefit halophilic microbes? Initially, we might think they provide an advantage to halobacteria for living in environments with very high salt concentrations. However, in 1972, American biochemist Thomas Langworthy discovered the same type of lipid in *T. acidophilum* and, two years later, in *S. acidocaldarius*. Consequently, it became evident that these lipids were characteristic of organisms being identified as "extremophiles". The concept of archaeobacteria proposed by Carl Woese suggested another hypothesis: could these distinctive lipids be a signature of archaeobacteria? If so, methanogens, though not extremophiles, should also possess this type of lipid. No one had previously considered studying the lipids of methanogens, assuming they were similar to those of classical bacteria. Alerted by Carl Woese to the significance of this question, Thomas Langworthy discovered these atypical lipids in the four methanogen strains studied at the time, even though three of them were neither halophilic nor thermophilic (Tornabene and Langworthy 1979). Thus, the lipids discovered by

2 The central carbon of glycerol, which forms the backbone of phospholipids, is indeed asymmetric.

Kates and his collaborators in the 1960s in *H. halobium* were indeed present in all archaeobacteria and could be considered a marker of this "primary kingdom".

1.4.2. *German biochemists: champions of the archaeobacteria concept*

It took some time before Carl Woese's discoveries were recognized and began to significantly influence the scientific community. A German microbiologist specializing in bacterial cell wall chemistry, Otto Kandler, played a major role in drawing the attention of some of his compatriots to the importance of archaea. In the mid-1970s, Otto Kandler made a surprising observation: unlike all the bacteria he had studied until then, the cell wall of methanogens did not contain peptidoglycan (Kandler and Hippe 1977). One of the four species of methanogens studied by Carl Woese, *Methanobacterium thermoautotrophicum*, did have a rigid wall composed of a peptidoglycan-type polymer, but it was chemically different from those previously studied in bacteria. When Otto Kandler, while visiting the United States, learned that Carl Woese had shown that methanogenic "bacteria" were not bacteria, he immediately recognized the significance of this discovery and became one of Carl Woese's first disciples. Back in Germany, Otto Kandler encouraged the community of microbiologists in his country to take an interest in archaeobacteria, which explains why, even today, German researchers are at the forefront of research on archaea.

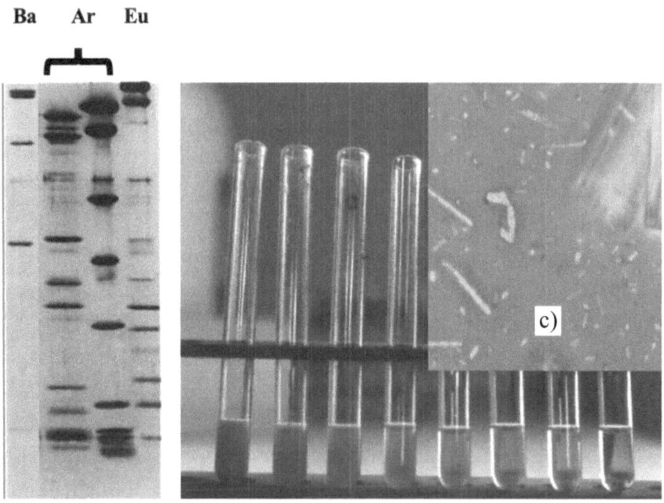

a) b)

Figure 1.4. *Revealing the evolutionary link between "archaeobacteria" and eukaryotes*

COMMENT ON FIGURE 1.4 – *(a) Electrophoresis gel (SDS-PAGE) of RNA polymerases from a bacterium (*Escherichia coli*),* two archaea (Ar) *Sulfolobus acidocaldarius and Thermoplasma acidophilum, and a eukaryote (Eu) Saccharomyces cerevisiae. Each band corresponds to a subunit (extracted from Figure 5 of Prangishvili et al. 1982). (b) Inhibition of Halobacterium halobium growth by increasing doses of aphidicolin (from left to right) (Forterre et al. 1984). The typical red color of these archaea disappears. (c) Pleomorphic cells of H. halobium (phase-contrast microscopy). Salt crystals can be seen in the upper left, indicating the high concentration of NaCl (4 M) in the culture medium.*

Otto Kandler succeeded in convincing Wolfram Zillig, a biochemist already internationally renowned for his work on RNA polymerases, to abandon bacteria and eukaryotes and devote himself exclusively to "archaeobacteria" (Albers et al. 2013). He also encouraged Karl Stetter, a young microbiologist who had completed his thesis under Kandler's supervision, to join Wolfram Zillig for his postdoctoral fellowship. In the late 1970s, these two researchers isolated and characterized the RNA polymerases of the two aerobic archaeobacteria available at the time, *H. halobium* and *S. acidocaldarius*. The results were spectacular; not only were these RNA polymerases very different from those of bacteria (e.g. they were resistant to rifampicin, an antibiotic used in the treatment of tuberculosis), but they were also much more complex (Zillig et al. 1979). In fact, the RNA polymerases of archaeobacteria resembled those of eukaryotes much more than those of bacteria. Bacterial RNA polymerases are formed by the association of four subunits, whereas those of archaea and eukaryotes are composed of about a dozen subunits (Figure 1.4(a)). Until then, it was thought that our RNA polymerases were more complex than those of bacteria because we (eukaryotes) were considered "higher" organisms compared to prokaryotes. However, here were archaea – prokaryotes – with RNA polymerases as complex as ours. This discovery heralded a paradigm shift.

1.4.3. *The evolutionary link between archaeobacteria and eukaryotes and the introduction of the term archaea*

Not only did the RNA polymerases of eukaryotes and archaeobacteria resemble each other in the number of subunits, but more importantly, they were evolutionarily closer. Indeed, by collaborating with André Sentenac's laboratory in Saclay near Paris, Wolfram Zillig demonstrated in 1983 that antibodies directed against the RNA polymerase of the yeast *S. cerevisiae* recognized the RNA polymerases of archaea but not those of bacteria (Huet et al. 1983). This was the first of a series of discoveries highlighting a close evolutionary relationship between the molecular mechanisms of archaea and those of eukaryotes. It was quickly realized that while archaeobacteria were resistant to most antibiotics effective against bacteria, they were sensitive to certain inhibitors previously considered specific to eukaryotes, such as diphtheria toxin, which inhibits translation, aphidicolin, an inhibitor of DNA polymerases (Figure 1.4(b)) and antitumor drugs acting on DNA topoisomerases (Kessel and Klink 1980; Forterre et al. 1984; Sioud et al. 1987).

The evolutionary proximity between "archaeobacteria" and eukaryotes led Carl Woese to propose in 1990 to replace the term "archaeobacteria" with "archaea" (see Chapter 2)[3]. With this name change, Carl Woese hoped that "archaeobacteria" would no longer be considered by most biologists as merely old bacteria or somewhat bizarre bacteria (extremophiles), but rather as representatives of a distinct "domain", with the term domain replacing that of "primordial kingdom".

By the late 1990s, the evolutionary link between archaea and eukaryotes was confirmed through the sequencing of the first archaeal genomes, specifically those of *Methanocaldococcus jannashii* and *Archaeoglobus fulgidus*, both isolated in the 1980s by Karl Stetter's team (see section 1.5.1). It became clear that for all major "informational" mechanisms (translation, transcription and DNA replication), archaea were much closer to eukaryotes than to bacteria (Forterre 1997; Woese and Olsen 1997). This link, which will be illustrated by numerous examples throughout this book, remains at the center of a scientific controversy regarding the topology of the universal tree of life: are archaea our sisters or our ancestors (see Chapter 3)?

This evolutionary connection between archaea and eukaryotes led to unexpected discoveries in laboratories studying archaea. For instance, the discovery and characterization of type II DNA topoisomerase in archaea led to the identification of the protein SPO11, which triggers genetic recombination during meiosis in eukaryotes by cutting parental chromosomes (Bergerat et al. 1997). Additionally, it led to the discovery of a topoisomerase responsible for the size of plant cells, and thereby the size of the plants themselves, by enabling the formation of polyploid giant cells (Sugimoto-Shirasu et al. 2002).

1.5. Archaea and high-temperature living

1.5.1. *The discovery of anaerobic hyperthermophilic archaea*

In the 1980s, while continuing their work on RNA polymerases and transcription in archaea, Wolfram Zillig and Karl Stetter embarked on a search for new families

3 In the remainder of this chapter and in the other two chapters authored by me, I will henceforth capitalize the names of the three domains.

of archaea, focusing on thermophiles. The first thermophilic archaea identified, *Thermoplasma* and *Sulfolobus*, were aerobic. Wolfram Zillig and Karl Stetter decided to search for thermophilic archaea that were anaerobic. Their reasoning was straightforward: if archaea were truly ancient, thermophilic archaea must have existed early on primitive Earth, which they envisioned as hot and anaerobic, since the atmosphere of primitive Earth lacked oxygen. They hoped that the descendants of these hypothetical archaea still inhabited the hot springs scattered across our planet. Wolfram Zillig and Karl Stetter chose to start their search in the numerous hot springs of Iceland (Figures 1.5(a) and (b)). From their first expedition in 1981, cultures conducted in the absence of oxygen allowed them to isolate four new anaerobic thermophilic microbes, which, unlike the previously known thermophilic archaea, grew at neutral pH (for reviews, see Stetter (2006, 2013) and Forterre (2007, 2016)). 16S rRNA analysis and RNA polymerase characterization of these anaerobic, neutrophilic thermophiles confirmed that they were indeed archaea.

The first isolated archaeon was a new methanogen, *Methanothermus fervidus*, which had an optimum growth temperature of 83°C and could grow up to 97°C, close to the boiling point of water (hence its Latin name, *fervidus*, meaning boiling) (Figure 1.5(c)). Next, they isolated two new archaea that were not methanogenic but were anaerobic. The first, rod-shaped and capable of forming branched structures, was named *Thermoproteus* in reference to the Greek god Proteus, who could take multiple forms (Figure 1.5(c)). Its optimum growth temperature was 88°C, higher than that of *Methanothermus*, and it could also grow up to 97°C. The second archaeon had spherical cells that could have initially been mistaken for *Sulfolobus*. However, instead of converting sulfur into sulfuric acid like *Sulfolobus*, it transformed it into hydrogen sulfide (H_2S). This archaeon was named *Desulfurococcus*, meaning "the coccus that reduces sulfur".

Finally, they isolated anaerobic thermophilic archaea that did not resemble either spheres or rods but had rather long and very thin filaments, which they named *Thermophilum*.

In the 1990s, due to the optimal growth temperatures of over 80°C exhibited by these new organisms, Karl Stetter proposed the cell and named them "hyperthermophile". Analysis of their 16S rRNAs and RNA polymerases revealed that *Thermoproteus*, *Desulfurococcus* and *Thermophilum* were phylogenetically closer to *Sulfolobus* than to other archaea known at the time. This marked the first indication of a fundamental division within archaea, with one lineage comprising *Sulfolobus*, *Thermoproteus* and *Desulfurococcus*, and another grouping *Thermoplasma* with methanogenic and halophilic archaea (Prangishvili et al. 1982).

One year later, during an expedition to the volcanic island of Vulcano in Italy, Karl Stetter and Wolfram Zillig discovered anaerobic hyperthermophilic archaea that were evolutionarily closer to certain methanogens (*Methanococcales* and *Methanobacteriales*) than to *Sulfolobus* and other anaerobic thermophiles (Zilig et al. 1983). These organisms included *Thermococcus celer* (literally, the swiftly moving thermal shell, from *celer*, meaning "speed" in Greek) and *Pyrococcus furiosus* (the furious burning shell). These two closely related organisms were grouped within the order Thermococcales. Thermococcales are currently associated with methanogenic archaea, *Thermoplasma*, and halophilic archaea within the phylum *Euryarchaeota* (from the Greek *Euryos*, meaning diversity), while *Sulfolobus* and its related hyperthermophiles are grouped within the phylum *Crenarchaeota* (from the Greek *Crenos*, meaning source/origin) (Woese et al. 1990) (see Chapter 2).

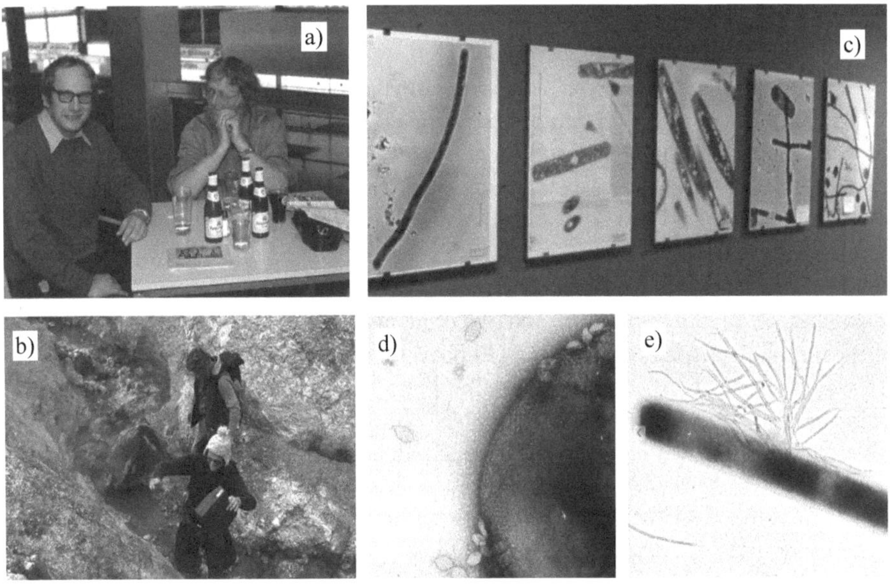

Figure 1.5. *The discovery of hyperthermophilic archaea and their viruses*

COMMENT ON FIGURE 1.5.– *(a) Karl Stetter (left) and Wolfram Zillig (right) at Reykjavik airport during their first expedition to Iceland in 1981. (b) The same individuals sampling a hot spring during their Icelandic expedition. (c) Photos displayed in the hallway of Wolfram Zillig's laboratory at the Max Planck Institute of Martinsried depicting hyperthermophilic archaea discovered in Iceland in 1981, with two photos of Methanothermus fervidus (left) and Thermoproteus* tenax *(right). (d) Sulfolobus shibatae cell surrounded by particles of the SSV1 virus.*

(e) Thermoproteus tenax cell surrounded by particles of the lipothrixvirus TTV1 (transmission electron microscopy).

From 1982 until the early 2000s, Karl Stetter, who became a professor at the University of Regensburg, continued his field research independently, exploring various volcanic areas around the globe with his students and postdoctoral researchers, and describing an extraordinary variety of hyperthermophilic archaea. Most of these new archaea belonged to the phylum *Crenarchaeota* (*Staphylothermus*, *Pyrobaculum*, *Pyrodictium*, *Pyrolobus* and *Ignicoccus*), but there were also some *Euryarchaeota* (*Archaeoglobus and Methanopyrus*). All these archaea were anaerobic, with a few exceptions, such as *Pyrobaculum aerophylum*, which was microaerobic. Karl Stetter and his team even discovered two hyperthermophilic bacteria, *Thermotoga maritima* and *Aquifex pyrophylus*, capable of dividing between 80 and 90°C (Stetter 2006, 2013).

One of the last fascinating discoveries of Karl Stetter's team was *Nanoarchaeum equitans*, a miniature archaeon living in symbiosis on the surface of a *Crenarchaeota* named *Ignicoccus hospitalis* (Huber et al. 2002). To Karl Stetter, *Nanoarchaeum equitans* seemed to be "riding" its "hospitable" host. This discovery foreshadowed the revelation of a whole group of miniature and parasitic archaea, the DPANN (see Chapter 2), which would be uncovered through metagenomic approaches 10 years later.

In the late 1990s, Karl Stetter was joined by other teams in the hunt for hyperthermophilic archaea. A Japanese team succeeded in isolating a strictly aerobic hyperthermophilic archaeon, *Aeropyrum pernix* (Sako et al. 1996). In France, Daniel Prieur's team in Brest specialized in isolating hyperthermophilic archaea from the walls of hydrothermal vents. These vents, or chimneys, appear black or white depending on the minerals in the plume and are located at faults formed by plate tectonics on the ocean floor. Sample collection was facilitated by the submarine Nautile owned by Ifremer (French Research Institute for Exploitation of the Sea). In 1993, Gael Erauso, a student of Daniel Prieur, isolated *Pyrococcus abyssi* from a sample taken from a smoker in the Pacific Ocean north of the Fiji Islands (Erauso et al. 1992). *P. abyssi* would later become a preferred model for the French community studying archaea (Forterre 2022a).

The discovery of hyperthermophilic archaea and bacteria has spurred biochemists to investigate the traits that enable these organisms to thrive at temperatures lethal to other life forms. This research led to the discovery of reverse gyrase in *Sulfolobus* in 1984, a type 1 DNA topoisomerase capable of introducing positive supercoils into DNA in vitro (Kikuchi and Asai 1984; Forterre et al. 1985), and to the identification of positively supercoiled DNA in a virus of this archaeon, the SSV1 virus (see Figure 1.5) (Nadal et al. 1986). Reverse gyrase has been found in all hyperthermophiles without exception, whether archaea or bacteria. Its exact role remains unclear, as some hyperthermophiles also possess gyrase, which negatively supercoils their DNA (Guipaud et al. 1997). Nevertheless, reverse gyrase appears to play a crucial role at high temperatures, as it is the only protein specific to hyperthermophiles (Forterre 2002; Catchpole and Forterre 2019). The discovery of archaea has particularly benefited research on DNA topoisomerases, especially in France, with the identification of new families and activities of these enzymes which manipulate DNA topology. Archaea have indeed become a veritable goldmine for researchers in this field (Garnier et al. 2021; Forterre 2022b).

1.5.2. *The race for the thermophilia record*

All the hyperthermophilic archaea previously mentioned had maximum growth temperatures between 80 and 100°C, except for *Pyrodictium* and *Pyrolobus*, which broke all records by growing at temperatures above the boiling point of water. How was this possible? In 1982, Karl Stetter isolated from underwater hot springs off the island of Vulcano, a Crenarchaeon that would hold the temperature record for nearly 10 years, *Pyrodictium occultum*, meaning "burning fingers" (Figure 1.6). The optimal growth temperature of this archaeon, whose cells resembled a hand with fingers, was 105°C, and it continued to divide at 110°C under pressure (Stetter 1982). It originated from hot springs located 10 m deep, where the water remains liquid above 100°C.

In 1989, Karl Stetter's team discovered a hyperthermophilic methanogenic archaeon (a Euryarchaeon) capable of dividing up to 110°C, which he named *Methanopyrus kandleri* in honor of his former thesis supervisor, Otto Kandler (Huber et al. 1989). In 1997, the same team isolated a Crenarchaeon from a sample of underwater smokers off the coast of Iceland, whose optimal growth temperature was 106°C and which continued to divide at 113°C, *Pyrolobus fumarii*, meaning "the burning lobe of smokers" (Blöchl et al. 1997). According to Karl Stetter, this archaeon still holds the temperature record.

Since then, however, two hyperthermophilic archaea with even higher maximum growth temperatures have been described. The first, in 2003, by an American laboratory, could divide up to 121°C (Kashefi and Lovley 2003). The second, described in 2008 by a Japanese team, was a new strain of *M. kandleri* that could divide up to 122°C, but only under pressure (Takai et al. 2008). However, no other laboratory has reproduced these results, which remain controversial.

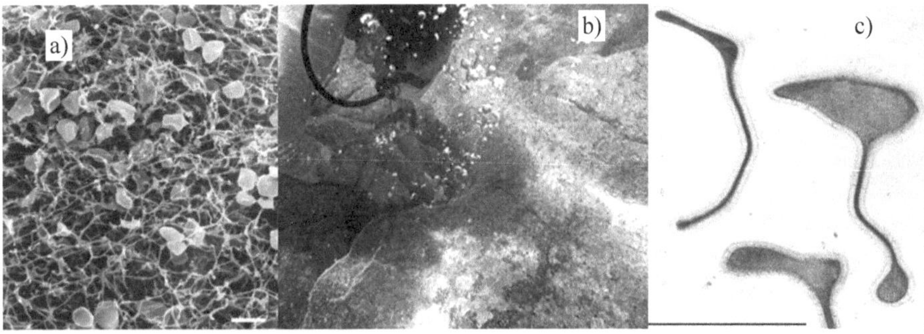

Figure 1.6. *Discovery of the first archaea with an optimal growth temperature above 100°C, Pyrodictium abyssi*

COMMENT ON FIGURE 1.6.– *Pyrodictium abyssi cells observed by scanning (a) and transmission (c) electron microscopy. (b) Karl Stetter sampling underwater hot springs off Vulcano beach in Italy. It was from these samples that he isolated this archaeon in 1982.*

It is true that some proteins are still active after being exposed for several tens of minutes at 120°C or even 140°C, but this is not the case for biological membranes. Vesicles formed with lipids of hyperthermophilic archaea quickly lose their impermeability to ions if the temperature exceeds 105°C, which corresponds to the optimal growth temperatures of *P. occultum* and *P. fumarii*, but also of the strain "121" and *M. kandleri* (Van de Vossenberg et al. 1995). It therefore seems difficult to imagine organisms capable of growing at temperatures well above 110°C. The maximum temperature at which an organism can divide on our planet, therefore, remains, in my opinion, an open question that has yet to be rigorously resolved.

The pursuit of the most thermophilic organism has sometimes led to errors. For instance, in 1982, the journal *Nature* published an article by two American researchers who claimed the discovery of bacteria capable of dividing at 250°C under 350 atmospheres in black underwater smokers (Baross and Deming 1983). This publication caused a media frenzy worldwide. However, it was what we would call today "fake news". Another American researcher, Jonathan Trent, identified all the flaws in this publication, but his rebuttal letter, also published in *Nature*, received much less attention (Trent et al. 1984). This episode, however, highlighted the issue of the stability of biological molecules at high temperatures. Specifically, the American biochemist Robert White published an article in 1984, again in *Nature*, which ended the controversy. Under conditions supposedly favorable to super-hyperthermophiles in black smokers – 300 atmospheres and 250°C – the typical molecules of living organisms rapidly degraded, with the ATP molecule hydrolyzing in a few seconds and proteins in a few minutes (White 1984).

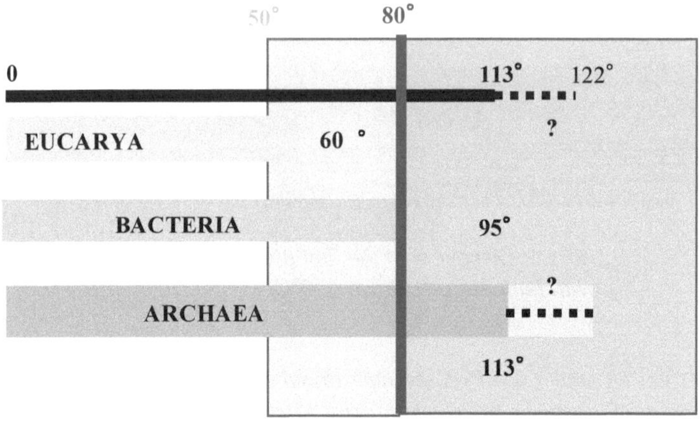

Figure 1.7. *Temperature ranges for organisms in the three domains of life. Hyperthermophiles are defined as organisms whose optimal growth temperature exceeds 80°C. The dotted lines indicate unconfirmed data*

While the upper temperature limit for life on Earth remains to be defined, it is already remarkable that this limit varies significantly across the three domains of life (Figure 1.7). Currently, no known eukaryote can divide above 60°C (a few fungi reach this limit). This is not for lack of trying. Karl Stetter himself, toward the end of his career, attempted to isolate hyperthermophilic eukaryotes. He only succeeded in isolating thermophilic protists capable of reproducing up to 55°C (De Jonckheere et al. 2011). Only archaea and bacteria seem capable of dividing above 60°C, and only archaea have surpassed the 100°C barrier. The difference in the upper viable temperature limits for archaea and bacteria (Figure 1.7) is likely due to differences in the composition of their membranes (see Chapter 2). The absence of hyperthermophilic eukaryotes could be explained by the lack of coupling between transcription and translation in eukaryotes, as messenger RNA is particularly fragile at high temperatures (Forterre 1995).

1.5.3. *The discovery of viruses in hyperthermophilic archaea*

While Karl Stetter specialized in searching for new hyperthermophilic archaea by investigating new metabolic types, Wolfram Zillig focused on the search for viruses infecting these microorganisms. Zillig aimed to isolate viruses to use as models for studying the molecular biology of archaea, like how pioneers of molecular biology in the 1950s–1970s used viruses infecting bacteria (bacteriophages or bacterioviruses). Zillig expected to find archaeal "phages" producing head-tail-like virions resembling those of bacteriophages. Indeed, the first archaeal virus discovered in the 1970s, even before Carl Woese's development of the concept of archaebacteria, produced classic head-tail virions. This virus infected a halophilic archaeon (Torsvik and Dundas 1974).

To his surprise, Zillig and his students managed to isolate several original viruses in Icelandic and Japanese hot springs that infected anaerobic hyperthermophilic archaea, such as *Thermoproteus* (Figure 1.5(e)), or aerobic thermoacidophiles, such as *Sulfolobus* (Figure 1.5d). Some of these viruses produced virions in the form of long rigid filaments (rudiviruses, or rigid viruses) or flexible filaments (lipothrixviruses, lipid filaments, as their virions contain lipids) (Zillig et al. 1996). These virions resembled those produced by some eukaryotic RNA viruses, such as the tobacco mosaic virus, but were DNA viruses. One of the new viruses isolated in the late 1980s, SSV1 (*Sulfolobus solfataricus* virus 1), produced lemon- or spindle-shaped virions (now classified in the *Fuselloviridae* family), a morphology never previously observed in viruses infecting bacteria or eukaryotes (Schleper et al. 1992) (Figure 1.5(d)). Subsequently, Mike Dyall-Smith in Australia and Claire Geslin in France discovered other lemon-shaped viruses that infected halophilic archaea and anaerobic hyperthermophilic archaea, respectively (Bath and Dyall-Smith 1998; Geslin et al. 2003). These atypical viruses are therefore present throughout the domain of archaea (see Chapter 5).

During one of these Icelandic expeditions in 1995, Wolfram was accompanied by a Georgian researcher, David Prangishvili, who had joined his team in 1992, following the fall of the USSR. David Prangishvili would go on to become the undisputed world specialist in archaeal viruses in the 2000s, discovering viruses producing virions with increasingly astonishing shapes: some resembling bottles or springs, others capable of changing shape after leaving the cell, and still others ending in claws used to grip the pili of the archaea they infect (see Chapter 2). The discovery of different viruses in archaea, distinct from those infecting bacteria and eukaryotes, supported Carl Woese's vision of a living world divided into three major domains, each with its specific set of viruses. David Prangishvili continued his studies on archaeal viruses in the unit I led at the Pasteur Institute of Paris from 2004 to 2019 (Prangishvili 2013; Prangishvili et al. 2017). His work is now being continued by one of our former postdoctoral researchers, Mart Krupovic, who heads a research unit on archaeal virology at this institute (Forterre 2022a).

1.6. Non-extremophilic archaea discovered by molecular ecology: a new vision of the third domain

The discovery of halophilic, thermo-acidophilic and then hyperthermophilic archaea had led most biologists to consider archaea as essentially extremophilic organisms. Methanogens could also, to some extent, be considered "extremophilic" due to their very strict anaerobiosis, with many of them being thermophilic or even hyperthermophilic, like *M. kandleri.* Thus, mesophilic archaea seemed to be an exception. The surprise was all the greater when the first molecular ecology studies, based on the identification of 16S rRNAs present in the environment, revealed new lineages of archaea present in large numbers in cold environments, the ocean, lakes and terrestrial soils (DeLong 1992; Furhman et al. 1992). This work was partly initiated in the early 1990s by the teams of Norman Pace (a former student of Carl Woese) and Edouard DeLong in the United States.

Although not culturable at the time, the sequence of their 16S rRNAs, their "phylotype", allowed the placement of these new archaeal lineages in a phylogenetic tree. A large number of them corresponded to new *Euryarchaeota*, but an equally large number seemed to correspond to *Crenarchaeota*, a phylum for which all cultivated organisms known at the time were either thermo-acidophilic or hyperthermophilic. This observation triggered a new "hunt for archaea", which remained unsuccessful for a long time. Analysis of genomic fragments from these organisms showed that they must be capable of oxidizing ammonia (Treusch et al. 2004). In 2005, after 15 years of unsuccessful trials in many laboratories, this observation enabled David Stahl's team in the United States to cultivate, for the first time, one of these new oceanic mesophilic archaea. It was a very small archaea that they named *Nitrosopumilus maritimus* (*pumilus* means "dwarf" in Latin) (Könneke et al. 2005). Curiously, this archaeon was not isolated from an oceanic water sample but from a sample taken from the Seattle Aquarium!

Subsequently, researchers demonstrated that these cold-dwelling archaea play a major role in the nitrogen cycle, a role previously attributed solely to bacteria. Today, thanks to genome sequencing, we know that *N. maritimus* and other mesophilic "*Crenarchaea*" are, in fact, very different from thermophilic *Crenarchaea*. They were eventually classified, along with some thermophiles, into a new phylum called *Thaumarchaea* (Brochier-Armanet et al. 2008). Additionally, *Thaumarchaea* were discovered in hot environments (see Chapter 2). One of the leading research teams studying *Thaumarchaea* is led by Christa Schleper, a former student of Wolfram Zillig, highlighting once again the importance of mentorship in scientific research[4].

1.7. Conclusion

The history of the discovery of archaea can be considered a triumph of the so-called "reductionist" approach in science (Forterre 2022a, 2022b). By using a single gene as an evolutionary marker to determine the relationships among microorganisms, Carl Woese was able to identify within "prokaryotes", a subgroup of microorganisms that constitute one of the three cellular lineages currently inhabiting our planet. This discovery was subsequently validated by genomics, but comprehensive approaches using hundreds and then thousands of genes did not provide fundamentally new insights compared to the concept proposed by Carl Woese. On the contrary, it was this concept that allowed scientists to make sense of the deluge of genomic and then metagenomic data which inundated the field of biology.

The concept of archaea faced significant resistance within the scientific community. Carl Woese encountered considerable opposition, particularly from American evolutionists such as Lynn Margulis and Ernst Mayr (Margulis and Guerrero 1991; Mayr 1998; Woese 1998; Quammen 2018). His exceptional contribution to 20th-century biology was only belatedly recognized with the prestigious Craaford Prize, awarded to him by the King of Sweden in 2003 (Figure 1.2(b)). Despite the considerable impact of his idea of characterizing microorganisms through molecular approaches, Carl Woese was never awarded the Nobel Prize, even though his work had profound implications not only in fundamental biology but also in medicine.

Even today, the concept of archaea is often ignored, or its importance underestimated by many biologists. Research continues to focus heavily on bacteria and eukaryotes; in the last 10 years, approximately 15,000 papers have been published on archaea, compared to 910,000 on bacteria and more than 7 million on eukaryotes.

However, history teaches us those researchers who ventured off the beaten path and took an early interest in archaea often had brilliant careers. This is a lesson for the younger generations, especially in an era where the pursuit of high-impact

4 An example worth mentioning is the discovery of the Asgard archaea (see Chapter 2) by Anja Spang, a former doctoral student of Christa Schleper, within the team of Thijs Ettema, a former doctoral student of Rolf Bernander, a pioneer in the study of the cell cycle in archaea.

factors and the influence of trends play an increasingly significant role in shaping research agendas and securing funding.

1.8. References

Albers, S.V., Forterre, P., Prangishvili, D., Schleper, C. (2013). The legacy of Carl Woese and Wolfram Zillig: From phylogeny to landmark discoveries. *Nat. Rev. Microbiol.*, 11, 713–719.

Baross, J.A. and Deming, J.W. (1983). Growth of 'black smoker' bacteria at temperature of at least 250°C. *Nature*, 303, 423–426.

Bath, C. and Dyall-Smith, M.L. (1998). His1, an archaeal virus of the *Fuselloviridae* family that infects *Haloarcula Hispanica*. *J. Virol.*, 72, 9392–9395.

Bergerat, A., de Massy, B., Gadelle, D., Varoutas, P.C., Nicolas, A., Forterre, P. (1997). An atypical topoisomerase II from Archaea with implications for meiotic recombination. *Nature*, 386, 414–417.

Blöchl, E., Rachel, R., Burggraf, S., Hafenbradl, D., Jannasch, H.W., Stetter, K.O. (1997). *Pyrolobus fumarii,* gen. and sp. nov., represents a novel group of archaea, extending the upper temperature limit for life to 113 degrees C. *Extremophiles*, 1, 14–21.

Bonen, L. and Doolittle, W.F. (1975). On the prokaryotic nature of red algal chloroplasts. *Proc. Natl. Acad. Sci. USA*, 72, 2310–2314.

Brochier-Armanet, C., Boussau, B., Forterre, P., Gribaldo, S. (2008). Mesophilic *Crenarchaeota*: Proposal for a third archaeal phylum, the *Thaumarchaeota*. *Nat. Rev. Microbiol.*, 6, 245–252.

Brock, T.D., Brock, K.M., Belly, R.T., Weiss, R.L. (1972). Sulfolobus: A new a new genus of sulfur-oxidizing bacteria living at low pH and high temperature. *Arch. Mikrobiol.*, 84, 54–68.

Catchpole, R.J. and Forterre, P. (2019). The evolution of reverse gyrase suggests a non-hyperthermophilic last universal common ancestor. *Mol. Biol. Evol.*, 36, 2737–2747.

Darland, G., Brock, T.D., Samsonoff, W., Conti, S.F. (1970). A thermophilic, acidophilic mycoplasma isolated from a coal refuse pile. *Science*, 170, 1416–1418.

De Jonckheere, J.F., Baumgartner, M., Eberhardt, S., Opperdoes, F.R., Stetter, K.O. (2011). *Oramoeba fumarolia* gen. nov., sp. nov., a new marine heterolobosean amoe-boflagellate growing at 54°C. *Eur. J. Protistol.*, 47, 16–23.

DeLong, E.F. (1992). Archaea in coastal marine environments. *Proc. Natl. Acad. Sci. USA*, 89, 5685–5689.

Erauso, G., Charbonnier, F., Barbeyron, T., Forterre, P., Prieur, D. (1992). Caractérisation préliminaire d'une archaebactérie hyperthermophile possédant un plasmide, isolée d'une source hydrothermale du bassin Nord Fidjien. *C.R. Acad. Sci.*, 314, 387–393.

Forterre, P. (1995). Thermoreduction, a hypothesis for the origin of prokaryotes. *C.R. Acad. Sci.*, 318, 415–422.

Forterre, P. (1997). Archaea: What can we learn from their sequences? *Curr. Opin. Genet. Dev.*, 7, 764–770.

Forterre, P. (2002). A hot story from comparative genomics: Reverse gyrase is the only hyperthermophile-specific protein. *Trends Genet.*, 18, 236–237.

Forterre, P. (2007). *Microbes de l'enfer*. Belin, Paris.

Forterre, P. (2016). *Microbes from Hell*. Chicago University Press, Chicago.

Forterre, P. (2022a). Carl Woese: Still ahead of our time. *mLife*, 1, 359–367.

Forterre, P. (2022b). Archaea: A goldmine for molecular biologists and evolutionists. *Methods Mol. Biol.*, 2522, 1–21. doi: 10.1007/978-1-0716-2445-6_1.

Forterre, P. (2022c). L'émergence d'une communauté "archée" en France. *Histoire de la recherche contemporaine*, 11, 29–40. doi: 10.4000/hrc.7830.

Forterre, P. (2022d). Les archées : une mine d'or pour l'étude des ADN topoisomérases. *Histoire de la recherche contemporaine*, 11, 17–27. doi: 10.4000/hrc.7956.

Forterre, P., Elie, C., Kohiyama, M. (1984). Aphidicolin inhibits growth and DNA synthesis in halophilic arachaebacteria. *J. Bacteriol.*, 159, 800–802.

Forterre, P., Mirambeau, G., Jaxel, C., Nadal, M., Duguet, M. (1985). High positive supercoiling in vitro catalyzed by an ATP and polyethylene glycol-stimulated topoisomerase from *Sulfolobus acidocaldarius*. *EMBO J.*, 4, 2123–2128.

Fox, G.E., Stackebrandt, E., Hespell, R.B., Gibson, J., Maniloff, J., Dyer, T.A., Wolfe, R.S., Balch, W.E., Tanner, R.S., Magrum, L.J. et al. (1980). *KashefiKashefi and Lo-velay* "The phylogeny of prokaryotes". *Science*, 209, 457–446.

Fuhrman, J.A., McCallum, K., Davis, A.A. (1992). Novel major archaebacterial group from marine plankton. *Nature*, 356, 148–149.

Garnier, F., Couturier, M., Débat, H., Nadal, M. (2021). Archaea: A gold mine for topoisomerase diversity. *Front Microbiol.*, 12, 661411.

Geslin, C., Le Romancer, M., Erauso, G., Gaillard, M., Perrot, G., Prieur, D. (2003). PAV1, the first virus-like particle isolated from a hyperthermophilic euryarchaeote. *Pyrococcus Abyssi*. *J. Bacteriol.*, 185, 3888–3894.

Guipaud, O., Marguet, E., Noll, K.M., de la Tour, C.B., Forterre, P. (1997). Both DNA gyrase and reverse gyrase are present in the hyperthermophilic bacterium *Thermotoga maritima*. *Proc. Natl. Acad. Sci. USA*, 94, 10606–10611.

Huber, R., Kurr, M., Jannash, H.W., Stetter, K.O. (1989). A novel group of abyssal methanogenic archaebacteria (*Methanopyrus*) growing at 110°C. *Nature*, 342, 833–834.

Huber, H., Hohn, M.J., Rachel, R., Fuchs.T., Wimmer, V.C., Stetter, K.O. (2002). A new phylum of Archaea represented by a nanosized hyperthermophilic symbiont. *Nature*, 417, 63–67.

Huet, J., Schnabel, R., Sentenac, A., Zillig, W. (1983). Archaebacteria and eukaryotes possess DNA-dependent RNA polymerases of a common type. *EMBO J.*, 2, 1291–1294.

Kandler, O. and Hippe, H. (1977). Lack of peptidoglycan in the cell walls of *Methanosarcina barkeri*. *Arch. Microbiol.*, 113, 57–60.

Kashefi, K. and Lovley, D.R. (2003). Extending the upper temperature limit for life. *Science*, 301, 934.

Kates, M., Yengoyan, L.S., Sastry, P.S. (1965). A diether analog of phosphatidyl glycerophosphate in *halobacterium cutirubrum*. *Biochim. Biophys. Acta*, 98, 252–268.

Kessel, M. and Klink, F. (1980). Archaebacterial elongation factor is ADP-ribosylated by diphtheria toxin. *Nature*, 287, 250–251.

Kikuchi, A. and Asai, K. (1984). Reverse gyrase – A topoisomerase which introduces positive superhelical turns into DNA. *Nature*, 309, 677–681.

Könneke, M., Bernhard, A.E., de la Torre, J.R., Walker, C.B., Waterbury, J.B., Stahl, D.A. (2005). Isolation of an autotrophic ammonia-oxidizing marine archaeon. *Nature*, 437, 543–546.

Lanyi, J.K. (1974). Salt-dependent properties of proteins from extremely halophilic bacteria. *Bacteriol. Rev.*, 38, 272–290.

Magrum, L.J., Luhersen, K.R., Woese, C.R. (1978). Are extreme halophiles actually "bacteria"? *J. Mol. Evol.*, 12, 1–8.

Margulis, L. (1975). The microbe contribution to evolution. *Biosystems*, 7, 266–292.

Margulis, L. and Guerrero, R. (1991). Kingdoms in turmoil. *New Scientist*, 1761, 46–50.

Mayr, E. (1998). Two empires or three? *Proc. Natl. Acad. Sci. USA*, 95, 9720–3722.

Nadal, M., Mirambeau, G., Forterre, P., Reiter, W.D., Duguet, M. (1986). Positively supercoiled DNA in a virus-like particle of an archaebacterium. *Nature*, 321, 256–258.

Olsen, G.J. and Woese, C.R. (1997). Archaeal genomics: An overview. *Cell*, 89, 991–994.

Prangishvili, D. (2013). The wonderful world of archaeal viruses. *Annu. Rev. Microbiol.*, 67, 565–585.

Prangishvilli, D., Zillig, W., Gierl, A., Biesert, L., Holz, I. (1982). DNA-dependent RNA polymerase of thermoacidophilic archaebacteria. *Eur. J. Biochem.*, 122, 471–477.

Prangishvili, D., Bamford, D.H., Forterre, P., Iranzo, J., Koonin, E.V., Krupovic, M. (2017). The enigmatic archaeal virosphere. *Nat. Rev. Microbiol.*, 15, 724–739.

Quammen, D. (2018). *The Tangle Tree: A Radical New History of Life*. Simon & Schuster, New York.

Sako, Y., Nomura, N., Uchida, A., Ishida, Y., Morii, H., Koga, Y., Hoaki, T., Maruyama, T. (1996). *Aeropyrum pernix* gen. nov., sp. nov., a novel aerobic hyperthermophilic archaeon growing at temperatures up to 100 degrees C. *Int. J. Syst. Bacteriol.*, 46, 1070–1077.

Sanger, F., Brownlee, G.G., Barrell, B.G. (1965). A two-dimensional fractionation procedure for radioactive nucleotides. *J. Mol. Biol.*, 13, 373–398.

Sapp, J. (2005). The prokaryote–eukaryote dichotomy: Meanings and mythology. *Microbiol. Mol. Biol. Rev.*, 69, 292–305.

Sapp, J. and Fox, G.E. (2013). The singular quest for a universal tree of life. *Microbiol. Mol. Biol. Rev.*, 77, 541–550.

Schleper, C., Kubo, K., Zillig, W. (1992). The particle SSV1 from the extremely thermophilic archaeon *Sulfolobus* is a virus: Demonstration of infectivity and of transfection with viral DNA. *Proc. Natl. Acad. Sci. USA*, 89, 7645–7649.

Sioud, M., Forterre, P., de Recondo, A.M. (1987). Effects of the antitumor drug VP16 (etoposide) on the archaebacterial Halobacterium GRB 1.7 kb plasmid in vivo. *Nucleic. Acids. Res.*, 15, 8217–8234.

Stanier, R.Y. and Van Niel, C.B. (1962). The concept of a bacterium. *Arch. Mikrobiol.*, 42, 17–35.

Stetter, K.O. (1982). Unltrathin mycelia-forming organisms from submarine volcanic areas having an optimal growth temperature of 105°C. *Nature*, 300, 258–260.

Stetter, K.O. (2006). History of discovery of the first hyperthermophiles. *Extremophiles*, 10, 357–362.

Stetter, K.O. (2013). A brief history of the discovery of hyperthermophilic life. *Biochem. Soc. Trans.*, 41, 416–420.

Sugimoto-Shirasu, K., Stacey, N.J., Corsar, J., Roberts, K., McCann, M.C. (2002). DNA topoisomerase VI is essential for endoreduplication in Arabidopsis. *Curr. Biol.*, 12, 1782–1786.

Takai, K., Nakamura, K., Toki, T., Tsunogai, U., Miyazaki, M., Miyazaki, J., Hirayama, H., Nakagawa, S., Nunoura, T., Horikoshi, K. (2008). Cell proliferation at 122 degrees C and isotopically heavy CH4 production by a hyperthermophilic methanogen under high-pressure cultivation. *Proc. Natl. Acad. Sci. USA*, 105, 10949–10954.

Tornabene, T.G. and Langworthy, T.A. (1979). Diphytanyl and dibiphytanyl glycerol ether lipids of methanogenic archaebacteria. *Science*, 203, 51–53.

Torsvik, T. and Dundas, I.D. (1974). Bacteriophage of *Halobacterium salinarium*. *Nature*, 248, 680–681.

Trent, J.D., Chastain, R.A., Yayanos, A.A. (1984). Possible artefactual basis for apparent bacterial growth at 250 degrees C. *Nature*, 307, 737–740.

Treusch, A.H., Kletzin, A., Raddatz, G., Ochsenreiter, T., Quaiser, A., Meurer, G., Schuster, S.C., Schleper, C. (2004). Characterization of large-insert DNA libraries from soil for environmental genomic studies of Archaea. *Environ. Microbiol.*, 6, 970–980.

Van de Vossenberg, J.L., Ubbink-Kok, T., Elferink, M.G., Driessen, A.J., Konings, W.N. (1995). Ion permeability of the cytoplasmic membrane limits the maximum growth temperature of bacteria and archaea. *Molecular Microbiology*, 18, 925–932.

White, R.H. (1984). Hydrolytic stability of biomolecules at high temperatures and its implication for life at 250 degrees C. *Nature*, 310, 430–432.

Whittaker, R.H. (1969). New concepts of kingdoms or organisms. Evolutionary relations are better represented by new classifications than by the traditional two kingdoms. *Science*, 163, 150–160.

Woese, C.R. (1981). Archaebacteria. *Scientific American*, 244, 98–125.

Woese, C.R. (1998). Default taxonomy: Ernst Mayr's view of the microbial world. *Proc. Natl. Acad. Sci. USA*, 95, 11043–11046.

Woese, C.R. and Fox, G.E. (1977). Phylogenetic structure of the prokaryotic domain: The primary kingdoms. *Proc. Natl. Acad. Sci. USA*, 74, 5088–5090.

Woese, C.R., Maniloff, J., Zablen, L.B. (1980). Phylogenetic analysis of the mycoplasmas. *Proc. Natl. Acad. Sci. USA*, 77, 494–498.

Woese, C.R., Kandler, O., Wheelis, M.L. (1990). Towards a natural system of organisms: Proposal for the domains Archaea, Bacteria, and Eucarya. *Proc. Natl. Acad. Sci. USA*, 87, 4576–4579.

Zablen, L.B., Kissil, M.S., Woese, C.R., Buetow, D.E. (1975). Phylogenetic origin of the chloroplast and prokaryotic nature of its ribosomal RNA. *Proc. Natl. Acad. Sci. USA*, 72, 2418–2422.

Zillig, W., Stetter, K.O., Janeković, D. (1979). DNA-dependent RNA polymerase from the archaebacterium *Sulfolobus acidocaldarius*. *Eur. J. Biochem.*, 96, 597–604.

Zillig, W., Schäfer, W., Reiter, W.D. (1983). The archaebacterium *Thermococcus celer* represents, a novel genus within the thermophilic branch of the archaebacteria. *Syst. Appl. Microbiol.*, 4, 88–94.

Zillig, W., Prangishvilli, D., Schleper, C., Elferink, M., Holz, I., Albers, S., Janekovic, D., Götz, D. (1996). Viruses, plasmids and other genetic elements of thermophilic and hyperthermophilic Archaea. *FEMS Microbiol. Rev.*, 18, 225–236.

2

Evolution of Archaea and Their Taxonomy

Patrick FORTERRE[1,2]
[1] *Institut Pasteur, Université Paris Cité, France*
[2] *Université Paris-Saclay, Gif-sur-Yvette, France*

2.1. Introduction

The discovery of archaea in the late 1980s (see Chapter 1) was made as part of a study aimed at determining kinship relationships among all microorganisms. From the outset, Carl Woese and his collaborators sought to understand the history of this domain and to base the taxonomy and nomenclature of archaea on their position in a phylogenetic tree that best represents their evolution.

Our knowledge of the diversity and evolution of archaea has continuously progressed since that time, primarily due to advancements in techniques. These include the analysis of oligonucleotide catalogs of 16S ribosomal RNAs (16S rRNAs), gene coding sequences for these RNAs, genome sequencing, reconstruction of genomes from metagenomes and the development of new models of phylogenetic analyses.

The taxonomy and nomenclature of the various lineages of archaea are still evolving, and there is currently no consensus among scientists on these issues. We hope that the historical presentation in this chapter will help readers understand the reasons for this situation and follow the developments and changes that are sure to occur in the future.

Biology of Archaea 1,
coordinated by Béatrice CLOUET-D'ORVAL, Bruno FRANZETTI and Philippe OGER.

2.2. One domain, three major branches and a few isolated "phyla"

2.2.1. *One domain, two phyla*

Soon after Carl Woese and his collaborators identified archaea as a group distinct from the "classical" bacteria, they proposed dividing the archaea (then still called archaebacteria) into two major branches. Early analyses of 16S rRNA oligonucleotide catalogs using the Sab method (see Chapter 1) demonstrated that the thermoacidophile *Sulfolobus acidocaldarius* was evolutionarily distant from methanogenic and halophilic archaea, which were more closely related to the thermoacidophile *Thermoplasma acidophilum*.

Subsequent analysis of their RNA polymerases and 16S rRNA sequences revealed that most hyperthermophilic archaea isolated by Wolfram Zillig and Karl Stetter in the 1980s–1990s (*Thermoproteus*, *Desulfurococcus*, *Pyrodictium*, *Pyrolobus* and *Pyrobaculum*) were grouped with *Sulfolobus*, except for the Thermococcales *Thermococcus* and *Pyrococcus*), which were found to be closer to certain methanogenic archaea (Prangishvili et al. 1982; Zilig et al. 1983). In the first universal tree published in 1986, the root of the archaea tree was surrounded by hyperthermophiles, with *Thermoproteus* and *Sulfolobus* on one side and *Thermococcus* on the other, at the base of a clade that included halophilic and methanogenic archaea (Woese and Olsen 1986) (Figure 2.1), reinforcing the idea of a thermophilic ancestor for all archaea.

In 1990, Woese et al. (1990) proposed formalizing these observations by dividing the archaea into two phyla, the *Euryarchaeota* and the *Crenarchaeota* (Figure 2.1). The *Euryarchaeota* included archaea with highly diverse phenotypes: methanogens, halophiles, mesophiles, thermoacidophiles and hyperthermophiles, hence the prefix "Eury", from the Greek *euryos*, meaning broad or spacious.

In contrast, the *Crenarchaeota* comprised only thermoacidophiles and hyperthermophiles, hence the prefix "Cren", from the Greek *crenos*, meaning spring, in the sense of origin. Carl Woese and Otto Kandler were indeed proponents of a hot origin of life. This proposal was published alongside their suggestion to replace "Archaebacteria" with "archaea" and to introduce the term "domain" as the highest rank in the taxonomy of the living world.

The division of archaea into Euryarchaea and Crenarchaea was also supported by biochemical data: for example, the presence in Euryarchaea of two eukaryotic-type proteins involved in chromosome structure and replication, which were not found in Crenarchaea. These proteins are histones (Sandman et al. 1990) and the RPA[1] (Kelly et al. 1998), as well as a DNA polymerase specific to Euryarchaea, DNA polymerase D (Cann and Ishino 1999).

These discoveries also indicated that the fundamental molecular mechanisms associated with chromosome structure and replication were more divergent within archaea than within the other two domains.

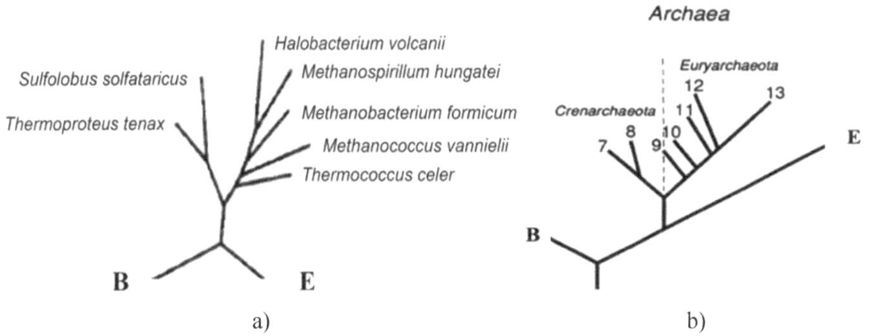

Figure 2.1. *First Universal Trees Based on 16S rRNAs*

COMMENT ON FIGURE 2.1.– *(a) A focus on archaebacteria in an unrooted tree (Woese and Olsen 1986). (b) A focus on archaea in the rooted tree published by Woese and his collaborators in 1990 to propose their new nomenclature (see Chapter 3 and Woese et al. 1990). The species used are the same in both trees (e.g. 9 corresponds to Thermococcus celer).*

2.2.2. *A first orphaned phylum, the korarchaeota*

In the 1990s, the rise of microbial molecular ecology (sequencing of genes coding for 16S rRNAs present in the environment) led to the discovery of many archaeal lineages that were still uncultivated at the time (and some of which remain uncultivated today) (DeLong 1992; Fuhrman et al. 1992).

All of these lineages, uniquely characterized by their phylotype, that is to say, by their position in the 16S rRNA tree, were attributed to the Euryarchaea or the Crenarchaea with one exception: two lineages of organisms detected in a

1 Replication Protein A (RPA) is part of the Single-Stranded DNA Binding proteins (SSB) that protect single-stranded DNA formed at the replication fork.

Yellowstone hot spring by Norman Pace's[2] team. These two lineages branched midway between the Crenarchaea and the Euryarchaea in the 16S rRNA tree (Figure 2.2).

Norman Pace and his collaborators proposed giving this lineage the status of a phylum under the name *Korarchaeota*, from the Greek *kouros*, meaning young man, again to reassert the hypothesis of a hot origin of life (Barns et al. 1996).

This new phylum, represented only by two environmental sequences, however, seemed quite atypical compared to the other two phyla represented by a wide variety of cultivated organisms.

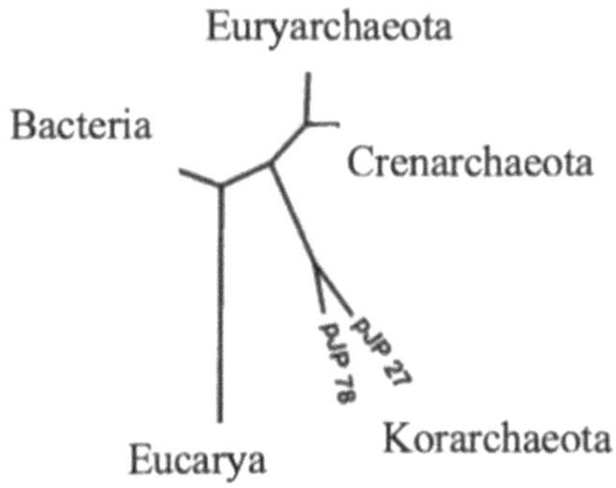

Figure 2.2. *Extract from the phylogeny of 16S rRNAs on which Norman Pace and his collaborators relied to propose the Korarchaeota phylum (Barns et al. 1996). Only the central part of the phylogeny is presented. PJP 27 and 76 correspond to two environmental 16S rRNA sequences*

2.2.3. *The first phylogenies based on conserved proteins*

From the 2000s, the sequencing of complete genomes made it possible to have protein sequences conserved for an increasing number of archaeal species, covering the entire domain known at the time. Phylogenies based on rRNA were gradually replaced by phylogenies based on the concatenation[3] of the sequences of ribosomal

2 Norman Pace, a former post-doctoral researcher of Carl Woese, was one of the pioneers in the use of 16S rRNA for the identification of microorganisms in the environment.

3 Concatenation refers to the end-to-end addition of all of the protein sequences used in a phylogenetic analysis present in a given species to form a single sequence characteristic of that species.

proteins conserved in all archaea (Matte-Taillez et al. 2002; Brochier et al. 2004, 2005a, 2005b). These phylogenies confirmed the division of the archaea into Euryarchaea and Crenarchaea (Figure 2.3).

The most hyperthermophilic organisms, the Thermococcales and/or the methanogenic archaea *Methanopyrus kandleri*, consistently appeared at the base of the Euryarchaea.

Conversely, mesophilic organisms were primarily found at the ends of the branches furthest from this base, with halophilic archaea close to the *Methanosarcinales* and *Methanomicrobiales*.

These results reinforced the idea that the common ancestor of all archaea, LACA (Last Archaeal Common Ancestor), lived at high temperatures, with mesophilic archaea adapting secondarily to increasingly cold environments.

The archaea at the base of the Euryarchaea and Crenarchaea were also all anaerobic, while the aerobic archaea were located at the ends of the branches of these two phyla. This confirmed the hypothesis that archaea had appeared in a world devoid of oxygen, with some subsequently adapting to its presence and eventually becoming strict aerobes capable of using oxygen for their energy needs.

Methanogenic archaea were divided into two groups: class 1 methanogens (*Methanobacteriales*, *Methanococcales* and *Methanopyrales*, close to Thermococcales) which included mesophilic, thermophilic, and hyperthermophilic organisms, and class 2 methanogens (*Methanosarcinales* and *Methanomicrobiales*) which included only mesophilic organisms.

Between these groups were the thermoacidophile *Thermoplasma acidophilum* and the hyperthermophile *Archaeoglobus fulgidus*. The phylogeny of the enzymes involved in methanogenesis overlapped with that of ribosomal proteins, indicating that these enzymes had evolved vertically (Bapteste et al. 2005).

It could be concluded that methanogenesis had appeared at the base of the Euryarchaea, and that non-methanogenic Euryarchaea, such as Haloarchaea, Thermoplasmatales and Archaeoglobales, had evolved from methanogenic archaea due to several independent losses of the enzymes and coenzymes involved in methanogenesis.

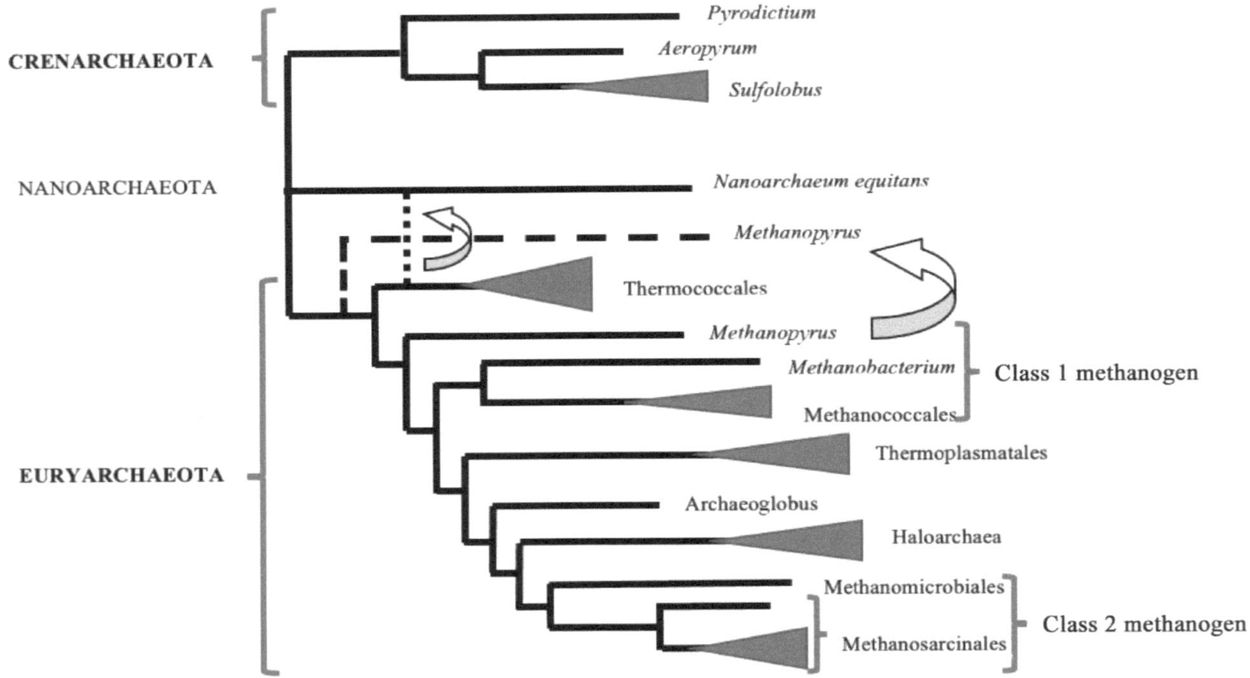

Figure 2.3. *Simplified phylogeny from Brochier et al. (2004, 2005a, 2005b), based on ribosomal proteins (solid lines). The positions of N. equitans in a concatenation limited to the small subunit of the ribosome and that of M. kandleri in a concatenation of RNA polymerase subunits are indicated by dotted lines. The arrows highlight the positional changes discussed in the text. For a color version of this figure, see www.iste.co.uk/clouet/archaea1.zip*

2.2.4. *The special case of Methanopyrus kandleri*

In 2005, Céline Brochier et al. (2005a) demonstrated that the phylogenies obtained from RNA polymerase subunits (transcription) were identical to those obtained from ribosomal proteins (translation) with one exception. Since translation and transcription are two mechanisms that could have evolved independently, this observation demonstrated the existence of a core set of proteins that evolved vertically, representing the species tree in archaea. The exception was the hyperthermophilic methanogenic archaeon *M. kandleri*, which appeared at the base of the Euryarchaea in the RNA polymerase (and 16S rRNA) tree, while it branched at the base of group I methanogens in the ribosomal protein tree (Figure 3.3). As the branch leading to *M. kandleri* was much longer in the RNA polymerase tree than in the ribosomal protein tree, it was evident that the correct position of *M. kandleri* was the one obtained with ribosomal proteins, and that its basal position in the RNA polymerase tree was due to the attraction of its long branch by that of the outgroup, the Crenarchaea[4].

In agreement with this conclusion, *M. kandleri* was closer to *Methanobacterium thermoautotrophicum* in the ribosomal protein tree than in the RNA polymerase tree (Figure 2.3). Notably, these two archaea are the only ones with a cell wall containing pseudomurein.

The long branch of *M. kandleri* in the RNA polymerase tree suggested an abnormally rapid evolution of this enzyme in this archaeon. This hypothesis was confirmed by the presence of an unusually high number of indels (insertions/deletions of amino acids) in the alignment of its sequence with those of other archaeal RNA polymerases. The authors noted 27 specific indels in the sequence alignment of *M. kandleri* RNA polymerase, whereas this number varied only from one to eight in other archaeal lineages (see Table 1 in Brochier et al. (2004)).

More generally, *M. kandleri* was known to exhibit several atypical traits, such as a high number of orphan genes in its genome and several genes split into two pieces. According to the authors of this study, the abnormally high rate of evolution in *M. kandleri* could be due to the absence of the transcription factor TFS, which is involved in transcription fidelity, a hypothesis that remains to be confirmed (Brochier et al. 2004).

---

4 This phenomenon, called long branch attraction (LBA), is due to the conservation of a greater number of ancestral characters in organisms with short branches. It was first described by Felsenstein (1978).

2.2.5. *The special case of Nanoarchaeum equitans*

In the early 2000s, Karl Stetter's team discovered a miniature archaeon (200–300 nm in diameter), which had the smallest genome known at that time for a cellular organism (489 kb) (Huber et al. 2002). This archaeon could only live and divide when attached to the surface of a "host" Crenarchaea called *Ignicoccus hospitalis*. It was, therefore, a mandatory ectosymbiont. Similar to the Korarchaea, this Nanoarchaea, named *Nanoarchaeum equitans* (small archaea riding on its host), branched halfway between the Euryarchaea and the Crenarchaea in phylogenies based on 16S rRNA. Karl Stetter and his collaborators proposed to consider *N. equitans* as the representative of a fourth phylum, the *Nanoarchaeota*.

However, this proposal was questioned by Patrick Forterre and his collaborators when they included this new archaeon in their analyses. *N. equitans* did branch between the Euryarchaea and the Crenarchaea in a phylogeny based on the concatenation of ribosomal proteins; however, it occupied this intermediate position only four times out of 27 in the individual phylogenies. Depending on the protein studied, *N. equitans* branched six times with the Crenarchaea and 17 times with the Euryarchaea (see Table 1 in Brochier et al. (2005b)). The phylogenies placing *N. equitans* in the middle of the Crenarchaea could be explained by gene transfers from the Crenarchaea *I. hospitalis*, the host of *N. equitans*, to the latter. In the 17 trees where *N. equitans* branched within the Euryarchaea, it was eight times the sister group of the Thermococcales. Thus, two opposing signals were observed, which could explain why *N. equitans* appeared between the Euryarchaea and the Crenarchaea when these signals were mixed by concatenation. Most of the signals placing *N. equitans* in the middle of the Crenarchaea corresponded to proteins of the large subunit of the ribosome. A phylogeny limited to the concatenation of the proteins of the small subunit placed *N. equitans* as the sister group of the Thermococcales (Brochier et al. 2005b). All of these data and reflections supported the idea that *N. equitans* did not represent a new phylum but was part of the Euryarchaea.

These early studies highlighted two phenomena: the attraction of long branches and gene transfers, which would continue to pose problems in studies aiming to establish the phylogeny of archaea or the topology of the universal tree of life (see Chapter 3). They established an important rule: the necessity to thoroughly analyze the individual phylogenies of the proteins used as markers before concatenating their sequences. This rule, as we will see, was not always followed subsequently.

2.2.6. *Thaumarchaea*

Some new lineages of archaea detected in the 1990s were found in cold and oxygenated environments, whether terrestrial, marine or lacustrine. Initially, some of these lineages seemed to be related to Crenarchaea, which contradicted the previously established mandatory thermophilic nature of Crenarchaea (Delong 1992; Fuhrman et al. 1992). The first genome of a psychrophilic Crenarchaea was published in 2006 and belonged to a marine archaea living in symbiosis with a sponge, *Candidatus Cenarchaeum symbiosum* (Hallam et al. 2006). Analysis of this genome allowed the determination of its evolutionary position in a phylogeny based on the concatenation of ribosomal proteins (Brochier-Armanet et al. 2008). Contrary to expectations, *Candidatus Cenarchaeum symbiosum* did not group with Crenarchaea but occupied an intermediate position between Euryarchaea and Crenarchaea. Additionally, unlike the situation observed previously with *N. equitans*, this intermediate position was consistently found in individual phylogenies. At the phenotypic level, the genome of *Ca. C. symbiosum* contained molecular characteristics of the eukaryotic type, previously specific to Euryarchaea[5], others previously specific to Crenarchaea[6], and some that had not been detected in archaea[7] before. The authors of this study proposed considering *Ca. C. symbiosum* as the representative of a new phylum, *Thaumarchaeota*, from the Greek *Thauma* (miracle) to signify their surprise at this result (Brochier-Armanet et al. 2008a). The validation of this new phylum was quickly confirmed by the analysis of the genomes of the first two Thaumarchaea that could be cultivated, a marine organism, *Nitrosopumilus maritimus*, and a terrestrial organism, *Nitrososphaera gargensis* (Spang et al. 2010), and the phylum *Thaumarchaeota* was soon accepted by the entire scientific community working on archaea.

Shortly after the identification of Thaumarchaea, Patrick Forterre and his collaborators published a phylogenetic tree based on the analysis of the first 100 sequenced archaeal genomes (Brochier-Armanet et al. 2011). This tree included the three main branches corresponding to the three proposed phyla and the relationships previously identified within Euryarchaea. Among Crenarchaea, the thermoacidophilic archaea of the genera Sulfolobales and Acidilobales formed a clade related to Desulfurococcales, with the clade formed by these three orders being

5 In particular, the replication protein RPA, eukaryote-type histones and DNA polymerase D.

6 For example, a protein that binds to single-stranded DNA and was known only in *Crenarchaeota* and eukaryotes (Wadsworth and White 2001) turned out to be present in *Thaumarchaeota*.

7 This was the case for the DNA topoisomerase IB family, present in all eukaryotes and previously unknown in archaea (Brochier et al. 2008b).

itself a sister group of Thermoproteales. The Korarchaea branched at the base of Crenarchaea (Figure 2.4). Remarkably, *N. equitans* now branched with Thermococcales in this concatenation, which grouped all of the ribosomal proteins and not just those of the small subunit. The addition of new genomes had thus counterbalanced the presence of Crenarchaea-like proteins in the large ribosomal subunit of *N. equitans*.

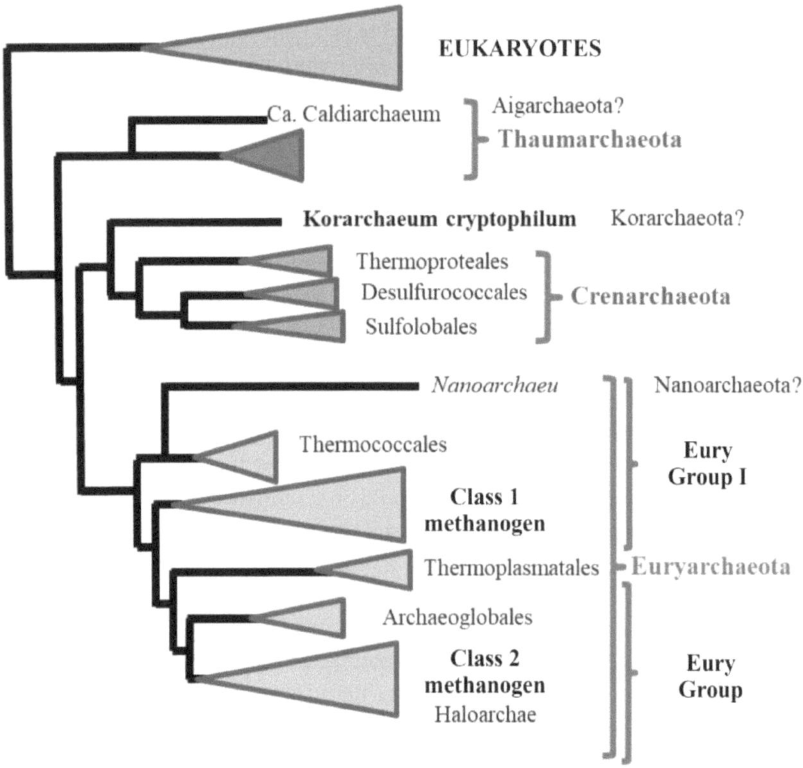

Figure 2.4. *Simplified phylogeny according to Brochier et al. (2011) (based on ribosomal proteins). The tree was rooted using eukaryotic homologous ribosomes as the outgroup. For a color version of this figure, see www.iste.co.uk/clouet/archaea1.zip*

The Thermococcales and *N. equitans* branched at the base of the Euryarchaea, followed by a clade grouping the class I methanogens, then a clade grouping the Archaeoglobus, the Thermoplasmatales, the Haloarchaea and the class II methanogens. This last clade was characterized by the systematic presence of a

DNA gyrase of bacterial origin that had been introduced at its base and which, according to phylogenetic analyses, had then co-evolved with this entire group of archaea.

Patrick Forterre and his collaborators therefore proposed dividing the Euryarchaea into two large clusters: group I and group II Euryarchaea. Group I (paraphyletic) Euryarchaea includes class I methanogens and Thermococcales (Figure 2.4), while group II (monophyletic) Euryarchaea includes class II methanogens, Archaeoglobus, Thermoplasmatales and Haloarchaea.

To analyze the position of *Ca. C. symbiosum*, Patrick Forterre and his collaborators constructed a tree of ribosomal proteins that for the first time included eukaryotic proteins homologous to those of archaea (Brochier-Armanet et al. 2008a) (Figure 2.4). This addition made it possible to root the archaea tree using eukaryotes as an external group (outgroup). In the resulting tree, the root of the archaea was no longer located as before between Euryarchaea and Crenarchaea, but between Thaumarchaea and the two other phyla of archaea. By the late 2000s, it thus seemed reasonable to divide the archaea into three large phyla: *Euryarcheota*, *Crenarchaeota* and *Thaumarchaeota*, with the positions of Nanoarchaea and Korarchaea in the archaeal tree remaining uncertain.

2.3. From phyla to superphyla

2.3.1. *Metagenomics and the explosion in the number of archaeal phyla*

The number of archaeal phyla, which had remained relatively limited until then (between three and five at most), exploded dramatically from the 2010s onwards. To date, there are several dozen archaeal phyla, more or less officially or unofficially recognized by the scientific community. Two main reasons can explain this sudden increase. The first was the development of metagenomic approaches (large-scale sequencing) that made it possible to reconstruct on a computer the genomes of the many lines of uncultivated archaea that were previously known only through their 16S rRNA phylotypes. Although these genomes, obtained from environmental DNA sequences, or more rarely isolated cells, are sometimes incomplete and may contain errors, they provide relevant information on the physiology, metabolism and evolutionary position of the underlying organisms. The new archaeal lines defined by these virtual genomes, called *MAGs* (metagenome-assembled genomes) or *SAGs* (single-cell amplified genomes), have been positioned in the archaeal tree not only based on rRNAs or ribosomal proteins but increasingly on the basis of the concatenation of groups of proteins common to most archaea.

The second reason that led to an explosion in the number of archaeal lineages is the proposal made by two teams in 2014 to use either the identity percentages of 16S rRNAs (Yarza et al. 2014) or the identity percentages between amino acids (AAI) between two genomes (Luo et al. 2014) to formally define bacterial and archaeal phyla. For Rosselo-Maura and his collaborators, two organisms belong to different phyla if the percentage identity of their 16S rRNAs is less than 75% (Yarza et al. 2014), while for Konstantidinis and his collaborators, two organisms belong to different phyla if their AAI is less than 45% (Luo et al. 2014). These less restrictive definitions have been adopted by researchers involved in the identification of new MAGs. They could therefore consider that each new line of archaea (or bacteria) they described allowed them to propose a new phylum. Thus, when Japanese researchers described in 2011 the MAG of an organism isolated from an underground hot spring that proved to be very close to the *Thaumarchaeota* in the phylogenies of 16S rRNA or proteins, they proposed to consider it not as a new *Thaumarchaeota*[8] but as the representative of a new phylum, the *Aigarchaeota* (from the Greek *aigi*, supposed to mean dawn in Greek[9]) (Nunoura et al. 2011). This organism, *Candidatus Caldiarchaeum Subterraneum*, coded for all eukaryote-type proteins characteristic of *Thaumarchaeota* (Brochier-Armanet et al. 2011; Nunoura et al. 2011). However, it had the particularity of coding for eukaryote-type[10] ubiquitins, which had never before been identified in other archaea (Nunoura et al. 2011).

The number of archaeal phyla also increased sharply with the reconstruction in the 2010s of numerous MAGs corresponding to lines of small archaea which, like *N. equitans*, have very small genomes (Narasingarao et al. 2012; Rinke et al. 2013; Castelle et al. 2015). Some of the organisms corresponding to these lines have been observed by electron microscopy, which has confirmed their small size (Comolli et al. 2009). As in the case of *N. equitans*, their genomes lack many genes coding for vital functions. These miniature archaea can therefore probably only live in symbiosis with other archaea, which has been demonstrated for several of them (see section 2.3.3.1). As a result, they had to evolve significantly to adapt to their hosts. Consequently, their rRNAs and proteins have greatly diverged from those of

8 In their initial description of *Thaumarchaeota*, the authors proposed including in this phylum all non-cultivable *Crenarchaeota* at the time, which formed a clade with *Candidatus Cenarchaeum Symbiosum* in 16S rRNA phylogenies, regardless of their mesophilic or thermophilic phenotypes (Brochier-Armanet et al. 2008a, 2012).

9 This term actually means "goat" in Greek, with the Greek name for "dawn" being "*aigu*".

10 Ubiquitins are small proteins that serve as markers for proteins to be eliminated by attaching to them.

other archaea, which probably explains why the numerous miniature archaeal lineages described between 2010 and 2020 were all assigned the status of phylum.

In parallel with this dramatic increase in the number of archaeal phyla, the last decade has seen an even more massive increase in the number of bacterial phyla, particularly with the discovery of a myriad of miniature bacteria very similar phenotypically to their miniature cousins in archaea (Brown et al. 2015). This general increase in the number of phyla in archaea and bacteria naturally led to the appearance of a higher ranking taxon, the superphylum.

2.3.2. *The superphylum TACK*

In 2011, Thijs Ettema and Lionel Guy proposed grouping *Thaumarchaeota, Aigarchaeota, Crenarchaeota and Korarchaeota* into a superphylum called *TACK*[11] (Guy and Ettema 2011). The existence of the TACK superphylum is now widely accepted by the microbiological community. However, it should be noted that this superphylum does not correspond to a monophyletic group (a clade) if the root of the archaeal tree is situated between *Thaumarchaeota* and *Aigarchaeota* on one side and all other archaea on the other, as was the case in the first publication describing *Thaumarchaeota* (Brochier-Armanet et al. 2008a).

This root, which was established using eukaryotes as an external group, was only valid within the framework of the classic Woese tree where archaea are monophyletic and share a common ancestor with eukaryotes (so-called 3D tree) (Figure 2.5). However, from the 2010s, most microbiologists adopted another perspective on the history of life, where eukaryotes emerge in the midst of archaea (tree with two primary domains, called 2D). In this case, archaea should be rooted using bacteria and not eukaryotes as the external group (Figure 2.5). This change makes phylogenetic analyses more difficult because the branch leading from bacteria to archaea is much longer than the one from eukaryotes to archaea, which can cause a phenomenon known as long branch attraction. We will return to this point in Chapter 3.

In the years following Guy and Ettema's proposal, the MAGs of seven archaeal lineages previously described as *Crenarchaeota* based on their 16S rRNAs were considered representatives of new phyla within the TACK: the *Geoarchaeota*

11 This acronym implied considering *Aigarchaeota* as a phylum, lacking a crucial vowel.

(Kozubal et al. 2013); *Bathyarchaeota*[12] (Meng et al. 2014); *Verstraetearchaeota*[13] (Vanwonterghem et al. 2016); *Geothermoarchaeota* (Jungbluth et al. 2017); *Marsarchaeota*[14] (Jay et al. 2018); *Nezhaarchaeota*[15] (Wang et al. 2019); and most recently the *Culexarchaeota*[16] (Kohtz et al. 2022). Until now, no archaea belonging to these lineages have been successfully cultivated.

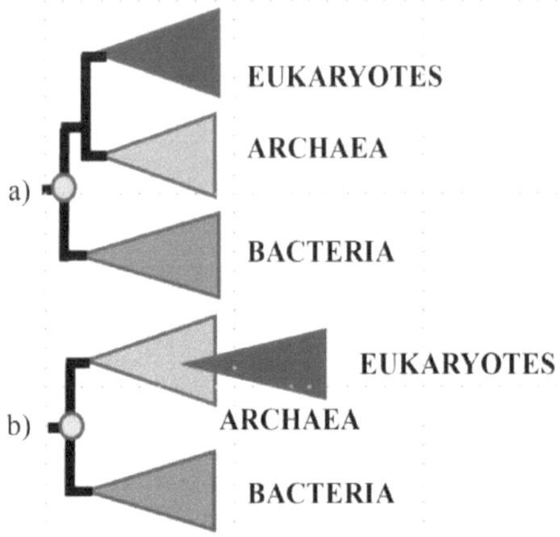

Figure 2.5. *The two possible rootings according to the topology of the universal tree. For a color version of this figure, see www.iste.co.uk/clouet/archaea1.zip*

COMMENT ON FIGURE 2.5.– *(a) Eukaryotes serve as an external group to root the archaeal tree because these two domains are sister groups (so-called 3D tree, or Woese tree, see Chapter 3). (b) Eukaryotes are a subgroup of archaea (so-called 2D tree), and bacteria must be used as an external group to root the archaeal tree.*

After the addition of these new phyla, the superphylum TACK can be divided into two large groups (Figure 2.6). The first group includes *Thaumarchaeota*, *Aigarchaeota*, and two new phyla, *Bathyarchaeota* and *Geothermoarchaeota*, the

12 From the Greek *bathys* meaning "deep", as the first MAGs of this lineage were obtained from deep environments and branch "deeply" at the base of *Thaumarchaeota* and *Aigarchaeota*.

13 In honor of Professor Willy Vestraerte, who played an important role in developing artificial ecosystems, particularly anaerobic digesters.

14 In reference to the red color of the planet Mars and their iron oxide-rich habitats.

15 Nezha is an immortal character in Chinese mythology.

16 Named after Culex Basin in Yellowstone National Park.

latter being very close to *Aigarchaeota*. These four phyla form a robust clade found in all the phylogenies published so far, both in 2D trees and in 3D trees, regardless of the proposed roots for the archaeal tree. Forterre and his collaborators have therefore proposed grouping these four phyla into a superphylum that would be called *BAT* (*Bathyarchaeota*, *Aigarchaeota*, *Thaumarchaeota*) (Gaïa et al. 2018). However, this terminology has not yet been validated by the community of researchers working on archaea, who remain attached to TACK.

Aigarchaeota and *Geothermoarchaeota* are primarily thermophiles found in hot and deep environments, while *Bathyarchaeota* and *Thaumarchaeota* include both mesophilic and thermophilic organisms. Phylogenetic analyses suggest that the ancestor of *Thaumarchaeota* was a thermophilic organism (Abby et al. 2020). Therefore, it is likely that the ancestor of the BAT superphylum was also a thermophilic organism. *Thaumarchaeota* have been extensively studied, and many species have been successfully cultivated from diverse marine and terrestrial environments. Most *Thaumarchaeota* species are capable of oxidizing ammonia, a metabolism they seem to have acquired while still thermophiles. *Thaumarchaeota*, which later adapted to cold or temperate environments, have retained this ability.

Bathyarchaeota, which are particularly diverse and abundant in anoxic environments, have been divided into numerous subgroups (Zhou et al. 2018). They correspond to a phylotype previously referred to as the Miscellaneous *Crenarchaeota* Group (MCG). Initially detected in the deep ocean, they have since been found in a wide variety of cold or hot anaerobic environments, typically marine or lake sediments, whether coastal or deep. Similar to *Aigarchaeota* and *Thaumarchaeota*, *Bathyarchaeota* proteomes contain many eukaryotic-like proteins that are absent in most archaea, except for *Asgardarchaeota* (section 2.3.4). Notably, *Bathyarchaeota* encode actins and tubulins that are very similar to those of eukaryotes (Da Cunha et al. 2022). Unexpectedly, the analysis of new TACK lineages revealed the presence of genes involved in methanogenesis in some *Bathyarchaeota* MAGs, particularly the *mcr* gene coding for methyl-coenzyme M reductase (Evans et al. 2015). The first two cultivated *Bathyarchaeota*, *Marisediminiarchaeum ligniniphilus* and *Bathyarchaeum tardum*, are strict anaerobic mesophiles (Khomyakova et al. 2022; Tian Tian et al. 2022).

The second group of TACK includes *Crenarchaeota* and four new lineages proposed as phyla: *Geoarchaeota*, *Verstraetearchaeota*, *Marsarchaeota*, *Nezhaarchaeota* and *Culexarchaeota* (Figure 2.6). The status of *Geoarchaeota* as a phylum is controversial, with Ettema and colleagues placing them within Crenarchaeota (Spang et al. 2014). *Marsarchaeota* are a sister group to *Crenarchaeota* and *Geoarchaeota* (Jay et al. 2018). These two archaeal lineages are the only ones lacking a family D DNA polymerase. As similar to thermophilic and acidophilic *Crenarchaeota*, *Geoarchaeota* and *Marsarchaeota* are found in hot and acidic terrestrial springs and can be either aerobic or anaerobic. *Verstraetearchaeota* are generally closely related to *Nezhaarchaeota* and *Culexarchaeota* in phylogenetic analyses (Kotz et al. 2022). *Culexarchaeota* have been detected in sediments from hot springs in Yellowstone[17].

Verstraetearchaea were detected by searching for new MAGs containing the *mcr* gene from DNA extracted from anaerobic digesters producing methane. They were subsequently found in anoxic environments where methane production is also observed (Vanwonterghem et al. 2016).

The discovery of methanogenesis in some Bathyarchaea and Verstraetearchaea demonstrated that the ability to produce methane is not confined to the Euryarchaea, as was long believed. Comparative genomic analysis of the metabolic pathways present in methanogenic Bathyarchaea and Verstraetearchaea suggests that these organisms utilize methylated compounds for methanogenesis (methylotrophy). The candidate class Methanomethylia has been proposed for methanogenic Verstraetearchaea. Nezhaarchaea are also likely methanotrophic. Their MAGs were similarly obtained using the *mcr* gene as a screening tool. The order Methanohydrogenales has been proposed for Nezhaarchaea whose MAGs have been reconstructed.

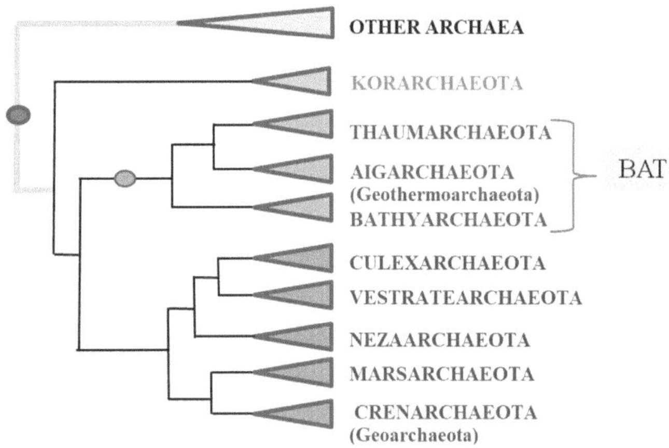

Figure 2.6. *Schematic phylogeny of the superphylum TACK. The phylum TACK forms a valid clade if the root of the archaea is outside this clade (blue circle). The phylum TACK is no longer valid if the root (pink circle) is within the BAT branch. For a color version of this figure, see www.iste.co.uk/clouet/archaea1.zip*

17 This lineage was named after Culex Basin in Yellowstone National Park, USA.

Environmental DNA sequencing has also enabled the acquisition of MAGs for numerous Korarchaea (McKay et al. 2019). These archaea consistently form a distinct, somewhat isolated group. According to the most recent phylogenies, Korarchaea are positioned at the base of the TACK superphylum (Liu and Li 2022). However, their position is challenging to pinpoint and varies among researchers, likely due to an unusually high rate of protein evolution. Consequently, indels (insertions/deletions) are found in their universal protein sequences in regions that are completely conserved in all other organisms, except those with rapid evolution (see Figure S38 in Da Cunha et al. 2017).

In universal phylogenies, Korarchaea often appear at the base of the archaea or eukaryotes, results likely influenced by the long-branch attraction phenomenon. Unexpectedly, the *mcr* gene has recently been identified in some of their MAGs, and the corresponding organisms, which may be new methanogens, have been named *Ca. Methanodesulfokores*.

2.3.3. *The DPANN*

The first lineages of miniature archaea resembling Nanoarchaea were identified in acidic terrestrial environments by Jill Banfield and her collaborators, who named them Arman[18] (Baker et al. 2010). These new archaea were later considered by Woyke and his collaborators to represent two new phyla, *Micrarchaeota* (ARMAN and 2) and *Parvarchaeota* (Arman 4 and 5) (Rinke et al. 2013). Simultaneously, another lineage of miniature archaea was discovered in hypersaline environments, the Nanohaloarchaea (Narasingarao et al. 2012).

Initially, Nanohaloarchaea were considered Euryarchaea, closely related to Haloarchaea based on 16S rRNA sequences (Narasingarao et al. 2012). They were later elevated to the status of a new phylum, *Nanohaloarchaeota*, as they grouped with *Nanoarchaea*, *Micrarchaeota* and *Parvarchaeota* in a phylogeny based on 38 conserved proteins in archaea and bacteria (Rinke et al. 2013). This clade also included two new phyla proposed by Woyke and his collaborators, *Aenigmarchaeota* and Diapherotrites (Rinke et al. 2013). These authors proposed grouping these new phyla into a superphylum called *DPANN* (*Diapherotrites*, *Parvarchaeota*, *Aenigmarchaeota*, *Nanoarchaeota*, *Nanohaloarchaeota*).

In 2015, Banfield and her collaborators reconstructed the MAGs of two new lineages of miniature archaea found in an aquifer, which branched within the

18 Arman is the acronym for Archaeal Richmond Mine Acidophilic Nanoorganisms.

DPANN in a phylogeny based on a subgroup of ribosomal proteins (Castelle et al. 2015). To honor the pioneers of archaeal research, these lineages were assigned to the phyla *Woesarchaeota* and *Pacearchaeota*. More recently, two new lineages of miniature archaea have been described from MAGs and included in DPANN as new phyla, *Huberarchaeota* (Probst et al. 2019) and *Undinarchaeota* (Dombrowski et al. 2020).

The hosts of these miniature archaea have been identified in three cases: a Nanohaloarchaea, which lives in salt-rich environments in symbiosis with the halophilic archaea *Halorubrum lacusprofundi* (Hamm et al. 2019); Micrarchaea, which live in acidic environments in symbiosis with thermo-acidophilic archaea of the order Thermoplasmatales (Golyshina et al. 2019; Krause et al. 2022); and a Huberarchaea, which lives in symbiosis with an archaea of the order Altiarchaeales (see section 2.3.3.2) (Schwank et al. 2019).

2.3.3.1. *The special case of Nanohaloarchaea*

In recent years, most published phylogenies have placed Nanohaloarchaea within the DPANN superphylum (Williams et al. 2017; Castelle et al. 2018, 2021; Schwank et al. 2019; Dombrowski et al. 2020; Aoud et al. 2022). This suggests that adaptation to high salinity conditions occurred independently in Nanohaloarchaea and Haloarchaea. Supporting this scenario, the adaptation of proteins to high salt concentrations differs between the two groups: Haloarchaea proteins are enriched in aspartic acid, whereas Nanohaloarchaea proteins are enriched in glutamic acid. This enrichment in acidic amino acids ensures the solubility of their proteins at salt concentrations near the cytoplasmic saturation of these organisms.

However, some researchers have proposed alternative positions for Nanohaloarchaea. By analyzing subgroups of universal proteins and/or excluding DPANN from their datasets, Peter Gogarten and his collaborators produced phylogenies in which Nanohaloarchaea appeared as a sister group to Haloarchaea (Feng et al. 2019; Rangel et al. 2020). Using methods designed to limit the long-branch[19] attraction phenomenon, Céline Brochier-Armanet and her colleagues

19 One method involves progressively enriching sequence alignments with slowly evolving positions, presumed to better reflect ancient divergences. Another method involves replacing letters corresponding to the 21 amino acids with a smaller number of letters (usually four or six) corresponding to groups of amino acids with equivalent physicochemical properties (referred to as recoding). This aims to eliminate rapidly evolving positions, as the replacement of one amino acid by another is generally much more frequent if the two amino acids have the same physicochemical properties.

obtained a phylogeny placing Nanohaloarchaea as a sister group to *Methanocellales*, within group II Euryarchaea (Aouad et al. 2018).

Two strong independent arguments, however, support the grouping of the Nanohaloarchaea with the DPANN superphylum. First, the Nanohaloarchaea and a subgroup of DPANNs share a monomeric DNA primase[20] whose sequences differ significantly from the dimeric primases present in other archaea (Raymann et al. 2014). This primase results from a fusion of the two subunits PriS and PriL, which are common to the primases of archaea and eukaryotes. Some archaeal plasmids also encode such monomeric primases (Gill et al. 2014). Therefore, it is likely that the ancestral primase of archaea was replaced by the primase of a mobile element at the base of the DPANN group, including the Nanohaloarchaea. It is difficult to imagine that the Nanohaloarchaea could have acquired this primase through gene transfer from a DPANN, as they are the only miniature archaea living in a saline environment, and their halophilic hosts possess the classic primase of archaea.

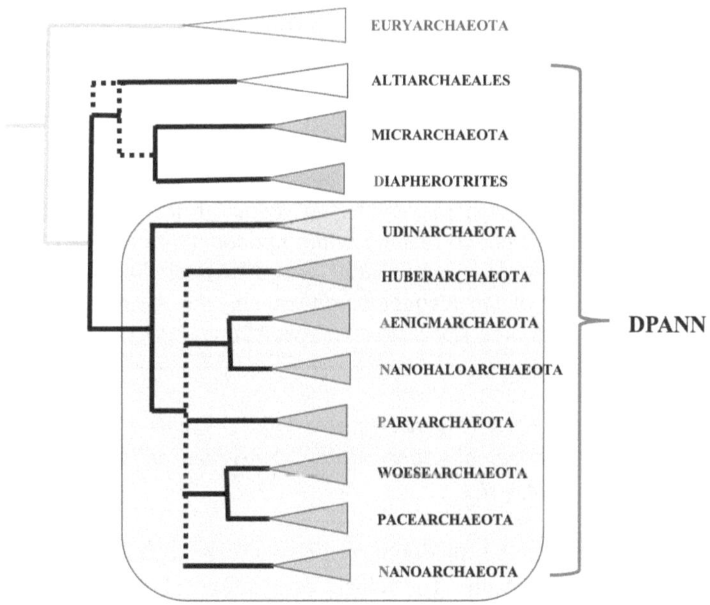

Figure 2.7. *Consensus schematic phylogeny of the DPANN Superphylum. The dotted lines indicate kinship relationships that are still not clearly resolved. Diapherotrites are also known as Iainarchaeota. For a color version of this figure, see www.iste.co.uk/clouet/archaea1.zip*

20 Primase is the enzyme that synthesizes RNA primers for Okazaki fragments during DNA replication.

The second argument concerns the presence of specific insertions in the sequences of five universal proteins, common to both the Nanohaloarchaea and the DPANN subgroup possessing the aforementioned monomeric primase. Three examples are shown in Figure 2.8. These regions are highly conserved among other archaea, bacteria and eukaryotes. One of these insertions specifically groups the Nanohaloarchaea with the *Aenigmarchaeota*, as observed in several published phylogenies.

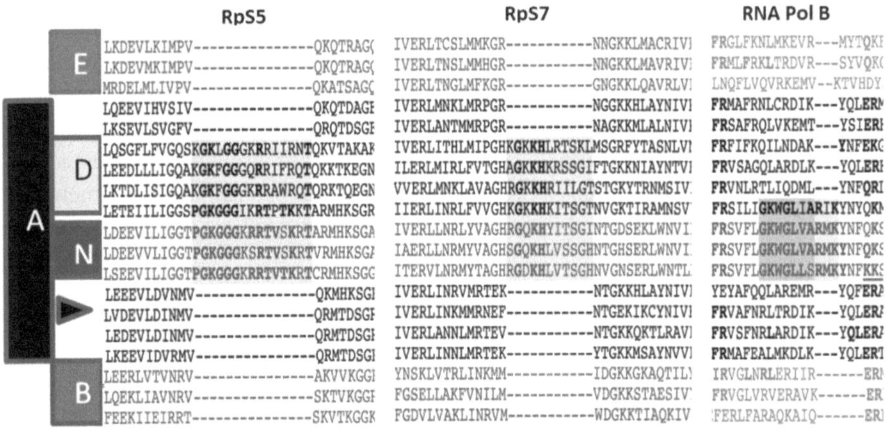

Figure 2.8. *Insertions common (highlighted in yellow) to the Nanohaloarchaea (N) and DPANN (D) observed in two regions of the ribosomal proteins RpS5 and Rps7, and a deletion followed by an insertion (gray and yellow) common to the B subunit of the RNA polymerases*[21] *of the Nanohaloarchaea and the Aenigmarchaeota. For a color version of this figure, see www.iste.co.uk/clouet/archaea1.zip*

COMMENT ON FIGURE 2.8.– *These insertions are not present in the Methanocellales (black triangle). These regions are highly conserved in other archaea (A), bacteria (B) and eukaryotes (E).*

These observations leave little doubt regarding the inclusion of Nanohaloarchaea in the DPANN superphylum. The close association of Nanohaloarchaea and Haloarchaea in some phylogenies could be due to gene transfers between

21 Author's observation. From top to bottom, the sequences correspond to a drosophila, an *Arabidopsis*, a trypanosome, a Bathyarchaea, a *Thermofilum*, four DPANN (a Parvarchaea, a Woesearchaea, a Pacearchaea and an Aenigmarchaea), three Nanohaloarchaea (a Nanohaloarchaeon, a Nanosalina and a Nanosalinarium), *Methanopyrus kandleri*, a Methanocella (black triangle), a Methanosarcina, *Acidoprofundum boonei*, *Bacillus subtilis*, *Escherichia coli* and *Thermotoga maritima*.

Nanohaloarchaea and their hosts (Haloarchaea), which may bias phylogenetic analyses. Such transfers are easily conceivable as the proteins of Nanohaloarchaea can function effectively in the highly saline cytoplasm of Haloarchaea and vice versa. Additionally, it can be noted that the close association between Nanohaloarchaea and Haloarchaea is often found in phylogenetic analyses where authors do not include other DPANN members, aiming to avoid the long-branch attraction phenomenon, which could artificially group Nanohaloarchaea within the DPANN. However, the absence of other DPANN members could eliminate many signals necessary for the correct placement of Nanohaloarchaea, thereby increasing the influence of signals related to gene transfers. Using methods intended to limit long-branch attraction could have a similar effect, as these methods also reduce the importance of the phylogenetic signal by removing rapidly evolving positions (Hernandez and Ryan 2021; Rangel and Fournier 2023).

The example of the Nanohaloarchaea clearly illustrates the difficulties in interpreting phylogenetic analyses, as reducing the overall signal (number of positions used) may be necessary in some cases to correctly position the species studied, while in other cases, it may distort the results. The quality of the signal (rather than its quantity) is therefore the determining factor in this type of analysis.

2.3.3.2. *DPANN phylogeny*

As seen in the cases of *M. kandleri*, Korarchaea and *N. equitans*, DPANNs are characterized by long branches in evolutionary trees, indicating an abnormally high rate of evolution. This accelerated rate of evolution is typical of parasitic organisms whose genomes evolve rapidly through reduction and whose protein sequences accumulate mutations at a higher rate compared to nonparasitic archaea. Consequently, the position of DPANNs is subject to the long-branch attraction phenomenon. In universal trees, DPANNs (archaea), like CPR (bacteria), appear grouped at the base of their respective domains (Castelle and Banfield 2018), which, in my opinion, confirms their susceptibility to the long-branch attraction phenomenon.

Recent phylogenetic analyses divide DPANN into two subgroups: the first includes *Micrarchaeota*, Diapherotrites and Altiarchaea, while the second groups all other DPANNs with *Undinarchaeota* at the base (Dombrowski et al. 2020; Castelle et al. 2021) (Figure 2.7). All DPANNs in this second group possess the monomeric primase and the specific insertions in certain universal proteins previously mentioned. These characteristics can therefore be considered synapomorphies[22] that support the grouping of these DPANNs into a single clade. Within this group, *Woesearchaeota* are the sister group of *Pacearchaeota*, and *Nanohaloarchaea* are the sister group of Aenigmarchaea (Figure 2.7).

The kinship relations between the group containing the monomeric primase and the *Micrarchaeota*, Diapherotrites and *Altiarchaeota* remain poorly defined. In fact, there is no clear marker proving the monophyly of DPANNs as a whole. If the position of these lineages in certain phylogenetic trees is due to long-branch attraction, they could form a paraphyletic group at the base of the Euryarchaea. This would explain early observations suggesting the grouping of Nanoarchaea with Thermococcales (Brochier et al. 2005b). Ultimately, the question remains whether DPANNs themselves should be considered a superphylum or if they owe this status to the accelerated evolution of their rRNAs and proteins.

2.3.3.3. *The special case of Altiarchaeota*

Altiarchaeota are a group of mesophilic anaerobic archaea discovered by Robert Huber from Karl Stetter's team in biofilms found at the bottom of marshes near their laboratory (Rudolph et al. 2001). These archaea could not be cultivated because they can only live in association with bacteria within particularly complex biofilms. They have harpoon-shaped appendages, called Hami, which allow them to form cross-linked networks within these biofilms (Mossi et al. 2005). Once their genome was sequenced, *Altiarchaeota* were initially defined as an order, Altiarchaeales, which branched at the base of the Euryarchaea (Probst et al. 2014; Adam et al. 2016). They were subsequently elevated to the status of phylum, *Altiarchaeota* and sometimes grouped with DPANN in the most recent phylogenies (Aouad et al. 2022).

2.3.4. *The Asgard archaea*

Several new lineages of archaea situated between the Euryarchaea and the TACK, first described in 2015 by Thijs Ettema and his collaborators, have garnered particular attention due to their possible kinship with eukaryotes (see Chapter 3) (Spang et al. 2015). The first three MAGs, obtained from DNA from sediments near an underwater hot spring called "Loki's Castle", were attributed to a new phylum

22 A synapomorphy corresponds to a derived character shared between two groups of organisms to the exclusion of all others, supporting the hypothesis that these two groups derive from a specific common ancestor and form a clade.

named *Lokiarchaeota*, after the fire god Loki from ancient Germanic mythology. A new lineage related to *Lokiarchaeota* was later detected in various marine sediments and attributed to a phylum named *Thorarchaeota*, after the Germanic god of thunder, Thor (Seitz et al. 2016). In 2017, Ettema and his collaborators described two new lineages that they named *Odinarchaeota* (Odin being the main god of the Norse pantheon) and *Heimdallarchaeota* (Heimdall being the guardian of the rainbow that separates the domain of the gods from the lower worlds) (Zaremba-Niedzwiedzka et al. 2017). Two of the three *Lokiarchaeota* lineages defined in 2015, Loki 2 and Loki 3, were then assigned to *Heimdallarchaeota*. The *Odinarchaeota* were found in terrestrial hot springs, while the other Asgard lineages had been detected in a wide variety of terrestrial, marine and lacustrine environments. These new Asgard lineages formed a monophyletic group in the phylogenies published by Ettema and his collaborators. They proposed grouping them into a superphylum named *Asgard*, the abode of the gods, equivalent to the Olympus of Greek gods. All Asgard archaea possess a highly conserved actin (Asgardactin) that is much closer to eukaryotic actins than to other archaeal actins (Crenactins), which can be considered a synapomorphy confirming the monophyly of Asgard archaea[23].

In 2019, Ken Takai and his collaborators succeeded in cultivating a Lokiarchaeon from an enrichment culture initiated 12 years earlier (Imachi et al. 2020). This Asgard archaeon, named *Candidatus Prometheoarchaeum syntrophicum*, could not be obtained in pure culture but was maintained in coculture with a sulfate-reducing bacterium and a methanogenic archaeon. In parallel, the authors conducted a comparative analysis of the genomes of Asgard archaea and observed the systematic absence of certain essential functions. According to them, it is possible that other Asgard archaea also rely on symbiotic associations for their development, which would explain the difficulty in cultivating them. The presence of numerous genes of bacterial origin in their genomes, coding for metabolic proteins, could be explained by such symbiotic relationships between Asgard archaea and bacteria (Filée et al. 2023).

In 2022, Schleper and his collaborators obtained a highly enriched culture of a second Lokiarchaeon, *Candidatus Lokiarchaeum ossiferum* (Rodrigues-Oliveira et al. 2022). Sulfate-reducing bacteria and a methanogenic archaeon were still present

---

23 In addition to this highly conserved actin, most *Asgardarchaeota* and some *Bathyarchaeota* possess more divergent actins. These Asgardactins and Bathyactins branch in phylogenetic trees between cytoplasmic actins and the various groups of actin-related proteins (ARPs) present in all eukaryotes, suggesting ancient transfers from proto-eukaryotes to certain archaea (Da Cunha et al. 2022).

in this culture. The two Lokiarchaea studied are characterized by numerous nanotube-like protuberances on their surface[24]. Schleper and his team demonstrated the presence of actin in these nanotubes and intracellular filamentous structures, suggesting the existence of both an endoskeleton and an exoskeleton in these archaea, related to the cytoskeleton of eukaryotic cells (Rodrigues-Oliveira et al. 2022).

We mentioned the presence in BAT genomes of several genes coding for eukaryotic-type proteins that had not been detected until then in other archaea. The genomes of the Asgard archaea were found to be even richer in these types of proteins, which were named ESPs (eukaryotic signature proteins) by Thijs Ettema and his collaborators. According to them, the presence of a very large number of ESPs among the Asgard archaea is explained by their evolutionary proximity to eukaryotes, who would be their direct descendants (see Chapter 3). This hypothesis has led to the search for new lineages belonging to this group. Consequently, the number of Asgard archaea phyla increased from 4 to 16 between 2019 and 2021 (Cai et al. 2020; Farag et al. 2021; Liu et al. 2021; Xi et al. 2021; reviewed in Da Cunha et al. 2022) (Figure 2.9). The new phyla continued to receive the names of Germanic gods, except for one, the phylum *Wukongarchaeota*, which was named after a character in Chinese mythology (see Table 2.1 for the origin of these names). This work led to the splitting of the previously defined *Heimdallarchaeota* into four phyla, with only one retaining the name *Heimdallarchaeota*, while the other three became new phyla: *Gerdarchaeota*, *Hodarchaeota* (corresponding to the former Loki 3) and *Kariarchaeota* (corresponding to the former Loki 2).

The phylogenies published by the various teams involved in this research, summarized in the phylogeny of Da Cunha et al. (2022), seem to divide the Asgard archaea into two main groups (Figure 2.9). The first group includes a clade formed by the *Lokiarchaea*, *Thorarchaea* and *Odinarchaea*, along with the new phyla *Sygnarchaeota*, *Helarchaeota*, *Baldarchaeota* and *Hermodarchaeota*. This group also includes a clade grouping *Borrarchaeota*, *Sifarchaeota* and *Freyrarchaeota*, and a third clade grouping *Jordarchaeaota* and a lineage (ALCG) that had not yet received a name in 2022. The second group comprises the four phyla corresponding to the old Heimdallarchaea plus two new phyla, *Njordarchaeota* and *Wukongarchaeota*.

The root of the *Asgardarchaeota* could lie between these two groups. However, in a recent publication, Thijs Ettema and his collaborators rooted the

24 It should be noted that some archaea, like Thermococcales, which do not possess actin, also form nanotubes (Marguet et al. 2013).

Asgardarchaeota within the first group, in the lineage leading to the *Jordarchaeota* (Eme et al. 2023). Their dataset, however, does not include the two subunits of RNA polymerase, which are considered the best phylogenetic markers (Martinez-Gutierrez and Aylward 2021). Additionally, archaea, particularly *Asgardarchaeota*, are significantly over-represented in the species dataset compared to the outgroup, which also poses a problem (Rangel and Fournier 2023). Eukaryotes have very short branches in most individual protein trees used for concatenation, and their position is variable and often unresolved (see the Figshare reference in Eme et al. (2003)).

LOKI	God of fire and trickery	(Spang et al. 2015)
SIGYN	God of victory	(Xi et al. 2021)
HEL	Sister of Loki, goddess of the underworld	(Seitz et al. 2017)
THOR	God of thunder	(Seitz et al. 2017)
BALD	God of light	(Liu et al. 2021)
HERMOD	Son of Odin	(Liu et al. 2021)
ODIN	Supreme god (Wotan)	(Zaremba-Niedzwiedzka et al. 2017)
JORD	Goddess of the Earth	(Sun et al. 2021)
BOR	Father of Odin	(Liu et al. 2021)
SIF	Goddess of the Earth	(Farag et al. 2021)
FREYR	God of peace	(Xie et al. 2021)
NJORD	God of the sea	(Xie et al. 2021)
WUKONG	Monkey king in Chinese literature	(Liu et al. 2021)
HOD	Hod, the blind warrior	(Sun et al. 2021)
GERD	Giantess goddess, wife of Freyr	(Cai et al. 2020)
HEIMDALL	Guardian of the rainbow bridge separating the underworld from the divine world	(Zaremba-Niedzwiedzka et al. 2017)
KARI	God of the wind	(Liu et al. 2021)

Table 2.1. *Meaning of the prefixes associated with Asgard archaeal phyla and publications proposing these names*

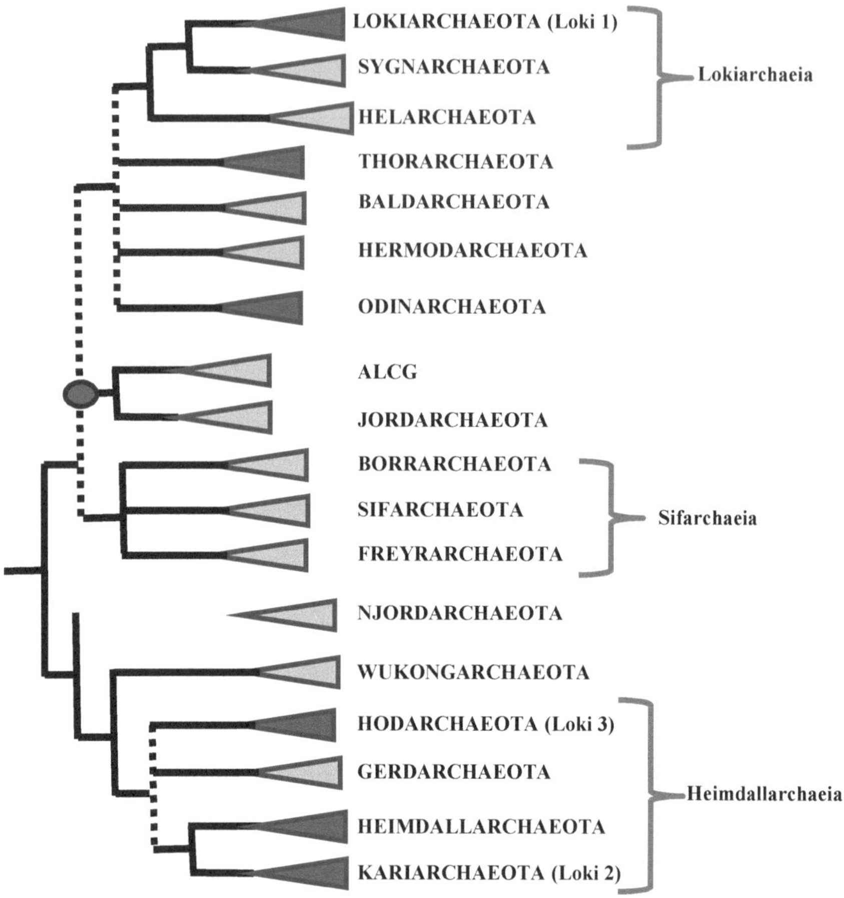

Figure 2.9. *Schematic phylogeny of the "superphylum" Asgard (reviewed in Da Cunha et al. (2022)). Dashed lines indicate unresolved kinship relationships. The positions of Loki 1, 2 and 3 from Spang et al. (2015) are shown. The root proposed by Ettema and his colleagues (Eme et al. 2023) is indicated by a red circle. For a color version of this figure, see www.iste.co.uk/clouet/archaea1.zip*

The rooting of the *Asgardarchaeota* remains an open and significant question, as it impacts the reconstruction of the last common ancestor of the *Asgardarchaeota* (LAsCA). Based on their root, Thijs Ettema and his collaborators suggest that this ancestor was hyperthermophilic because it possessed reverse gyrase (Eme et al. 2023), an enzyme characteristic of organisms living at high temperatures (Catchpole and Forterre 2019) (see section 2.5.2). However, contrary to this suggestion, their phylogenetic analysis shows that reverse gyrase was introduced several times independently among the *Asgardarchaeota* from other groups of archaea (Eme et al. 2023, Figure 32 in the supplementary information).

It is expected that new lineages of *Asgardarchaeota* will be described in the coming years. Fortunately, mythological resources seem inexhaustible. Conversely, in contrast to the division of *Asgardarchaeota* into multiple phyla, we will see later in the discussion on the definition of phyla that Phil Hugenholtz and his collaborators propose considering all *Asgardarchaeota* as a single phylum (Rinke et al. 2021). In their new taxonomy, the previously proposed phyla are relegated to the rank of order or class (Figure 2.9 and Table 2.2).

2.3.5. *Stygia/Hadarchaeota*

Other lineages described through metagenomics seem to occupy an intermediate position between the Euryarchaea, the Asgard and the TACK superphyla. These include the Hadesarchaea (Baker et al. 2016), proposed as a phylum under the name *Hadarchaeota* (Chuvochina et al. 2019), and the Persephonarchae, initially named MSBL-1 and later renamed after Persephone, the wife of Hades in Greek mythology (Mwirichia et al. 2016; Adams et al. 2017). These two lineages form a monophyletic clade. Hadesarchaea are present in many anoxic terrestrial or marine environments, both thermophilic and mesophilic. One of the MAGs from Hadesarchaea contains the *mcr* gene and could therefore correspond to methanogenic or methanotrophic organisms (Hua et al. 2019). Persephonarchaea are mainly found in mesophilic, anoxic, salty environments and may correspond to halophilic organisms.

In Greek mythology, the god Hades and his wife Persephone rule over the underworld traversed by the river Styx. The Hadesarchaea and Persephonarchaea form a robust clade that Simonetta Gribaldo and her collaborators have proposed to name *Stygia*. The taxonomic rank corresponding to Stygia has not been specified. It could correspond to *Hadarchaeota* if Persephonarchaea are included in this group.

2.3.6. *Hydrothermarchaeota*

Hydrothermarchaeota were detected from hydrothermal fluids in the hot deep biosphere (Jungbluth et al. 2017) at the same time as the *Geothermarchaeota*. Their position in phylogenies is imprecise and they are sometimes considered to be a sister group to the Stygia (Rinke et al. 2021), sometimes placed at the base of the Euryarchaea (Aouad et al. 2022), or as a sister group to group I Euryarchaea (Dombrowski et al. 2021) (Figure 2.10). In fact, the relative positions of Asgard archaea, Stygia, *Hydrothermarchaeota*, DPANN and group I Euryarchaea remain imprecise and vary from one phylogeny to another. This highlights the limitations of current analyses and the need to test other approaches, such as searching for significant insertions, to resolve these deep nodes in the archaeal tree.

2.4. Euryarchaea

As their name suggests, the Euryarchaea exhibit considerable phenotypic diversity. They are the oldest archaea to have been described and encompass the largest number of cultivated species. As previously discussed, they have been divided into two subgroups by Patrick Forterre and his colleagues: group I Euryarchaea, which includes Thermococcales and class 1 methanogens, and group II Euryarchaea, encompassing Archaeoglobales, Thermoplasmatales, Haloarchaea and class 2 methanogens (Forterre et al. 2015). I will retain this nomenclature here, although it has not been adopted by the scientific community due to ongoing debates over the root of the archaeal tree. Gribaldo, Brochier-Armanet and their collaborators proposed further division of each group into two classes: *Acherontia* and *Methanomada* for group I Euryarchaea, and *Diaphorachaea* and *Methanotecta* for group II Euryarchaea (Figure 2.9) (Brochier-Armanet et al. 2015; Adam et al. 2017). The latter were subsequently referred to by these authors as Gaiarchaea (Aouad et al. 2022) (see section 2.4.1). While the monophyly of group II Euryarchaea is consistently supported across published phylogenies, the relationships of group I Euryarchaea with Hydrothermarchaea, Stygia and DPANN remain poorly defined (Figure 2.10).

The diversity of Euryarchaea, notable since their initial description, has expanded significantly over the past decade with the identification of new lineages from environmental DNA. From early molecular ecology studies, new Euryarchaea, distinct from those previously known, were identified in the deep ocean, such as *Marine Group II Euryarchaea* (DeLong 1992)[25]. Subsequent analyses of environmental 16S rRNA revealed additional marine lineages (*Marine Group III, IV*), as well as novel lineages inhabiting terrestrial environments, often in anoxic conditions.

Over the last decade, numerous MAGs and/or SAGs corresponding to these lineages have been reconstructed. Metabolic analyses have led to the identification of new methanogenic Euryarchaea and the discovery of methanotrophic Euryarchaea, such as *Anaerobic Methanotrophs* (ANME), some of which engage in symbiotic relationships with sulfate-reducing bacteria within biofilms. These

25 The Marine Group I described in 1992 by Ed Delong corresponded to *Thaumarchaeota*.

Euryarchaea utilize a pathway of "reverse methanogenesis", reversing the typical methane production pathway. Concurrently, several new species of group II Euryarchaea have been cultured, including Methanomassilicoccales, Methanonatronoarchales, Thermophilic acidophiles and Aciduliprofundales.

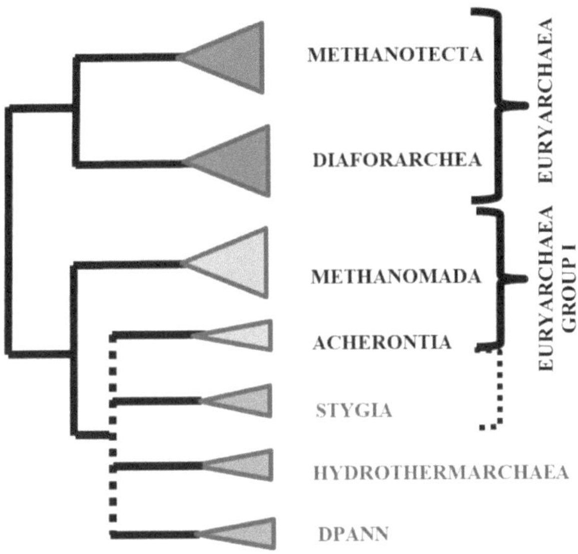

Figure 2.10. *Schematic phylogeny of the Euryarchaea (in black) highlighting their debated kinship relationships (dashed lines with adjacent groups (in red)). For a color version of this figure, see www.iste.co.uk/clouet/archaea1.zip*

2.4.1. *Group I Euryarchaea*

The cultivated species of group I Euryarchaea include Thermococcales and class 1 methanogens (*Methanococcales*, *Methanobacteriales* and *Methanopyrales*). The Thermococcales (genera *Thermococcus*, *Pyrococcus Paleococcus*), all hyperthermophiles, form a robust clade with three unique marine archaeal lineages characterized solely by MAGs, which have been tentatively classified as possible classes: Theinoarchaea (Lazar et al. 2017); *Methanofastidiosa* (Nobu et al. 2016); and *Nuwarchaeia*[26] (Wang et al. 2021) (Figure 2.11). The Thermococcales (Thermococci class) emerge at the base of this clade. Simonetta Gribaldo and her collaborators proposed naming this clade *Acherontia*, after the river Acheron that

26 Nuwa is the mother of Earth in Chinese mythology.

bathes the entrance to Hades in Greek mythology (Adam et al. 2017). Based on analysis of their MAGs, *Methanofastidiosa* and *Nuwarchaeales* are methanogens capable of reducing methylated groups. It has been suggested to classify them as class 3 methanogens (Wang et al. 2021).

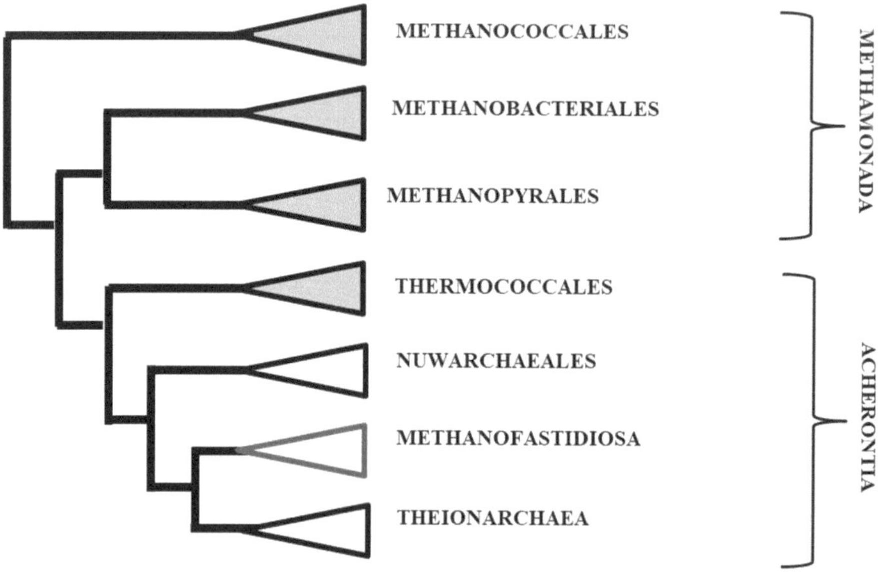

Figure 2.11. *Schematic phylogeny of group I Euryarchaea. Groups comprising cultivated species are depicted in beige, while those uniquely defined by MAGs are in white. For a color version of this figure, see www.iste.co.uk/clouet/archaea1.zip*

Céline Brochier and her collaborators proposed grouping *Methanococcales*, *Methanobacteriales* and *Methanopyrales* into the class *Methanomada* (Petitjean et al. 2015) (where *omada* means group in Greek).

However, not all authors are convinced of the monophyly of this grouping, and some phylogenies place *Methanococcales* at the base of group II Euryarchaea (Dombrowski et al. 2021). The relative positioning of these three groups varies among different phylogenies.

In some phylogenies, *Methanopyrales* form a clade with *Methanobacteriales* (Adam et al. 2017; Dombrowski et al. 2021). This association is highly likely because, as previously noted, these are the only two archaeal lineages with a cell wall composed of pseudomurein.

2.4.2. *Group II Euryarchaea*

Group II Euryarchaea, which includes hyperthermophilic Archaeoglobales, universally possess a bacterial-derived DNA gyrase that was already present in the last common ancestor of this group and subsequently evolved vertically within the group (Raymann et al. 2015; Villain et al. 2022). The DNA gyrase introduces negative supercoiling into the DNA, a change in chromosome topology that likely had significant consequences on the physiology and molecular biology of these archaea. This acquisition may have contributed to the evolutionary success of group II Euryarchaea, enabling them to thrive in mesophilic environments rich in bacteria (Forterre and Gadelle 2009). Subsequently, the gyrase was transferred from group II Euryarchaea to a few DPANN and Asgard archae, but almost never to other archaea (Villain et al. 2022). This bias could be explained by symbiotic relationships between group II Euryarchaea, DPANN and Asgard archaea. As discussed in section 2.3, such relationships have been observed between Micrarchaea and Thermoplasmatales, Nanohaloarchaea and Haloarchaea, and Lokiarchaea and methanogenic Euryarchaea.

2.4.2.1. *Diaforarchaea*

Diaforarchaea (from the Greek *diaphora*, meaning "diverse") encompass thermo-acidophiles (Thermoplasmatales, Acidoprofondales), mesophilic methanogens (Methanomassilicoccales) and MAGs corresponding to DNA extracted from various environments (Figure 2.12). Thermoplasmatales have diversified significantly, currently including the genera *Acidiplasma*, *Thermoplasma*, *Picrophilus*, *Cuniculiplasma* and *Ferroplasma*. Many species in this group have been successfully cultivated. All of these organisms are characterized by the absence of an S-layer, making them highly fragile in laboratory settings. Aciduliprofondales are also thermo-acidophilic and have primarily been identified in underwater hydrothermal springs, representing a phylotype of 16S rRNA previously termed DHVE2 for *deep-sea hydrothermal vent Euryarchaeota 2*. The only cultivated organism in this group remains *Aciduliprofundum boonei* (Reysenbach et al. 2006).

For a long time, methanogens were classified into six genera corresponding to classes 1 and 2. Methanomassilicoccales (Gorlas et al. 2012), from which several members have been isolated, cultured and characterized, were initially described as the seventh order of methanogens (Borrel et al. 2013). Methanomassilicoccales are distinguished by their double membrane. They inhabit various terrestrial marshy environments as well as in the digestive tracts of animals. Some species coexist in our intestines alongside *Methanobacteriales* (Gorlas et al. 2012; Borrel et al. 2013).

Methanomassilicoccales have the capacity to utilize amine compounds such as mono-, di- or trimethylamine to produce methane and may play a crucial role in detoxifying these compounds, some of which are carcinogenic to humans (Brugère et al. 2014). Thermoplasmatales, Aciduliprofondales and Methanomassilicoccales form a clade proposed by Gribaldo and colleagues to be classified as a class named Thermoplasmata (Adams et al. 2017).

The Diaforarchaea encompass multiple uncultivated marine archaeal lineages. The two most abundant lineages are Pontarchaeales and Posidoniales, which form a clade in recently published phylogenies (Rinke et al. 2019; Dombrowski et al. 2021) (Figure 2.12). Pontarchaeales correspond to the previously named *Marine Group III*. They dominate in the deep ocean but can also be found near the surface. Posidoniales (*Ca. Poseidonaceae* and *Ca. Thalassarchaeaceae*) correspond to the previously named *Marine Group II a and b* (Rinke et al. 2019). They exhibit high diversity and are particularly abundant in the illuminated regions of marine environments. Surface-dwelling Posidoniales and Pontarchaeales possess genes encoding proteorhodopsin[27], also found in planktonic bacteria and Haloarchaea, where it facilitates non-chlorophyll photosynthesis, suggesting these organisms are photoheterotrophs (Rinke et al. 2019; Tully et al. 2019). Interestingly, these two archaeal groups lack the key gene involved in archaeal-type lipid synthesis, glycerol-1 phosphate dehydrogenase, but possess the bacterial phospholipid-type gene, glycerol-3 phosphate dehydrogenase, indicating they may have atypical lipids, combining archaeal and bacterial characteristics (Villanueva et al. 2017; Rinke et al. 2019).

A highly abundant lineage of Diaforarchaea in marine sediments, corresponding to the MBG-D phylotype (*Marine Benthic Group D*), has been described either as a class, *Izemarchaea* (Adams et al. 2017), or as an order, *Thermoprofundales* (Zhou et al. 2019). According to different studies, Izemarchaea branch either at the base of Thermoplasmata, Diapherotrites (Adams et al. 2017) or as a sister group to Pontarchaeales (Zhou et al. 2018). Izemarchaea (Thermoprofundales) may engage in symbiotic relationships with Lokiarchaea and Hadesarchaea (Zhou et al. 2019). Other Diaforarchaea lineages closely related to Methanomassilicoccales have been described as orders, *Lunaplasmatales* (Zinke et al. 2019) and *Proteinoplasmatales* (Yin et al. 2022).

Further characterization of other Diaforarchaea lineages is ongoing, with some potentially representing new groups of methanogens (Wang et al. 2021).

27 Proteorhodopsin is a proton pump that uses retinal as a chromophore to capture photons. Its presence indicates life in a highly sunlit environment.

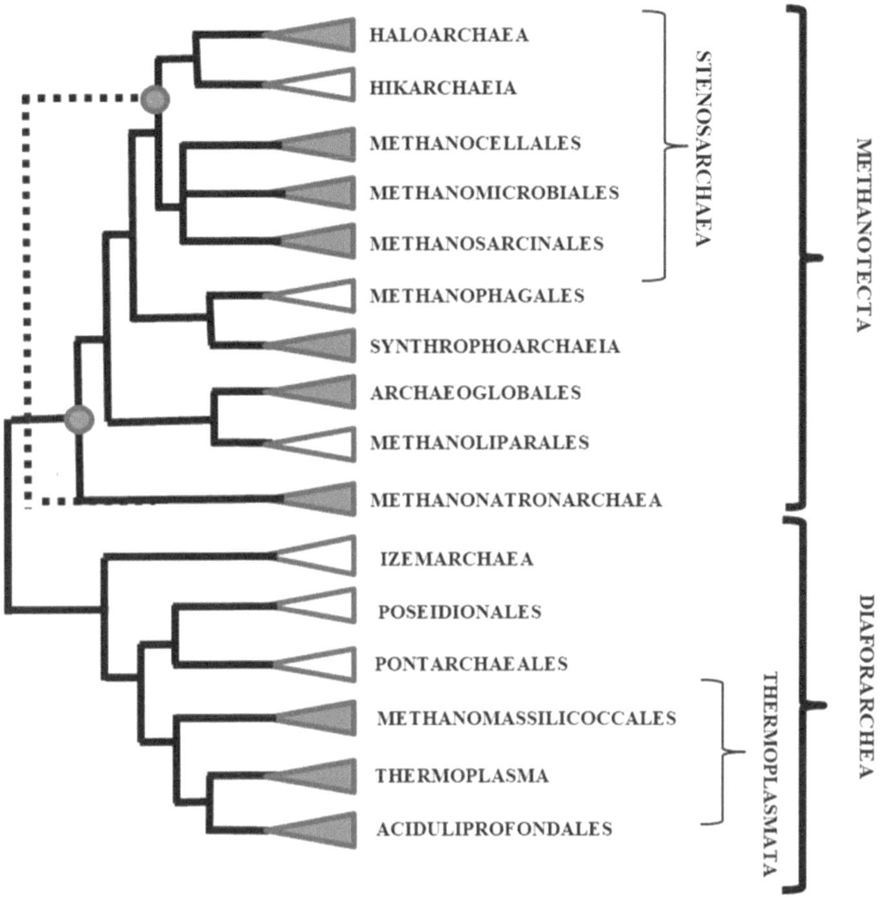

Figure 2.12. *Schematic phylogeny of Group II Euryarchaea. Groups containing cultivated species are depicted in brown, while those defined solely by MAGs are in white. Brown circles indicate the two proposed positions for the Methanonatroarchaea. The dotted line denotes the less probable position based on current published data. For a color version of this figure, see www.iste.co.uk/clouet/archaea1.zip*

2.4.2.2. *Methanotecta*

The name Methanotecta has been proposed to describe a large clade that includes Haloarchaea, Archaeoglobales, class 2 methanogens (*Methanosarcinales*, *Methanomicrobiales* and *Methanocellales*) and a group of recently cultivated halophilic methanogens, *Methanonatronarchaeales* (Sorokin et al. 2018). The

Methanotecta also encompass numerous methanogenic or methanotrophic archaea identified in enrichment cultures or defined by MAGs. Recent phylogenetic analyses suggest that the root of Methanotecta separates Methanonatronarchaeales from the other group members (see section 2.4.2.3). The Methanotecta most likely descended from a methanogenic ancestor, which is reflected in their name (from the Greek *tiktomai*, to have been born).

All published phylogenies identify a clade formed by Haloarchaea and class 2 methanogens, which Céline Brochier and her collaborators proposed to name *Stenosarchaea*, from the Greek *stenos*, meaning to associate/close (Aouad et al. 2018). The relative positions of *Methanomicrobiales*, *Methanosarcinales* and *Methanocellales* within the Stenosarchaea vary according to recent phylogenies. The Stenosarchaea include the lineage corresponding to the ANME-2 and ANME-3 lineages that branch within the *Methanosarcinales*.

Among the lineages that branch outside the Stenosarchaea are the *Methanophagales*, corresponding to ANME-1, which form a clade with another methanotrophic archaeal lineage, the *Syntrophoarchaeales* (Figure 2.12). Syntrophoarchaeales can oxidize various alkanes (propane, butane, etc.) and include several species obtained in enrichment cultures (Laso-Pérez et al. 2016). These archaea live in symbiosis with sulfate-reducing bacteria. Two other archaeal lineages capable of degrading alkanes, the *Alkanophagales* and the *Santabarbaracales*, appear to be closely related to the Syntrophoarchaeales. Xiao and his colleagues grouped them with the Syntrophoarchaeales in the class Syntropharchaeia (Wang et al. 2021).

In some phylogenies, Archaeoglobales form a clade with *Methanoliparales* (Dombrowski et al. 2021). Methanoliparales are capable of both methanogenesis and alkane oxidation (Meyerdierks et al. 2010). Notably, uncultivated methanotrophic Archaeoglobales lineages have recently been identified (Boyd et al. 2019). As previously discussed, *Archaeoglobales* (genera *Archaeoglobus*, *Ferroglobus* and *Geoglobus*) were for a long time positioned alone at the base of group II Euryarchaea and later at the base of Methanotecta. This basal position is now claimed by the Methanonatronarchaeales. However, the exact position of the Methanonatronarchaeales has sparked a controversy that we will discuss below.

2.4.2.3. *The origin of Haloarchaea*

The position of Haloarchaea within the Stenosarchaea suggests that these halophilic and aerobic archaea emerged from methanogens, which are non-halophilic and anaerobic. This dramatic phenotypic change may have accelerated the evolution of proteins and rRNAs in Haloarchaea, explaining their typically long branches in phylogenies. The emergence of Haloarchaea from methanogenic archaea likely involved intermediate stages. To identify these intermediates, Thijs Ettema and his collaborators reconstructed the MAGs of several organisms from a lineage previously known as *Marine Group IV archaea*, which branched close to Haloarchaea in 16S rRNA phylogenies. This analysis led to the identification of a new lineage of marine aerobic archaea named *Hikarchaeia*[28] (Martijn et al. 2020). Phylogenetic analyses confirmed the close relationship between Hikarchaeia and Haloarchaea, and proteomic analysis showed that Hikarchaeia had already lost the ability to produce methane and had become aerobic. Additionally, they had acquired many UV resistance and/or DNA repair genes.

William Martin's team had previously proposed that the appearance of Haloarchaea occurred in a single step through the acquisition of approximately 1,000 bacterial genes (Nelson-Sathi et al. 2012, 2017). However, these results were challenged by Vincent Daubin and his collaborators, who, through more robust phylogenetic analyses, demonstrated that the number of gene transfers estimated by Martin's team had been greatly overestimated (Groussin et al. 2016). Furthermore, an analysis based on a much larger number of Haloarchaea genomes also supported the scenario of multiple successive transfers (Becker et al. 2014). The work of Ettema's team confirmed this scenario and revealed a complex evolutionary history involving the emergence of new genes and gene transfers of both bacterial and archaeal origin (Martijn et al. 2020).

In 2017, van Loosdrecht and his collaborators isolated poly-extremophilic archaea from salt lakes that live in hot, hypersaline, and alkaline environments, known as *Methanonatronarchaeales*[29] (Sorokin et al. 2017). These archaea accumulate large concentrations of salt in their cytoplasm, similar to Haloarchaea and Nanohaloarchaea. Thus, Methanonatronarchaeales were considered good candidates as intermediates in the transition from methanogens to halophiles. Indeed, they were identified as a sister group to Haloarchaea in the phylogeny published by van Loosdrecht and his collaborators (before the discovery of Hikarchaeia). However, this position was questioned by Gribaldo and his collaborators, who obtained phylogenies positioning Methanonatronarchaeales either at the base of Methanotecta or within Methanotecta (Aouad et al. (2019); see Sorokin et al. (2019) for the response).

---

28 Hika means "sea" in the language of the Yamanas of Tierra del Fuego, near the sites where samples containing DNA of Hikarchaeia were collected.

29 Natron is a mineral composed of sodium carbonate found in large quantities in certain environments.

By focusing on the Methanotecta and using different approaches to construct their phylogenies, Thijs Ettema and his collaborators consistently found that Hikarchaeia are the sister group to Haloarchaea, with Methanonatronarchaeales either positioned at the base of a halophilic clade grouping them with Hikarchaeia and Haloarchaea, or more likely, within or at the base of the Methanotecta (Martijn et al. 2020). In the latter case, adaptation to halophilic conditions would have occurred independently in Hikarchaeia and Haloarchaea, on the one hand, and in Methanonatronarchaeales, on the other hand. If Nanohaloarchaea belong to the DPANN, this suggests that adaptation to halophilic conditions occurred at least three times independently during the evolution of archaea. It is also important to note that adaptation to halophilic conditions occurred independently in bacteria as well.

2.5. A new nomenclature for the taxonomy of archaea?

After describing the various phyla and superphyla of archaea according to the "traditional" nomenclature, it is important to note that the term superphylum remains arbitrary, as no precise definition has been proposed for this new taxonomic rank. Moreover, defining phyla based on the percentage identity of 16S rRNA or proteomes (AAI) introduces a significant bias by not accounting for differences in evolutionary rates between organisms and their proteins. For instance, if two species, A and B, have evolved over a time T (e.g. 500 million years), they could be placed in the same phylum if their 16S rRNA evolved at rate X, but in two different phyla if it evolved at rate 2X. This problem is particularly acute for archaeal lineages that appear to have evolved at markedly different rates, as seen with *M. kandleri*, DPANN and Korarchaea.

This criticism was addressed by Phil Hugenholtz and his collaborators, who proposed a new definition of phyla (and other taxonomic ranks) based on the phylogeny of a set of 122 conserved proteins found in all archaea (Rinke et al. 2021). In the phylogenetic tree they constructed by concatenating these proteins, they recovered most of the previously described monophyletic groups. Based on this phylogeny, they calculated a parameter of "relative evolutionary divergences" (RED) to delineate boundaries between different taxonomic ranks. Their goal was to propose a new classification of archaea and bacteria, avoiding the issues posed by varying evolutionary rates of rRNAs that artificially inflate the number of phyla in rapidly evolving species, and to break from the historical paradigm that maintained, for example, a single phylum for the Euryarchaea despite their remarkable diversity.

It should be noted immediately that the approach used by Phil Hugenholtz and his colleagues (Rinke et al. 2021) could not eliminate the problem of differences in evolutionary rates between "phyla" because proteins can also evolve at very different speeds depending on the lineage. Indeed, their analysis has not yet corrected the bias introduced by the rapid evolution of the DPANN, as this group is still divided into seven phyla in their new nomenclature (Table 2.2). In contrast, the *Euryarchaeota* are divided into only three phyla, while the TACK and Asgard superphyla are each reduced to a single phylum (Table 2.2). The reduction of the 16 Asgard phyla to a single phylum suggests that the rRNAs of these archaea also evolve faster than the average for other archaea, but slower than those of the DPANN. Moreover, the results obtained by Phil Hugenholtz and his colleagues (Rinke et al. 2021) are valid only for a specific set of proteins, and the inclusion of the DPANN in their phylogeny necessarily biased the topology. Finally, the absence of an external group does not allow the tree to be rooted with certainty. The authors chose to root their tree between the DPANN and the other archaea, which is arbitrary (see Chapter 3). The definition of some phyla depends on the chosen root for the archaeal tree. For example, the phylum *Thermoproteota*, which corresponds to the superphylum TACK, is not valid if the root is located within this "phylum" (Da Cunha et al. 2017).

Based on their results, Phil Hugenholtz and his collaborators proposed a new nomenclature for all taxa defined by their approach. This was incorporated into a database used for the identification of MAGs, the *Genome Taxonomy DataBase* (GTDB). The new names proposed for the various taxa diverge significantly from the "historical" names used in this chapter and still commonly used by many authors (see Table 2.2 for correspondences between the different nomenclatures). The Methanotecta, Diapherotrites and group I *Euryarchaeota* have become phyla called *Halobacteriota*, *Thermoplasmatota* and *Methanobacteriota*, respectively (Table 2.2). The Asgard phyla become orders, sometimes grouped into classes (see Figure 2.9). The *Thaumarchaeota* and *Crenarchaeota* phyla and all their relatives disappear, with all members of the TACK superphylum being grouped into a single phylum, the *Thermoproteota*. The *Thaumarchaeota* become Nitrososphaeria and the *Crenarchaeota* become Thermoprotei (classes) (for complete correspondence, see Table 1 in Rinke et al. (2021)).

This new nomenclature is already contested (Lloyd and Tahon 2022; Panda et al. 2022) (for an ICSP response, see Sutcliffe et al. (2021)). The name changes are such that they will complicate bibliographic searches for future researchers who will not necessarily be aware of the correspondences between the old and new nomenclatures. Furthermore, Phil Hugenholtz and his collaborators based the names of the different taxa on the name of the first organism described in that taxon. Consequently, several of these names have a phenotypic connotation that can be confusing. For example, most *Thermoproteota* and *Thermoplasmatota* are not thermophilic, most *Halobacteriota* are not halophilic, and some groups of the phylum *Methanobacteriota* are not methanogens. The names of the phyla *Methanobacteriota* and *Halobacteriota*, inherited from the genus names *Methanobacterium* and *Halobacterium*, two historically significant archaea long mistaken for bacteria, can also be confusing for archaea phyla.

GTDB	"Traditional" Nomenclature (GTDB)	Superphyla and Groups
Halobacteriota	Methanotecta	Group II Eury (Gaiarchaea)
Thermoplasmatota	Diaphorarchaea	Group II Eury (Gaiarchaea)
Methanobacteriota	Methamonada	*Eury group I*
Hydrothermarchaeota	Hydrothermarchaea	–
Hadarchaeota	*Hadesarchaeota* *Persephonarchaeota*	Stygia
Thermoproteota	*Bathyarchaeota* *Aigarchaeota* *Thaumarchaeota* (Nitrososphaeria) *Crenarchaeota* (Thermoprotei) *Marsarchaeota* *Verstraetearchaeota*	(BAT) TACK (Proteoarchaea)
Asgardarchaeota	*Asgardarchaeota*	ASGARD
Nanoarchaeota	*Nanoarchaeota* *Parvarchaeota* Woesearchaeota *Pacearchaeota*	DPANN
Nanohaloarchaeota	*Nanohaloarchaeota*	DPANN
Aenigmarchaeota	*Aenigmarchaeota*	DPANN
Micrarchaeota	*Micrarchaeota*	DPANN
Iainarchaeota	Diapherotrites	DPANN
Altarchaeota	*Altiarchaeota*	DPANN
Huberarchaeota	*Huberarchaeota*	DPANN

Table 2.2. *Phylum-level correspondence between the Genome Taxonomy Database (GTDB) Nomenclature and the traditional nomenclature still used by the NCBI and/or in this chapter. The right column gives the correspondence with the previously proposed superphyla and/or groups*

In fact, several phylum names ending in "archaea" have disappeared, whereas this suffix had been systematically added to the names of archaeal phyla in 1990 by Carl Woese and his collaborators to clearly distinguish archaea from bacteria (Panda et al. 2022). It can be noted that the names proposed by the ICSP (*International Committee on Systematics of Prokaryotes*) are based on the rules set by the *International Code of Nomenclature of Prokaryotes* (ICNP), whose very names reflect an attachment to the "prokaryote/eukaryote" paradigm established before the discovery of archaea (Forterre 2022).

It can therefore already be predicted that the traditional nomenclature will survive alongside the new one, causing considerable confusion. Sharma and his collaborators propose returning to a division of the archaea into three phyla: *Euryarchaeota*, *Crenarchaeota* and *Thaumarchaeota*, which reverts to the proposal made by Forterre and his collaborators at the time of the identification of these groups (Brochier-Armanet et al. 2008; Panda et al. 2022). In my opinion, it seems premature today to propose a definitive nomenclature for *archaea* as long as there is no consensus on the root of this domain and researchers have not agreed on how to evaluate the relative evolutionary speeds of proteins from different organisms.

2.6. Reconstructing the last common ancestor of archaea: LACA

2.6.1. *Rooting the archaeal tree*

Figure 2.13 presents a tree grouping all archaea based on the phylogeny published in 2022 by Céline Brochier-Armanet and her collaborators, who concatenated the sequences of 72 proteins conserved in most archaea (Aouad et al. 2022). This tree is quite similar to the one obtained by Rinke and his collaborators, which was used to propose their new nomenclature. Many nodes in the tree are still under discussion (indicated by dashed lines). While there is consensus on the monophyly of BAT, Asgard archaea, and group II Euryarchaea, and the monophyly of DPANN is likely, the monophyly of other groups and their relative positions remain debated.

The most important unresolved question, in my opinion, is the position of the root, which remains controversial. This point is particularly crucial because the position of the root in the archaeal tree corresponds to that of their last common ancestor, LACA. Therefore, it is essential to determine the position of the root of the archaea to reconstruct LACA.

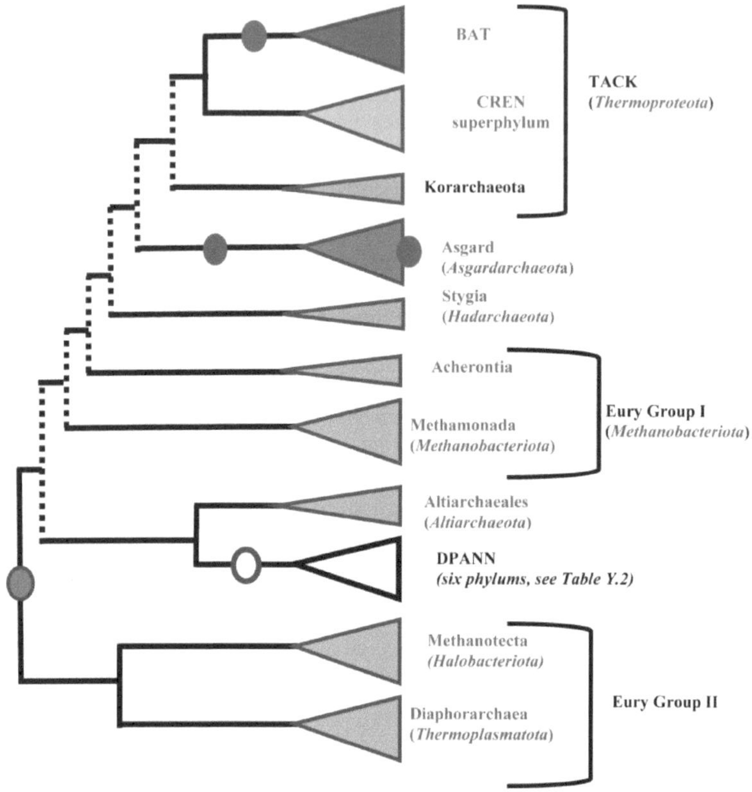

Figure 2.13. *Global schematic phylogeny of archaea based on the publication of Brochier-Armanet et al. (Aouad et al. 2022). For a color version of this figure, see www.iste.co.uk/clouet/archaea1.zip*

COMMENT ON FIGURE 2.13.– *Branch lengths are arbitrary. The colored circles indicate the different positions proposed for the root of the archaea tree (see Figure 2.14). The red circle corresponds to the positions proposed for eukaryotes in a 2D tree. The names in red correspond to the GTDB nomenclature.*

The rooting of the archaeal tree requires the use of an external group; however, this choice depends on the topology chosen for the universal tree of life, which is itself controversial (Figure 2.4) (Chapter 3).

In the case of a tree with three domains (3D), eukaryotes must serve as the external group because they are much closer to archaea than bacteria are. On the other hand, in the case of a tree with two domains (2D) where eukaryotes are a subgroup of archaea, bacteria must be used as the external group.

Using bacteria as an external group, Moreira and his collaborators obtained a root located between the Euryarchaea (including some DPANNs) and the TACK, which they proposed to rename *Proteoarchaea* (Petitjean et al. 2014) (Figure 2.14(a)). In contrast, Gribaldo and his collaborators, using a different dataset without the DPANNs, found a root located between group I and group II Euryarchaea (Raymann et al. 2015) (Figure 2.14(b)). Gribaldo, Brochier-Armanet and their collaborators recently confirmed this result with a much larger dataset including DPANNs (corresponding to the tree in Figure 2.13). They proposed naming the clade (cluster II) grouping group II *Euryarchaea as* Gaiarchaea and the clade (cluster I) grouping *all other archaea as* Ouransarchaea (Figure 2.14(b)). Still using bacteria as an external group, Martinez-Gutierrez and Aylward (2021) positioned the root between the DPANNs and the other archaea (Figure 2.14(c)). They used a protein-enriched dataset with a strong signal. However, they highlighted the possibility of a long-branch attraction phenomenon between DPANNs and bacteria.

As part of a 3D model, Forterre and his collaborators found the root located between the BAT clade (Bathyarchaea, Aigarchaea, Thaumarchaea) and all other archaea, using eukaryotes as the external group and either a concatenation of the two RNA polymerase subunits or a selection of six proteins yielding a 3D tree (Da Cunha et al. 2017) (Figure 2.14(d)) (see Chapter 3). These 3D trees were obtained by removing rapidly evolving archaea (DPANN and *M. kandleri*, the korarchae) from the dataset. Conversely, using a tree built with 35 proteins, the majority of which yielded a 2D tree, they positioned the root of *archaea* in the branch leading to the Asgard archaea. The 2D trees were obtained with smaller proteins, and the topology of archaea in these trees varied significantly from tree to tree, though the root in the BAT branch seemed more robust to them. Brochier-Armanet and his collaborators also positioned the root in the branch leading to Asgard archaea by concatenating the sequences of 64 proteins common to archaea and eukaryotes (Aouad et al. 2022) (red circle of Figure 2.14). They achieved the same result by removing DPANN, group II Euryarchaea and Methanomada from their dataset to avoid the phenomenon of long-branch attraction.

However, the variability of individual trees was not discussed. In the context of the 2D model, they interpreted this root as representative of the emergence of eukaryotes within the archaea, as a sister group to the Asgard archaea.

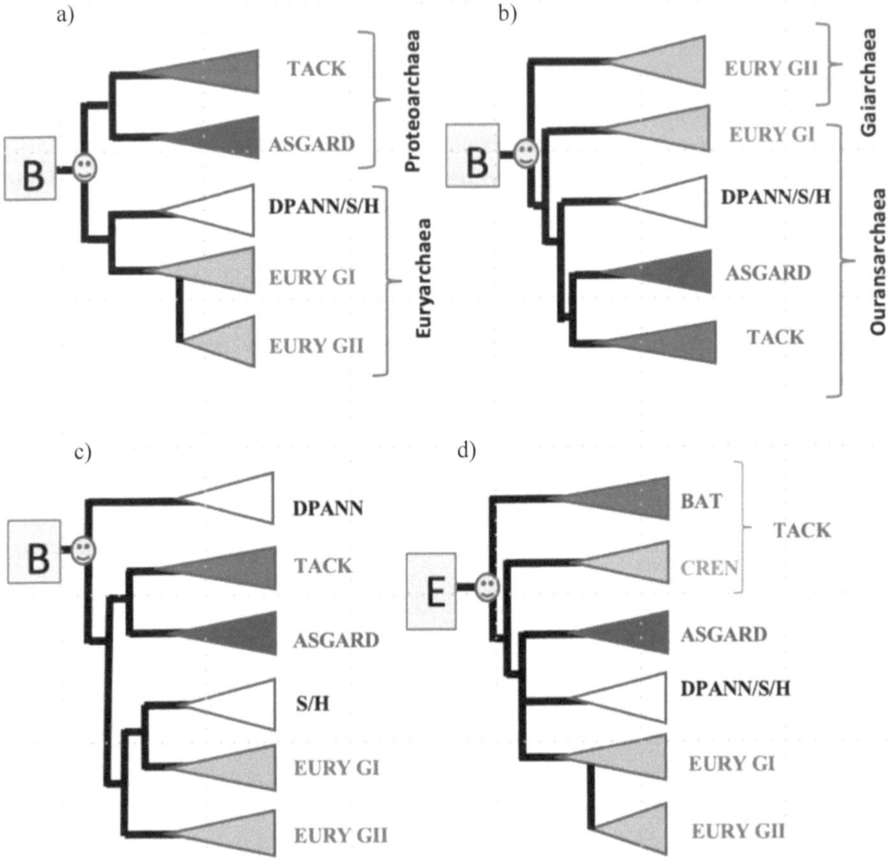

Figure 2.14. *Trees of archaea with different roots obtained using bacteria as the external group (a, b, c) in the 2D scenario or eukaryotes in the 3D scenario (d). S: Stygia; H: Hydrothermarchaeota. The yellow circle corresponds to LACA. For a color version of this figure, see www.iste.co.uk/clouet/archaea1.zip*

To avoid the problems posed by using an external group, Embley and collaborators employed an approach based on modeling the evolutionary process from a very large number of individual trees, including gene transfers, losses and gains (Williams et al. 2017). They obtained a root located between the DPANN and the other archaea (Figure 2.13(b)). However, the validity of this approach, called "reconciliation", is difficult to assess. Individual trees can be affected by the phenomenon of long-branch attraction, and the large number of transfers between these symbiotic organisms complicates this type of analysis (Wade 2020).

2.6.2. *Two opposing visions of LACA: simple or complex*

Several authors have attempted to reconstruct the history of archaea and the nature of LACA from their phylogeny and the comparison of their genomes to determine gene losses and gains during their evolution, as well as gene transfers between archaea lineages and between archaea and bacteria. An initial study based on the analysis of 29 genomes and a root positioned at the level of the Thaumarchaea (between the Crenarchaea and the Euryarchaea) concluded that LUCA was at least as complex, if not more so, than modern archaea, implying a rather reductive evolution within this domain (Csurös and Miklós 2009). This result aligned with a previous observation that the ribosome of LACA must have been more complex than that of modern archaea (Lecompte et al. 2002). It suggested a trend toward the simplification of major molecular mechanisms during the evolution of archaea.

Conversely, using the root obtained between the DPANN and the other archaea, Embley and his collaborators proposed that LACA had a smaller genome than modern archaea, implying a general trend toward increased complexity (Williams et al. 2017). The positioning of the root at the level of the DPANNs indeed favors this type of scenario, and the same could be predicted for a root positioned within the Euryarchaea. On the other hand, if the root is located between the Euryarchaea and the other archaea, or in the branch leading to the BAT, Crenarchaea or Asgard archaea, the situation is more open, allowing for both simplification and complexity trends.

The debate between a simple or complex LACA overlaps with the discussion on the relationship between eukaryotes and archaea and the 2D/3D models, referring to the problem of *eukaryotic signature proteins* (ESPs). ESPs are more abundant in Asgard archaea, BAT and Crenarchaea than in Euryarchaea, particularly in the Asgard archaea (Zaremba-Niedzwiedzka et al. 2017; Liu et al. 2021). In the 2D model, this uneven distribution among different groups of archaea is explained by the gradual emergence of these proteins (absent in LACA) during the evolution of archaea (see Chapter 3). In the 3D model, some of these proteins were already present in LACA before being lost in Euryarchaea and, to a lesser extent, in BAT and Crenarchaea. The situation is complex, as in both scenarios, some ESPs could have been transferred from proto-eukaryotes to archaea (see Chapter 3). Nonetheless, it can be stated that LACA already possessed a strongly "eukaryotic" character, as the vast majority of eukaryotic-type proteins in archaea involved in replication, transcription and translation, as well as membrane proteins such as those in the sec secretion system, are present in all archaea.

In any case, the general trend toward complexity or simplification has not been uniform. Indeed, during the evolution of archaeal lineages, gene losses, series of duplications and gene gains by transfer are observed. These transfers occurred between lineages of archaea living in the same environments (e.g. between acidophilic Euryarchaea and acidophilic Crenarchaea) but also between archaea and bacteria. Transfers appear to have been particularly numerous from bacteria to Euryarchaea and DPANN (perhaps via Euryarchaea hosts). These transfers have played a very important role in the evolution of archaea, for example, by promoting their adaptation to life at low temperatures or in the presence of oxygen (see many examples in Lopez-Garcia et al. 2015). Moreira and colleagues detected a negative correlation in archaea between the hyperthermophilic character and the percentage of genes of bacterial origin present in their genomes. These transfers often involve proteins known to facilitate adaptation to cold environments.

DNA transfers between archaea and bacteria could also explain the striking resemblances between their mobilomes. Plasmids, transposons and some viruses, especially *Caudoviricetes*, which produce head-tail virions, are very similar in archaea and bacteria and very different from plasmids, transposons, retrotransposons and eukaryotic viruses. In particular, archaeal conjugative plasmids appear to be of bacterial origin (Guglielmini et al. 2013). Antiviral defense mechanisms, such as CRISPR and restriction enzymes, are common to archaea and bacteria, as are toxin–antitoxin systems.

2.6.3. *The likely hyperthermophilic nature of LACA*

The presence of thermophiles and hyperthermophiles in almost all groups of archaea has long suggested the idea that LACA was a hyperthermophile.

This hypothesis could explain both the presence in all archaea of lipids particularly well adapted to living conditions in terrestrial or underwater hot springs and the absence of RNA viruses in archaea, as RNA is a particularly fragile molecule at high temperature (Forterre 1996). The hyperthermophilic nature of LACA seemed to be confirmed by the reconstruction of its proteome by Manolo Gouy's team in 2008 based on a phylogeny obtained by concatenating several universal proteins (Boussau et al. 2008). LACA's proteome was indeed enriched with amino acids characteristic of high-temperature living. Finally, the hyperthermophilic nature of LACA seemed to be validated by phylogenetic analysis of reverse gyrase. This enzyme is the only protein specific to hyperthermophiles (Forterre 2002). It is systematically present in all hyperthermophiles (archaea and bacteria), in some thermophiles, but never in mesophiles (Catchpole and Forterre 2019). For a long time, the phylogeny of archaeal reverse gyrases could be superimposed on that of archaea based on ribosomal proteins, indicating that LACA was either a thermophile or a hyperthermophile (Brochier-Armanet and Forterre 2007).

The situation has become more complex today. The reconstruction of the LACA proteome has not been updated by incorporating all of the genomes obtained over the last 10 years, and the updated phylogeny of reverse gyrase can no longer guarantee the presence of this enzyme in LACA (Catchpole and Forterre 2019). In this phylogeny, an early divergence of the reverse gyrases of the Euryarchaea and the Crenarchaea is observed (see Figure 3.8 and section 3.5.3 of Chapter 3), but the position of these two groups and that of the groups of archaea lacking reverse gyrase relative to LACA is no longer a consensus. It all depends, once again, on the position of the root.

The example of Acherontia illustrates how the discovery of new lineages can call previous conclusions into question. For a long time, it seemed evident that the ancestor of the Thermococcales, located at the base of the Euryarchaea, was a hyperthermophile. However, the discovery of uncultivated mesophilic lineages, such as the Nuwarchaeales, the Theinoarchaea and the Methanofastidiosa, sister groups of the Thermococcales, raises questions. In the case of Thaumarchaea, phylogenetic analyses indicate that their ancestor was most likely a thermophilic organism (Abby et al. 2020a). For the Asgard archaea, a mesophilic ancestor can be assumed since they include only one thermophilic lineage, the Odinarchaea and their reverse gyrase appears to have been acquired from the Thermococcales (Author's observation). For all other major groups of archaea, the question remains unresolved because, like the Acherontia, they include a mixture of mesophilic and thermophilic organisms.

An approach aimed at progressively reconstructing the proteomes of increasingly distant ancestors (thus increasingly close to LACA) and determining whether their amino acid compositions indicate a mesophilic or thermophilic nature seems promising. Using this approach, Brochier-Armanet and colleagues showed that the *Methanococcales*, which can divide over a particularly wide temperature range with optimal growth temperatures from 25 to 95°C, originated from a hyperthermophilic ancestor (Lecocq et al. 2021). Among the other Methamonada, it can be noted that the *Methanopyrales* are all extreme hyperthermophiles with optimal growth temperatures of 106°C and that the *Methanobacteriales* also divide within a temperature range of 25–95°C. The ancestor of the Methamonada was therefore most likely a hyperthermophile. However, the relative position of the appearance of the Thermococcales and Methamonada will change depending on the position of the root, leading to different scenarios for the transition between thermophilic and mesophilic characteristics.

The only certainty for the moment is that several initially thermophilic archaeal lineages (Thaumarchaea and *Methanococcales*) have adapted to mesophilia, while the reverse has not yet been observed. If this observation becomes generalized in the future, it could be seen as a strong argument in favor of the thermophilic or hyperthermophilic nature of LACA, consistent with the results of Manolo Gouy and his collaborators, as well as the absence of RNA viruses in archaea and the stability of their lipids at high temperatures.

2.6.4. *The possibility of a methanogenic LACA*

It is also difficult to determine with certainty whether LACA's metabolism was due, once again, to the lack of consensus on the root and the ease with which genes encoding enzymes involved in metabolic pathways can be transferred between evolutionarily distant lineages. However, it can be agreed that LACA was most likely an anaerobic organism, as it must have existed long before the accumulation of oxygen in the planet's atmosphere.

Was LACA a methanogen? At the time of the identification of the first archaea, all of which were methanogens, this phenotype seemed perfectly adapted to life on the early Earth, in an atmosphere rich in CO_2 and hydrogen. This was one of the major arguments used by Carl Woese to propose the name "Archaebacteria" (see Chapter 1). The discovery of other lifestyles among archaea, the absence of methanogens among the Crenarchaea, and the historical positioning of the root of the archaea between the Euryarchaea and the Crenarchaea had called into question the hypothesis of a methanogenic LACA and led to the belief that methanogenesis appeared among the Euryarchaea. The discovery, thanks to metagenomics, of methanogenic archaea outside the Euryarchaea has revived the hypothesis of a methanogenic LACA. This hypothesis is supported by the partial overlap between the archaeal tree and that of the gene coding for a methanogenesis enzyme, methyl-coenzyme M (CoM) reductase (MCR) (Garcia et al. 2022; Wang et al. 2022). In methanogens, this enzyme reduces methyl-coenzyme A to methane. MCR is also present in methanotrophic archaea, where it catalyzes the formation of methyl-coenzyme M from methane.

The MCR enzyme is present in all major groups of archaea, except (for now) for DPANN (Wang et al. 2022). Phylogenetic analysis of MCRs highlights four clades corresponding to classes 1, 2, and 3 methanogens and one clade corresponding to the TACK superphylum (Garcia et al. 2022). The topology of this tree overlaps with that of the archaea, with a few exceptions likely due to gene transfers. If the root of the archaea is located at the DPANN level, it is indeterminable whether MCR was present in LACA. Conversely, if the root of the archaea lies outside the DPANN, MCR was probably already present in LACA before being lost in all archaea that are neither methanogenic nor methanotrophic. The distribution of methanotrophs in archaea (which is more limited than that of methanogens) suggests that methanotrophs appeared after methanogens, likely due to the accumulation of electron acceptors such as sulfates or nitrates, which are presumed to have appeared later in Earth's history (Wang et al. 2022). If methanogenesis preceded reverse methanogenesis, the probable presence of the MCR protein in LACA strongly suggests that LACA was indeed a methanogen, and therefore an anaerobic organism.

There are different types of methanogens depending on the molecules used to produce methane, raising the question of which type of methanogens were present in LACA. The direct use of CO_2 depends on an enzymatic complex present only in classes 1 and 2 methanogens (Wang et al. 2021), except for Nezhaarchaea, which are part of the BAT superphylum. These latter organisms appear to have acquired this complex through gene transfer from class 2 methanogens. If the root of the archaea is located between the Euryarchaea of groups I and II, LACA could have utilized this type of metabolism. For all other roots, LUCA could have used methylated compounds in the presence of hydrogen, which corresponds to the metabolism (methylotrophy) observed today in Acherontia, Bathyarchaea and Hadesarchaea. According to Xiao and colleagues, this scenario would align with the presumed abundance of methylated compounds on early Earth. The ability to directly use CO_2 would have appeared in class 1 methanogens, where it predominates. Class 2 methanogens, which have been highly successful evolutionarily, would have retained this capability while developing the ability to use acetate and various methylated compounds to produce methane.

The loss of methanogenesis in many archaea lineages would have led to the diversification of their metabolism, resulting in the emergence of new modes of aerobic and/or anaerobic respiration and the appearance of many heterotrophic lineages. Gene exchanges between archaea and bacteria likely played a major role in this process, contributing to the remarkable metabolic diversity common to these two domains.

2.7. Conclusion

Our knowledge of the diversity of archaea and their evolution has made significant progress in recent years, largely due to the development of metagenomic approaches. However, many unresolved questions remain regarding the taxonomy and phylogeny of archaea. Resolving these questions will largely depend on the probable discovery of new lineages of archaea in the future and greater attention to the various biases that can compromise phylogenetic analyses. It is particularly important to analyze and understand the origin of the conflicting signals observed between different individual trees. The use of insertions as synapomorphies for clade identification, coupled with phylogenetic analysis, is expected to play a more important role in establishing a robust phylogeny of archaea in the future.

The cultivation of archaea belonging to groups currently represented only by MAGs is now one of the main priorities of archaea research. The development of metagenomic approaches has led in recent years to an explosion of new phyla, posing numerous coherence problems and sometimes leading different authors to propose different names for the same groups. The very definition and scope of phyla are debated. However, there is a risk that the nomenclature proposed by the GTDB will be increasingly accepted by researchers working on archaea without being specialists in evolution, potentially becoming a paradigm that temporarily hinders research in this field. One of the major objectives for the future remains to harmonize the taxonomy of archaea with that of bacteria, based on phylogeny and a more precise understanding of organism physiology, particularly the relative rate of evolution of protein markers used in phylogenetic analyses.

2.8. References

Abby, S.S., Kerou, M., Schleper, C. (2020). Ancestral reconstructions decipher major adaptations of ammonia-oxidizing archaea upon radiation into moderate terrestrial and marine environments. *mBio*, 11(5), e02371–20. doi: 10.1128/mBio.02371-20.

Adam, P.S., Borrel, G., Brochier-Armanet, C., Gribaldo, S. (2017). The growing tree of Archaea: New perspectives on their diversity, evolution and ecology. *ISME J.*, 11, 2407–2425.

Aouad, M., Taib, N., Oudart, A., Lecocq, M., Gouy, M., Brochier-Armanet, C. (2018). Extreme halophilic archaea derive from two distinct methanogen Class II lineages. *Mol. Phylogenet. Evol.*, 127, 46–54.

Aouad, M., Borrel, G., Brochier-Armanet, C., Gribaldo, S. (2019). Evolutionary placement of Methanonatronarchaeia. *Nat. Microbiol.*, 4, 558–559.

Aouad, M., Flandrois, J.P., Jauffrit, F., Gouy, M., Gribaldo, S., Brochier-Armanet, C. (2022). A divide-and-conquer phylogenomic approach based on character supermatrices resolves early steps in the evolution of the Archaea. *BMC Ecol. Evol.*, 22(1), 1. doi: 10.1186/s12862-021-01952-0.

Baker, B.J., Comolli, L.R., Dick, G.J., Hauser, L.J., Hyatt, D., Dill, B.D., Land, M.L., Verberkmoes, N.C., Hettich, R.L, Banfield, J.F. (2010). Enigmatic, ultrasmall, uncultivated Archaea. *Proc. Natl. Acad. Sci. USA*, 107, 8806–8811.

Baker, B.J., Saw, J.H., Lind, A.E., Lazar, C.S., Hinrichs, K.U., Teske, A.P., Ettema, T.J. (2016). Genomic inference of the metabolism of cosmopolitan subsurface Archaea Hadesarchaea. *Nat. Microbiol.*, 15(1), 16002. doi: 10.1038/nmicrobiol.2016.2.

Bapteste, E., Brochier, C., Boucher, Y. (2005). Higher-level classification of the Archaea: Evolution of methanogenesis and methanogens. *Archaea*, 1, 353–363.

Barns, S.M., Delwiche, C.F., Palmer, J.D., Pace, N.R. (1996). Perspectives on archaeal diversity, thermophily and monophyly from environmental rRNA sequences. *Proc. Natl. Acad. Sci. USA*, 93, 9188–9193.

Borrel, G., O'Toole, P.W., Harris, H.M., Peyret, P., Brugère, J.F., Gribaldo, S. (2013). Phylogenomic data support a seventh order of methylotrophic methanogens and provide insights into the evolution of methanogenesis. *Genome Biol. Evol.*, 5, 769–1780.

Borrel, G., Brugère, J.F., Gribaldo, S., Schmitz, R.A., Moissl-Eichinger, C. (2020). The host-associated archaeome. *Nat. Rev. Microbiol.*, 18, 622–636.

Boussau, B., Blanquart, S., Necsulea, A., Lartillot, N., Gouy, M. (2008). Parallel adaptations to high temperatures in the Archaean eon. *Nature*, 456, 942–945.

Brochier, C., Forterre, P., Gribaldo, S. (2004). Archaeal phylogeny based on proteins of the transcription and translation machineries: Tackling the *Methanopyrus kandleri* paradox. *Genome Biol.*, 5(3), R17. doi: 10.1186/gb-2004-5-3-r17.

Brochier, C., Forterre, P., Gribaldo, S. (2005a). An emerging phylogenetic core of Archaea: Phylogenies of transcription and translation machineries converge following addition of new genome sequences. *BMC Evol. Biol.*, 5(36). doi: 10.1186/1471-2148-5-36.

Brochier, C., Gribaldo, S., Zivanovic, Y., Confalonieri, F., Forterre, P. (2005b). Nanoarchaea: Representatives of a novel archaeal phylum or a fast-evolving euryarchaeal lineage related to Thermococcales? *Genome Biol.*, 6(5), R42. doi: 10.1186/gb-2005-6-5-r42.

Brochier, C., Forterre, P., Gribaldo, S. (2011). Phylogeny and evolution of the Archaea: One hundred genomes later. *Curr. Opin. Microbiol.*, 14, 274–281.

Brochier-Armanet, C. and Forterre, P. (2007). Archaea. Widespread distribution of archaeal reverse gyrase in thermophilic bacteria suggests a complex history of vertical inheritance and lateral gene transfers. *Archaea*, 2, 83–93.

Brochier-Armanet, C., Boussau, B., Forterre, P., Gribaldo, S. (2008a). Mesophilic *Crenarchaeota*: Proposal for a third archaeal phylum, the *Thaumarchaeota*. *Nat. Rev. Microbiol.*, 6, 245–252.

Brochier-Armanet, C., Gribaldo, S., Forterre, P. (2008b). A DNA topoisomerase IB in *Thaumarchaeota* testifies for the presence of this enzyme in the last common ancestor of Archaea and Eucarya. *Biol Direct.*, 23(3), 54. doi: 10.1186/1745-6150-3-54.

Brochier-Armanet, C., Gribaldo, S., Forterre, P. (2012). Spotlight on the *Thaumarchaeota*. *ISME J.*, 6, 227–230.

Brown, C.T., Hug, L.A., Thomas, B.C., Sharon, I., Castelle, C.J., Singh, A., Wilkins, M.J., Wrighton, K.C., Williams, K.H., Banfield, J.F. (2015). Unusual biology across a group comprising more than 15% of domain bacteria. *Nature*, 523, 208–211.

Brugère, J.F., Borrel, G., Gaci, N., Tottey, W., O'Toole, P.W., Malpuech-Brugère, C. (2014). Archaebiotics: Proposed therapeutic use of archaea to prevent trimethylaminuria and cardiovascular disease. *Gut Microbes*, 5, 5–10.

Cai, M., Liu, Y., Yin, X., Zhou, Z., Friedrich, M.W., Richter-Heitmann, T., Nimzyk, R., Kulkarni, A., Wang, X., Li, M. et al. (2020). Diverse Asgard archaea including the novel phylum *Gerdarchaeota* participate in organic matter degradation. *Sci. China Life Sci.*, 63, 886–897.

Cann, I.K. and Ishino, Y. (1999). Archaeal DNA replication: Identifying the pieces to solve a puzzle. *Genetics*, 152, 1249–1267.

Castelle, C.J. and Banfield, J.F. (2018). Major new microbial groups expand diversity and alter our understanding of the tree of life. *Cell*, 172, 1181–1197.

Castelle, C.J., Wrighton, K.C., Thomas, B.C., Hug, L.A., Brown, C.T., Wilkins, M.J., Frischkorn, K.R., Tringe, S.G., Singh, A., Markillie, L.M. et al. (2015). Genomic expansion of domain archaea highlights roles for organisms from new phyla in anaerobic carbon cycling. *Curr. Biol.*, 25, 690–701.

Castelle, C.J., Brown, C.T., Anantharaman, K., Probst, A.J., Huang, R.H., Banfield, J.F. (2018). Biosynthetic capacity, metabolic variety and unusual biology in the CPR and DPANN radiations. *Nat. Rev. Microbiol.*, 16, 629–645.

Castelle, C.J., Méheust, R., Jaffe, A.L., Seitz, K., Gong, X., Baker, B.J., Banfield, J.F. (2021). Protein family content uncovers lineage relationships and bacterial pathway maintenance mechanisms in DPANN archaea. *Front Microbiol.*, 12, 660052. doi: 10.3389/fmicb.2021.660052.

Catchpole, R.J. and Forterre, P. (2019). The evolution of Reverse gyrase suggests a non-hyperthermophilic last universal common ancestor. *Mol. Biol. Evol.*, 36, 2737–2747.

Chuvochina, M., Rinke, C., Parks, D.H., Rappé, M.S., Tyson, G.W., Yilmaz, P., Whitman, W.B., Hugenholtz, P. (2019). The importance of designating type material for uncultured taxa. *Syst. Appl. Microbiol.*, 42, 15–21.

Comolli, L.R., Baker, B.J., Downing, K.H., Siegerist, C.E., Banfield, J.F. (2009). Three-dimensional analysis of the structure and ecology of a novel, ultra-small archaeon. *The ISME Journal*, 3, 159–167.

Csurös, M. and Miklós, I. (2009). Streamlining and large ancestral genomes in Archaea inferred with a phylogenetic birth-and-death model. *Mol. Biol. Evol.*, 26, 2087–2095.

Da Cunha, V., Gaia, M., Ogata, H., Jaillon, O., Delmont, T.O., Forterre, P. (2022). Giant viruses encode actin-related proteins. *Mol. Biol. Evol.*, 39(2), msac022. doi: 10.1093/molbev/msac022.

DeLong, E.F. (1992). Archaea in coastal marine environments. *Proc. Natl. Acad. Sci. USA*, 89, 5685–5689.

Dombrowski, N., Williams, T.A., Sun, J., Woodcroft, B.J., Lee, J.H., Minh, B.Q., Rinke, C., Spang, A. (2020). *Undinarchaeota* illuminate DPANN phylogeny and the impact of gene transfer on archaeal evolution. *Nat. Commun.*, 11(1), 3939. doi: 10.1038/s41467-020-17408-w.

Dombrowski, N., Mahendrarajah, T., Gross, S.T., Eme, L., Spang, A. (2021). Archaea. In *Practical Handbook of Microbiology*, Green, L.H. and Goldman, E. (eds). CRC Press, London/New York.

Elkins, J.G., Podar, M., Graham, D.E., Makarova, K.S., Wolf, Y., Randau, L., Hedlund B.P., Brochier-Armanet, C., Kunin, V., Anderson, I. et al. (2008). A korarchaeal genome reveals insights into the evolution of the Archaea. *Proc. Natl. Acad. Sci. USA*, 105, 8102–8107.

Eme, L., Tamarit, D., Caceres, E.F., Stairs, C.W., De Anda, V., Schön, M.E., Seitz, K.W., Dombrowski, N., Lewis, W.H., Homa, F. et al. (2023). Inference and reconstruction of the heimdallarchaeial ancestry of eukaryotes. *Nature*, 618, 992–999.

Evans, P.N., Parks, D.H., Chadwick, G.L., Robbins, S.J., Orphan, V.J., Golding, S.D., Tyson, G.W. (2015). Methane metabolism in the archaeal phylum *Bathyarchaeota* revealed by genome-centric metagenomics. *Science*, 350, 434–438.

Felsenstein, J. (1978). Cases in which parsimony or compatibility methods will be positively misleading. *Systematic Biology*, 27, 401–410.

Feng, Y., Neri, U., Gosselin, S., Louyakis, A.S., Papke, R.T., Gophna, U., Gogarten, J.P. (2021). The evolutionary origins of extreme halophilic archaeal lineages. *Genome Biol. Evol.*, 13(8), evab166. doi: 10.1093/gbe/evab166.

Filée, J., Becker, H.F., Mellottee, L., Eddine, R.Z., Li, Z., Yin, W., Lambry, J.C., Liebl, U., Myllykallio, H. (2023). Bacterial origins of thymidylate metabolism in Asgard archaea and Eukarya. *Nat. Commun.*, 14(1), 838. doi: 10.1038/s41467-023-36487-z.

Forterre, P. (1996). A hot topic: The origin of hyperthermophiles. *Cell*, 85, 789–792.

Forterre, P. (2002). A hot story from comparative genomics: Reverse gyrase is the only hyperthermophile-specific protein. *Trends Genet.*, 18, 236–237.

Forterre, P. (2022). Carl Woese: Still ahead of our time. *mLife*, 1(4), 359–367. doi: 10.1002/mlf2.12049.

Forterre, P. and Gadelle, D. (2009). Phylogenomics of DNA topoisomerases: Their origin and putative roles in the emergence of modern organisms. *Nucleic. Acids. Res.*, 37, 679–692.

Forterre, P., Krupovic, M., Raymann, K., Soler, N. (2014). Plasmids from *Euryarchaeota. Microbiol Spectr.*, 2(6). doi: 10.1128/microbiolspec.PLAS-0027-2014.

Fuhrman, J.A., McCallum, K., Davis, A. (1992). Novel major archaebacterial group from marine plankton. *Nature*, 356, 148–149.

Gaia, M., Da Cunha, V., Forterre, P. (2018). The tree of life. In *Molecular Mechanisms of Microbial Evolution*, Rampelotto, P.H. (ed.). Springer, Cham. doi: 10.1007/978-3-319-69078-0_3.

Garcia, P.S., Gribaldo, S., Borrel, G. (2022). Diversity and evolution of methane-related pathways in Archaea. *Annu. Rev. Microbiol.*, 76, 727–755.

Gill, S., Krupovic, M., Desnoues, N., Béguin, P., Sezonov, G., Forterre, P. (2014). A highly divergent archaeo-eukaryotic primase from the *Thermococcus nautilus* plasmid, pTN2. *Nucleic. Acids. Res.*, 42, 3707–3719.

Golyshina, O.V., Bargiela, R., Golyshin, P.N. (2019). Cuniculiplasmataceae, their ecogenomic and metabolic patterns, and interactions with ARMAN. *Extremophiles*, 23, 1–7.

Gorlas, A., Robert, C., Gimenez, G., Drancourt, M., Raoult, D. (2012). Complete genome sequence of *Methanomassiliicoccus luminyensis*, the largest genome of a human-as-sociated Archaea species. *J. Bacteriol.*, 194(17), 4745. doi: 10.1128/JB.00956-12.

Guglielmini, J., de la Cruz, F., Rocha, E.P. (2013). Evolution of conjugation and type IV secretion systems. *Mol. Biol. Evol.*, 30, 315–331.

Guy, L. and Ettema, T.J. (2011). The archaeal 'TACK' superphylum and the origin of eukaryotes. *Trends Microbiol.*, 19, 580–587.

Guy, L., Spang, A., Saw, J.H., Ettema, T.J. (2014). 'Geoarchaeote NAG1' is a deeply rooting lineage of the archaeal order Thermoproteales rather than a new phylum. *ISME J.*, 2, 1353–1357.

Hallam, S.J., Konstantinidis, K.T., Putnam, N., Schleper, C., Watanabe, Y., Sugahara, J., Preston, C., de la Torre, J., Richardson, P.M., DeLong, E.F. (2006). Genomic analysis of the uncultivated marine crenarchaeote Cenarchaeum symbiosum. *Proc. Natl. Acad. Sci. USA*, 103, 18296–18301.

Hamm, J.N., Erdmann, S., Eloe-Fadrosh, E.A., Angeloni, A., Zhong, L., Brownlee, C., Williams, T.J., Barton, K., Carswell, S., Smith, M.A. et al. (2019). Unexpected host dependency of Antarctic *Nanohaloarchaeota. Proc. Natl. Acad. Sci. USA*, 116, 14661–14670.

Hernandez, A.M. and Ryan, J.F. (2021). Six-state amino acid recoding is not an effective strategy to offset compositional heterogeneity and saturation. *Phylogenetic Analyses. Syst. Biol.*, 70, 1200–1212.

Hua, Z.S., Wang, Y.L., Evans, P.N., Qu, Y.N., Goh, K.M., Rao, Y.Z., Qi, Y.L., Li, Y.X., Huang, M.J., Jiao, J.Y. et al. (2019). Insights into the ecological roles and evolution of methyl-coenzyme M reductase-containing hot spring Archaea. *Nat. Commun.*, 10(1), 4574. doi: 10.1038/s41467-019-12574-y.

Huber, H., Hohn, M.J., Rachel, R., Fuchs, T., Wimmer, V.C., Stetter, K.O. (2002). A new phylum of Archaea represented by a nanosized hyperthermophilic symbiont. *Nature*, 417, 63–67.

Imachi, H., Nobu, M.K., Nakahara, N., Morono, Y., Ogawara, M., Takaki, Y., Takano, Y., Uematsu, K., Ikuta, T., Ito, M. et al. (2020). Isolation of an archaeon at the prokaryote-eukaryote interface. *Nature*, 577, 519–525.

Jay, Z.J., Beam, J.P., Dlakić, M., Rusch, D.B., Kozubal, M.A., Inskeep, W.P. (2018). *Marsarchaeota* are an aerobic archaeal lineage abundant in geothermal iron oxide microbial mats. *Nature Microbiol.*, 3, 732–740.

Jungbluth, S.P., Amend, J.P., Rappé, M.S. (2017). Metagenome sequencing and 98 microbial genomes from Juan de Fuca Ridge flank subsurface fluids. *Sci. Data.*, 4, 170037. doi: 10.1038/sdata.2017.37.

Kelly, T.J., Simancek, P., Brush, G.S. (1998). Identification and characterization of a single-stranded DNA-binding protein from the archaeon *Methanococcus jannaschii*. *Proc. Natl. Acad. Sci. USA*, 95, 14634–14639.

Khomyakova, M.A., Merkel, A.Y., Mamiy, D.D., Klyukina, A.A., Slobodkin, A.I. (2023). Phenotypic and genomic characterization of *Bathyarchaeum tardum* gen. nov., sp. nov., a cultivated representative of the archaeal class *Bathyarchaeia*. *Front Microbiol.*, 14, 1214631. doi: 10.3389/fmicb.2023.1214631.

Kohtz, A.J., Jay, Z.J., Lynes, M.M., Krukenberger, V., Hatzenpichler, R. (2022). Culexar-chaeia, a novel archaeal class of anaerobic generalists inhabiting geothermal environments. *ISME Commun.*, 2, 86. doi: 10.1038/s43705-022-00175-8.

Könneke, M., Bernhard, A.E., de la Torre, J.R., Walker, C.B., Waterbury, J.B., Stahl., D.A. (2005). Isolation of an autotrophic ammonia-oxidizing marine archaeon. *Nature*, 437, 543–546.

Kozubal, M.A., Romine, M., Jennings, R.D., Jay, Z.J., Tringe, S.G., Rusch, D.B., Beam, J.P., McCue, L.A., Inskeep, W.P. (2013). *Geoarchaeota*: A new candidate phylum in the Archaea from high-temperature acidic iron mats in Yellowstone National Park. *ISME J.*, 7, 622–664.

Krause, S., Gfrerer, S., von Kügelgen, A., Reuse, C., Dombrowski, N., Villanueva, L., Bunk, B., Spröer, C., Neu, T.R., Kuhlicke, U. et al. (2022). The importance of biofilm formation for cultivation of a Micrarchaeon and its interactions with its Thermoplasmatales host. *Nat Commun.*, 13(1), 1735. doi: 10.1038/s41467-022-29263-y.

Laso-Pérez, R., Hahn, C., van Vliet, D.M., Tegetmeyer, H.E., Schubotz, F., Smit, N.T., Pape, T., Sahling, H., Bohrmann, G., Boetius, A. et al. (2019). Anaerobic degradation of non-methane alkanes by "*Candidatus* Methanoliparia" in hydrocarbon seeps of the Gulf of Mexico. *mBio*, 10, e01814–19. doi: 10.1128/mBio.01814-19.

Lazar, C.S., Baker, B.J., Seitz, K.W., Teske, A.P. (2017). Genomic reconstruction of multiple lineages of uncultured benthic archaea suggests distinct biogeochemical roles and ecological niches. *ISME J.*, 11, 1118–1129.

Lecocq, M., Groussin, M., Gouy, M., Brochier-Armanet, C. (2021). The molecular determinants of thermoadaptation: *Methanococcales* as a case study. *Mol. Biol. Evol.*, 38, 1761–1776.

Lecompte, O., Ripp, R., Thierry, J.C., Moras, D., Poch, O. (2002). Comparative analysis of ribosomal proteins in complete genomes: An example of reductive evolution at the domain scale. *Nucleic. Acids. Res.*, 30, 5382–5390.

Liu, Y. and Li, M. (2022). The unstable evolutionary position of *Korarchaeota* and its relationships with other TACK and Asgard Archaea. *mLife*, 1, 218–222.

Liu, Y., Makarova, K.S., Huang, W.-C., Wolf, Y.I., Nikolskaya, A.N., Zhang, X., Cai, M., Zhang, C.J., Xu, W., Luo, Z. et al. (2021). Expanded diversity of Asgard archaea and their relationships with eukaryotes. *Nature*, 593, 553–557.

Lloyd, K.G. and Tahon, G. (2022). Science depends on nomenclature, but nomenclature is not science. *Nat. Rev. Microbiol.*, 20, 123–124.

López-García, P., Zivanovic, Y., Deschamps, P., Moreira, D. (2015). Bacterial gene import and mesophilic adaptation in archaea. *Nat. Rev. Microbiol.*, 13, 447–456.

Luo, C., Rodriguez-R, L.M., Konstantinidis, K.T. (2014). MyTaxa: An advanced taxonomic classifier for genomic and metagenomic sequences. *Nucleic. Acids. Res.*, 42, e73. doi: 10.1093/nar/gku169.

Marguet, E., Gaudin, M., Gauliard, E., Fourquaux, I., le Blond du Plouy, S., Matsui, I., Forterre, P. (2013). Membrane vesicles, nanopods and/or nanotubes produced by hyperthermophilic archaea of the genus *Thermococcus*. *Biochem. Soc. Trans.*, 41, 436–442.

Martijn, J., Schön, M.E., Lind, A.E., Vosseberg, J., Williams, T.A., Spang, A., Ettema, T.J.G. (2020). Hikarchaeia demonstrate an intermediate stage in the methanogen-to-halophile transition. *Nat. Commun.*, 11(1), 5490. doi: 10.1038/s41467-020-19200-2.

Martinez-Gutierrez, C.A. and Aylward, F.O. (2020). Genome size distributions in bacteria and archaea are strongly linked to evolutionary history at broad phylogenetic scales. *PLoS Genet.*, 18(5), e1010220. doi: 10.1371/journal.pgen.1010220.

Martinez-Gutierrez, C.A. and Aylward, F.O. (2021). Phylogenetic signal, congruence, and uncertainty across bacteria and archaea. *Mol. Biol. Evol.*, 38, 5514–5527.

Matte-Tailliez, O., Brochier, C., Forterre, P., Philippe, H. (2002). Archaeal phylogeny based on ribosomal proteins. *Mol. Biol. Evol.*, 19, 631–639.

McKay, L.J., Dlakić, M., Fields, M.W., Delmont, T.O., Eren, A.M., Jay, Z.J., Klingelsmith, K.B., Rusch, D.B., Inskeep, W.P. (2019). Co-occurring genomic capacity for anaerobic methane and dissimilatory sulfur metabolisms discovered in the *Korarchaeota*. *Nat. Microbiol.*, 4, 614–622.

Meng, J., Xu, J., Qin, D., He, Y., Xiao, X., Wang, F. (2014). Genetic and functional properties of uncultivated MCG archaea assessed by metagenome and gene expression analyses. *ISME J.*, 8, 650–659.

Meyerdierks, A., Kube, M., Kostadinov, I., Teeling, H., Glöckner, F.O., Reinhardt, R., Amann, R. (2010). Metagenome and mRNA expression analyses of anaerobic methanotrophic archaea of the ANME-1 group. *Environ. Microbiol.*, 12, 422–439.

Moissl, C., Rachel, R., Briegel, A., Engelhardt, H., Huber, R. (2005). The unique structure of archaeal 'hami', highly complex cell appendages with nano-grappling hooks. *Mol. Microbiol.*, 6, 361–370.

Moody, E.R.R., Mahendrarajah, T.A., Dombrowski, N., Clark, J.W., Petitjean, C., Offre, P., Szöllősi, G.J., Spang, A., Williams, T.A. (2022). An estimate of the deepest branches of the tree of life from ancient vertically evolving genes. *Elife*, 11, e66695. doi: 10.7554/eLife.66695.

Murray, A.E., Freudenstein, J., Gribaldo, S., Hatzenpichler, R., Hugenholtz, P., Kämpfer, P., Konstantinidis, K.T., Lane, C.E., Papke, R.T., Parks, D.H. et al. (2020). Roadmap for naming uncultivated Archaea and bacteria. *Nat. Microbiol.*, 5, 987–994.

Mwirichia, R., Alam, I., Rashid, M., Vinu, M., Ba-Alawi, W., Kamau, A., Ngugi, D., Göker, M., Klenk, H.P., Bajic, V., Stingl, U. (2016). Metabolic traits of an uncultured archaeal lineage – MSBL1– From brine pools of the Red Sea. *Sci. Rep.*, 6, 19181. doi: 10.1038/srep19181.

Narasingarao, P., Podell, S., Ugalde, J.A., Brochier-Armanet, C., Emerson, J.B., Brocks, J.J., Heidelberg, K.B., Banfield, J.F., Allen, E.E. (2012). *De novo* metagenomic assembly reveals abundant novel major lineage of Archaea in hypersaline microbial communities. *ISME J.*, 6, 81–93.

Nobu, M.K., Narihiro, T., Kuroda, K., Mei, R., Liu, W.T. (2016). Chasing the elusive *Euryarchaeota* class WSA2: Genomes reveal a uniquely fastidious methyl-reducing methanogen. *ISME J.*, 10, 2478–2487.

Nunoura, T., Takaki, Y., Kakuta, J., Nishi, S., Sugahara, J., Kazama, H., Chee, G.J., Hattori, M., Kanai, A., Atomi, H. et al. (2011). Insights into the evolution of Archaea and eukaryotic protein modifier systems revealed by the genome of a novel archaeal group. *Nucleic. Acids. Res.*, 39, 3204–3223.

Panda, A., Islam, S.T., Sharma, G. (2022). Harmonizing prokaryotic nomenclature: Fixing the fuss over phylum name flipping. *mBio.*, 13(3). doi: 10.1128/mbio.00970-22.

Petitjean, C., Deschamps, P., López-García, P., Moreira, D. (2014). Rooting the domain archaea by phylogenomic analysis supports the foundation of the new kingdom *Proteoarchaeota*. *Genome Biol. Evol.*, 7, 191–204.

Petitjean, C., Deschamps, P., López-García, P., Moreira, D, Brochier-Armanet, C. (2015). Extending the conserved phylogenetic core of archaea disentangles the evolution of the third domain of life. *Mol. Biol. Evol.*, 32, 1242–1254.

Prangishvilli, D., Zillig, W., Gierl, A., Biesert, L., Holz, I. (1982). DNA-dependent RNA polymerase of thermoacidophilic archaebacteria. *Eur. J. Biochem.*, 122, 471–477.

Probst, A.J. and Moissl-Eichinger, C. (2015). Altiarchaeales: Uncultivated archaea from the subsurface. *Life*, 5, 1381–1395.

Probst, A.J., Weinmaier, T., Raymann, K., Perras, A., Emerson, J.B., Rattei, T., Wanner, G., Klingl, A., Berg, I.A., Yoshinaga, M. et al. (2014). Biology of a widespread uncultivated archaeon that contributes to carbon fixation in the subsurface. *Nat. Commun.*, 5, 5497. doi: 10.1038/ncomms6497.

Rangel, L.T. and Fournier, G. (2023). Fast-evolving sites are highly informative for reconstructions of deep tree of life phylogenies. *Microorganisms*, 11(10), 2499. doi: 10.3390/microorganisms11102499.

Rangel, L.T., Soucy, S.M., Setubal, J.C., Gogarten, J.P., Fournier, G.P. (2021). An efficient, nonphylogenetic method for detecting genes sharing evolutionary signals in phylogenomic data sets. *Genome Biol. Evol.*, 13(9), evab187. doi: 10.1093/gbe/evab187.

Raymann, K., Forterre, P., Brochier-Armanet, C., Gribaldo, S. (2014). Global phylogenomic analysis disentangles the complex evolutionary history of DNA replication in archaea. *Genome Biol. Evol.*, 6, 192–212.

Raymann, K., Brochier-Armanet, C., Gribaldo, S. (2015). The two-domain tree of life is linked to a new root for the Archaea. *Proc. Natl. Acad. Sci. USA*, 112, 6670–6675.

Reysenbach, A.L., Liu, Y., Banta, A.B., Beveridge, T.J., Kirshtein, J.D., Schouten, S., Tivey, M.K., Von Damm, K.L., Voytek, M.A. (2006). A ubiquitous thermoacidophilic archaeon from deep-sea hydrothermal vents. *Nature*, 442, 444–447.

Rinke, C., Schwientek, P., Sczyrba, A., Ivanova, N.N., Anderson, I.J., Cheng, J.F., Darling, A., Malfatti, S., Swan, B.K., Gies, E.A. et al. (2013). Insights into the phylogeny and coding potential of microbial dark matter. *Nature*, 499, 431–437.

Rinke, C., Rubino, F., Messer, L.F., Youssef, N., Parks, D.H., Chuvochina, M., Brown, M., Jeffries, T., Tyson, G.W., Seymour, J.R., Hugenholtz, P. (2019). A phylogenomic and ecological analysis of the globally abundant Marine Group II archaea (*Ca.* Poseidoniales ord. nov.). *ISME J.*, 13, 663–675.

Rinke, C., Chuvochina, M., Mussig, A.J., Chaumeil, P.A., Davín, A.A., Waite, D.W., Whitman, W.B., Parks, D.H., Hugenholtz, P. (2021). A standardized archaeal taxonomy for the genome taxonomy database. *Nat. Microbiol.*, 6, 946–959.

Rodrigues-Oliveira, T., Wollweber, F., Ponce-Toledo, R.I., Xu, J., Rittmann, S.K., Klingl, A., Pilhofer, M., Schleper, C. (2022). Actin cytoskeleton and complex cell architecture in an Asgard archaeon. *Nature*, 613, 332–339. doi: 10.1038/s41586-022-05550-y.

Rudolph, C., Wanner, G., Huber, R. (2001). Natural communities of novel archaea and bacteria growing in cold sulfurous springs with a string-of-pearls-like morphology. *Appl. Environ. Microbiol.*, 67, 2336–2344.

Sandman, K., Krzycki, J.A., Dobrinski, B., Lurz, R., Reeve, J.N. (1990). HMf, a DNA-binding protein isolated from the hyperthermophilic archaeon *Methanothermus fervidus*, is most closely related to histones. *Proc. Natl. Acad. Sci. USA*, 87, 5788–5791.

Schwank, K., Bornemann, T.L.V., Dombrowski, N., Spang, A., Banfield, J.F., Probst, A.J. (2019). An archaeal symbiont-host association from the deep terrestrial subsurface. *ISME J.*, 13, 2135–2139.

Seitz, K.W., Lazar, C.S., Hinrichs, K.U., Teske, A.P., Baker, B.J. (2016). Genomic reconstruction of a novel, deeply branched sediment archaeal phylum with pathways for acetogenesis and sulfur reduction. *ISME J.*, 10, 1696–1705.

Seitz, K.W., Dombrowski, N., Eme, L., Spang, A., Lombard, J., Sieber, J.R., Teske, A.P., Ettema, T.J.G., Baker, B.J. (2019). Asgard archaea capable of anaerobic hydrocarbon cycling. *Nat Commun.*, 10(1), 1822. doi: 10.1038/s41467-019-09364-x.

Sorokin, D.Y., Makarova, K.S., Abbas, B., Ferrer, M., Golyshin, P.N., Galinski, E.A., Ciordia, S., Mena, M.C., Merkel, A.Y., Wolf, Y.I. et al. (2017). Discovery of extremely halophilic, methyl-reducing Euryarchaea provides insights into the evolutionary origin of methanogenesis. *Nat. Microbiology*, 2, 17081. doi: 10.1038/nmicro-biol.2017.81.

Sorokin, D.Y., Merkel, A.Y., Abbas, B., Makarova, K.S., Rijpstra, W.I.C., Koenen, M., Sinninghe Damsté, J.S., Galinski, E.A., Koonin, E.V., van Loosdrecht, M.C.M (2018). *Methanonatronarchaeum thermophilum* gen. nov., sp. nov. and '*Candidatus* Methanohalarchaeum *thermophilum*', extremely halo(natrono)philic methylreducing methanogens from hypersaline lakes comprising a new Euryarchaeal class Methanonatronarchaeia classis nov. *International Journal of Systematic and Evolutionary Microbiology*, 68, 2199–2208.

Sorokin, D.Y., Makarova, K.S., Abbas, B., Ferrer, M., Golyshin, P.N., Galinski, E.A., Ciordia, S., Mena, M.C., Merkel, A.Y., Wolf, Y.I. et al. (2019). Reply to evolutionary placement of Methanonatronarchaeia. *Nat. Microbiol.*, 4, 60–561.

Spang, A., Hatzenpichler, R., Brochier-Armanet, C., Rattei, T., Tischler, P., Spieck, E., Streit, W., Stahl, D.A., Wagner, M., Schleper, C. (2010). Distinct gene set in two different lineages of ammonia-oxidizing archaea supports the phylum *Thaumarchaeota. Trends Microbiol.*, 18, 331–340.

Spang, A., Saw., J.H., Jørgensen, S.L., Zaremba-Niedzwiedzka, K., Martijn, J., Lind, A.E., van Eijk, R., Schleper, C., Guy, L., Ettema, T.J.G. (2015). Complex archaea that bridge the gap between prokaryotes and eukaryotes. *Nature*, 521, 173–179.

Sutcliffe, I.C., Arahal, D.R., Göker, M., Oren, A. (2022). ICSP response to 'Science depends on nomenclature, but nomenclature is not science'. *Nat. Rev. Microbiol.*, 20, 249–250.

Tully, B.J. (2019). Metabolic diversity within the globally abundant Marine Group II Euryarchaea offers insight into ecological patterns. *Nat. Commun.*, 10(1), 271. doi: 10.1038/s41467-018-07840-4.

Vanwonterghem, I., Evans, P.N., Parks, D.H., Jensen, P.D., Woodcroft, B.J., Hugenholtz, P., Tyson, G.W. (2016). Methylotrophic methanogenesis discovered in the archaeal phylum *Verstraetearchaeota. Nat. Microbiol.*, 1, 16170. doi: 10.1038/nmicrobiol.2016.170.

Villain, P., Catchpole, R., Forterre, P., Oberto, J., da Cunha, V., Basta, T. (2022). Expanded dataset reveals the emergence and evolution of DNA gyrase in archaea. *Mol. Biol. Evol.*, 39(8), msac155. doi: 10.1093/molbev/msac155.

Villanueva, L., Schouten, S., Damsté, J.S. (2017). Phylogenomic analysis of lipid biosynthetic genes of Archaea shed light on the 'lipid divide'. *Environ. Microbiol.*, 19, 54–69.

Wade, T., Rangel, L.T., Kundu, S., Fournier, G.P., Bansal, M.S. (2020). Assessing the accuracy of phylogenetic rooting methods on prokaryotic gene families. *PLoS One*, 15(5), e0232950. doi: 10.1371/journal.pone.0232950.

Wadsworth, R.I. and White, M.F. (2001). Identification and properties of the crenarchaeal single-stranded DNA binding protein from *Sulfolobus solfataricus*. *Nucleic. Acids. Res.*, 15, 914–920.

Wang, Y., Wegener, G., Hou, J., Wang, F., Xiao, X. (2019). Expanding anaerobic alkane metabolism in the domain of Archaea. *Nat. Microbiol.*, 4, 595–602.

Wang, Y., Wegener, G., Williams, T.A., Xie, R., Hou, J., Tian, C., Zhang, Y., Wang, F., Xiao, X. (2021). A methylotrophic origin of methanogenesis and early divergence of anaerobic multicarbon alkane metabolism. *Sci. Adv.*, 7(27), eabj1453. doi: 10.1126/sciadv.abj1453.

Williams, T.A., Szöllősi, G.J., Spang, A., Foster, P.G., Heaps, S.E., Boussau, B., Ettema, T.J.G., Embley, T.M. (2017). Integrative modeling of gene and genome evolution roots the archaeal tree of life. *Proc. Natl. Acad. Sci. USA*, 114, 4602–4611.

Woese, C.R. (1981). Archaebacteria. *Scientific American*, 244, 98–125.

Woese, C.R. (1998). Default taxonomy: Ernst Mayr's view of the microbial world. *Proc. Natl. Acad. Sci. USA*, 95, 11043–11046.

Woese, C.R. and Fox, G.E. (1977). Phylogenetic structure of the prokaryotic domain: The primary kingdoms. *Proc. Natl. Acad. Sci. USA*, 74, 5088–5090.

Woese, C.R. and Olsen, G.J. (1986). Archaebacterial phylogeny: Perspectives on the urkingdoms. *Syst. Appl. Microbiol.*, 7, 161–717.

Woese, C.R., Maniloff, J., Zablen, L.B. (1980). Phylogenetic analysis of the mycoplasmas. *Proc. Natl. Acad. Sci. USA*, 77, 494–498.

Woese, C.R., Kandler, O., Wheelis, M.L. (1990). Towards a natural system of organisms: Proposal for the domains Archaea, Bacteria, and Eucarya. *Proc. Natl. Acad. Sci. USA*, 87, 4576–4579.

Xie, R., Wang, Y., Huang, D., Hou, J., Li, L., Hu, H., Zhao, X., Wang, F. (2021). Expanding Asgard members in the domain of Archaea sheds new light on the origin of eukaryotes. *Sci. China Life Sci.*, 65, 818–829. doi: 10.1007/s11427-021-1969-6.

Yarza, P., Yilmaz, P., Pruesse, E., Glöckner, F.O., Ludwig, W., Schleifer, K.H., Whitman, W.B., Euzéby, J., Amann, R., Rosselló-Móra, R. (2014). Uniting the classification of cultured and uncultured bacteria and archaea using 16S rRNA gene sequences. *Nat. Rev. Microbiol.*, 12, 635–645.

Yin, X., Zhou, G., Cai, M., Zhu, Q.Z., Richter-Heitmann, T., Aromokeye, D.A., Liu, Y., Nimzyk, R., Zheng, Q., Tang, X., Friedrich, M.W. et al. (2022). Catabolic protein degradation in marine sediments confined to distinct archaea. *ISME J.*, 16, 1617–1626.

Yu, T., Hu, H., Zeng, X., Wang, Y., Pan, D., Wang, F. (2022). Cultivation of widespread *Bathyarchaeia* reveals a novel methyltransferase system utilizing lignin-derived aromatics. *mLife*, 2(3), 272–282. doi: 10.1101/2022.10.23.513425.

Zablen, L.B., Kissil, M.S., Woese, C.R., Buetow, D.E. (1975). Phylogenetic origin of the chloroplast and prokaryotic nature of its ribosomal RNA. *Proc. Natl. Acad. Sci. USA*, 72, 2418–2422.

Zaremba-Niedzwiedzka, K., Caceres, E.F., Saw, J.H., Bäckström, D., Juzokaite, L., Vancaester, E., Seitz, K.W., Anantharaman, K., Starnawski, P., Kjeldsen K.U. et al. (2017). Asgard archaea illuminate the origin of eukaryotic cellular complexity. *Nature*, 41, 353–358.

Zhang, X., Huang, Y., Liu, Y., Xu, W., Pan, J., Zheng, X., Du, H., Zhang, C., Lu, Z., Zou, D. et al. (2022). An ancient respiratory system in the widespread sedimentary archaea Thermoprofundales. *Mol. Biol. Evol.*, 39(10), msac213. doi: 10.1093/molbev/msac213. 2022.

Zhou, Z., Pan, J., Wang, F., Gu, J.D., Li, M. (2018). *Bathyarchaeota*: Globally distributed metabolic generalists in anoxic environments. *FEMS Microbiol. Rev.*, 42, 639–655.

Zhou, Z., Liu, Y., Lloyd, K.G., Pan, J., Yang, Y., Gu, J.D., Li, M. (2019). Genomic and transcriptomic insights into the ecology and metabolism of benthic archaeal cosmopolitan, Thermoprofundales (MBG-D archaea). *ISME J.*, 13, 885–901.

Zillig, W., Stetter, K.O., Janeković, D. (1979). DNA-dependent RNA polymerase from the archaebacterium *Sulfolobus acidocaldarius*. *Eur. J. Biochem.*, 96, 597–604.

Zillig, W., Schäfer, W., Reiter, W.D. (1983). The archaebacterium *Thermococcus celer* represents, a novel genus within the thermophilic branch of the archaebacteria. *Syst. Appl. Microbiol.*, 4, 88–94.

Zillig, W., Prangishvilli, D., Schleper, C., Elferink, M., Holz, I., Albers, S., Janekovic, D., Götz, D. (1996). Viruses, plasmids and other genetic elements of thermophilic and hyperthermophilic Archaea. *FEMS Microbiol. Rev.*, 18, 225–236.

Zinke, L.A., Evans, P.N., Santos-Medellin, C., Schroeder., A.L., Parks, D.H., Varner, R.K., Rich, V.I., Tyson, G.W., Emerson, J.B. (2021). Evidence for non-methanogenic metabolisms in globally distributed archaeal clades basal to the *Methanomassiliicoccales*. *Environ. Microbiol.*, 23, 340–357.

3

Archaea and the Tree of Life

Patrick FORTERRE[1,2]
[1] *Institut Pasteur, Université Paris Cité, France*
[2] *Université Paris-Saclay, Gif-sur-Yvette, France*

3.1. Introduction

The discovery of archaea at the end of the last century (see Chapter 1) immediately rekindled the interest of evolutionists in studying the early stages of life on our planet. Since then, various, often contradictory hypotheses have been proposed regarding the nature of the last universal common ancestor (LUCA)[1] and the relationships between the three domains of life, leading to numerous disputes among scientists.

For the past 10 years, the debate on the origin of eukaryotes and their relationship with archaea has been at the forefront. Currently, the dominant hypothesis is that eukaryotes emerged from the association of an Asgard archaea and one or more bacteria.

However, the data supporting this hypothesis are contested by some authors, including myself. A reexamination of recently published phylogenetic analyses suggests that the Woese tree remains valid. As in Chapter 2, I will adopt a historical approach, which, in my view, is essential for fully understanding the stakes in the ongoing debates.

1 The acronym LUCA, which stands for Last Universal Common Ancestor, was proposed in 2016 during a symposium organized by the Fondation des Treilles (Forterre 1996a).

Biology of Archaea 1,
coordinated by Béatrice CLOUET-D'ORVAL, Bruno FRANZETTI and Philippe OGER.

3.2. The progenote concept

In their early publications on archaea, Carl Woese and George Fox suggested that the three primordial kingdoms (urkingdoms) they had identified diverged almost simultaneously from a much simpler universal common ancestor than modern organisms, which they called the progenote (Woese and Fox 1977a, 1977b; Woese 1981). According to them, this progenote still had an RNA genome and a primitive translation system which did not yet allow the production of proteins with high fidelity. For Carl Woese, this could explain why the evolution of RNA sequences occurred much faster at that time than today. Over roughly a billion years, Earth's history saw the emergence of life and its evolution to LUCA with the development of RNA and the ribosome. Subsequently, ribosomal RNA (rRNA) sequences gradually stabilized in their current state between the progenote, which lived between 3.5 and 4 billion years ago, and the last common ancestors of archaea, bacteria and eukaryotes (the LACA, LBCA and LECA)[2], which lived between 3 and 3.5 billion years ago.

The three versions of rRNA for each domain have since evolved very little over the last three billion years. The same observation can be made today for the ribosome as a whole (Bowman et al. 2020) and for all the features unique to each domain: bacteria remained bacteria, archaea remained archaea and eukaryotes remained eukaryotes. To use a term employed by Carl Woese, the period between LUCA and the three ancestors experienced a slowdown in the "tempo" of evolution.

To propose that the progenote was a much simpler organism than modern organisms, Carl Woese and George Fox based their idea on the significant differences revealed by molecular biologists in the 1970s between the translation mechanisms in bacteria and eukaryotes. For example, the mechanisms of translation initiation involved the recognition of the Shine–Dalgarno sequence of messenger RNAs by 16S rRNA in bacteria and the recognition of the cap of messenger RNAs by the ribosome in eukaryotes. According to them, the mechanisms of replication, transcription and translation had refined independently in the three cellular lineages. Cells had thus progressed from the "progenote" stage to the "prokaryote" stage. Carl Woese and George Fox proposed to name the now-extinct prokaryotes that gave rise to eukaryotes, before the appearance of the nucleus, "Urkaryotes" (Woese and Fox 1977a).

2 Last Archaeal Common Ancestor, Last Bacterial Common Ancestor, Last Eukaryotic Common Ancestor.

3.3. Archaea: prokaryotes related to eukaryotes

The rapid development of research on archaea by biochemists in the 1980s led to a completely unexpected discovery. Alongside characteristics unique to them, such as the structure of their lipids, archaea exhibited numerous traits that brought them closer to eukaryotes (see Chapters 1 and 2). What is the evolutionary significance of this molecular resemblance between prokaryotic organisms, the archaea and eukaryotes?

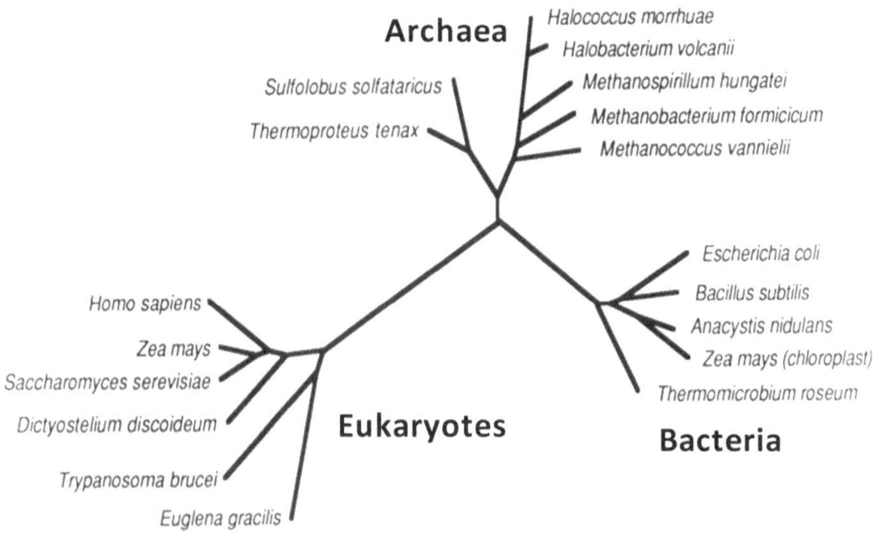

Figure 3.1. *First universal tree based on 16S rRNA. In this tree, archaea appear closer to bacteria than to eukaryotes*

In the first universal tree based on the phylogenetic analysis of complete 16S rRNA sequences, published in 1986 by Carl Woese and his colleagues, archaea were not closer to eukaryotes than to bacteria (Pace et al. 1986) (Figure 3.1).

However, when researchers were able to construct universal trees based on protein sequences, archaea appeared very close to eukaryotes, confirming the results obtained with RNA polymerases (for a review, see Gaia et al. (2018)) (see the example of ribosomal proteins, Figure 3.2). Notably, this contradiction between trees based on 16S rRNA and those based on proteins is still not fully understood today.

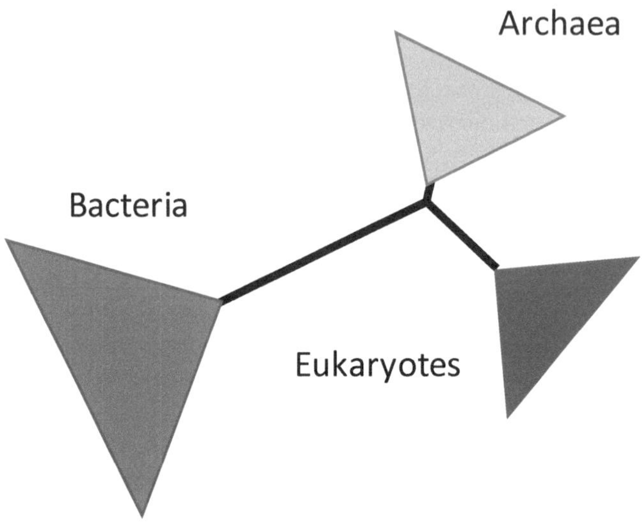

Figure 3.2. *Universal schematic tree based on the concatenation of 29 universal ribosomal proteins (branch lengths are respected) (adapted from Fournier and Gogarten (2010)). For a color version of this figure, see www.iste.co.uk/clouet/archaea1.zip*

COMMENT ON FIGURE 3.2.– *In this tree, archaea are much closer to eukaryotes than to bacteria. With the addition of an ever-increasing number of archaeal sequences, eukaryotes would now be found within archaea (2D tree, see section 3.6).*

3.4. Rooting the universal tree

In 1989, two teams used paralogous universal protein pairs[3], which had diverged by gene duplication before LUCA, to root the universal tree (Gogarten et al. 1989; Iwabe et al. 1989). These were, on the one hand, the translation elongation factors, called EFTu and EFG in bacteria and EF1 and EF2 in archaea and eukaryotes; and, on the other hand, the two catalytic subunits of bacterial and archaeal ATP synthases and the vacuolar ATPase of eukaryotes. For each pair of paralogous proteins, two universal trees were obtained, each of which could serve as an outgroup to root the other. In both cases, the root was located on the branch leading to bacteria (Figure 3.3 shows the result obtained with the elongation factors).

3 Paralogous proteins are homologous proteins that diverged not by speciation but through gene duplication.

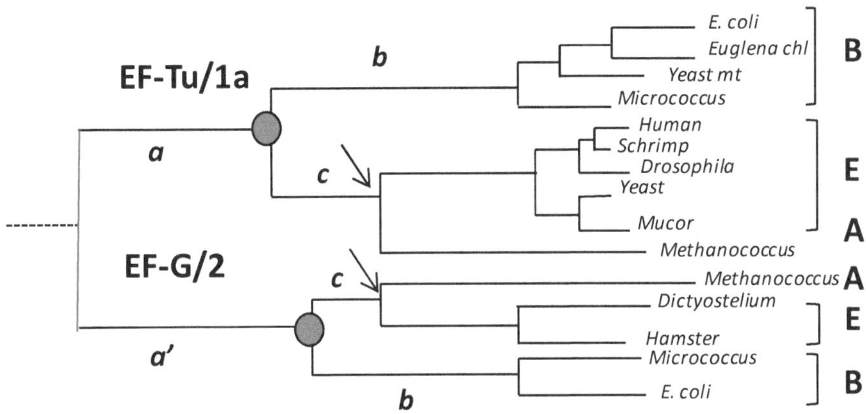

Figure 3.3. *Universal trees of elongation factors EF1(Tu) and EF2(G) (Iwabe et al. 1989)*

COMMENT ON FIGURE 3.3.– *The gray circles indicate the position of the root, which in both cases is located in the bacterial branch. The arrows indicate the positions of the common ancestor to archaea and eukaryotes. As in Figure 3.2, archaea are much closer to eukaryotes than to bacteria. At that time, the sequences of these proteins were known for only one archaeon, Methanococcus. Since then, numerous EF2 paralogs have been discovered in eukaryotes and archaea, and the overall phylogeny of this elongation factor has become extremely complex (Narrowe et al. 2018).*

By positioning this root on the 16S rRNA tree (which could not be rooted directly due to the absence of an outgroup), a tree was obtained in which bacteria were separated from archaea by this root (Woese et al. 1990) (Figure 3.4). For Carl Woese, this separation justified abandoning the term "Archaebacteria" in favor of "Archaea" to permanently sever the umbilical cord between bacteria and archaea (see Chapter 2).

In this tree, often referred to today as the Woese tree, archaea and eukaryotes diverged from a specific common ancestor. We thus obtained a tree with two major branches: one leading to bacteria and the other leading to an ancestor common to archaea and eukaryotes. The clade corresponding to the grouping of archaea and eukaryotes did not receive a name, likely because Carl Woese preferred to maintain a three-domain (3D) system (Forterre 2022a, 2022b). I proposed to name this clade Arcarya, grouping archaea and eukaryotes (Forterre 2015), but this term has not yet been adopted in the scientific literature.

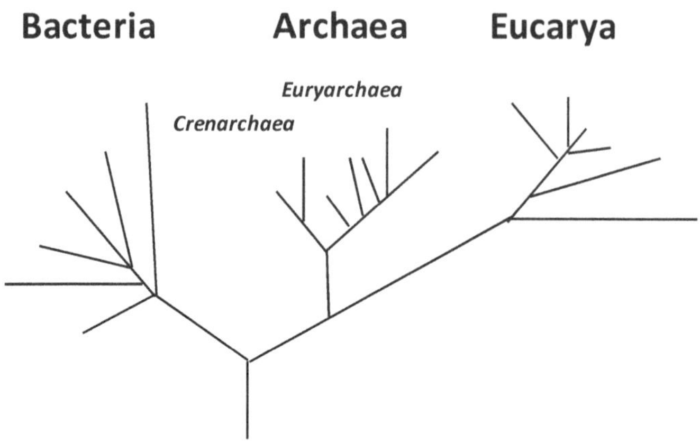

Figure 3.4. *The Woese tree. The 16S rRNA tree from Figure 3.1, rooted between bacteria and an archaea/eukaryotic clade using data obtained from two pairs of universal paralogous proteins (Gogarten et al. 1989; Iwabe et al. 1989)*

In the late 1990s, Hervé Philippe and I criticized the Woese tree, rooted in the "bacterial branch", due to significant differences between the lengths of the bacterial branches (b) in Figure 3.3 and those of archaea and eukaryotes (c) in the paralogous protein trees used to position the root (Forterre and Philippe 1999; Philippe and Forterre 1999). It was impossible to dismiss the hypothesis that the long bacterial branches of the two paralogous proteins were attracted to the long branch separating the two paralogous proteins (a + a'), a phenomenon known as "Long Branch Attraction" (LBA), artificially placing the root within these branches[4]. This issue remains unresolved and is unlikely to be solved using this approach because the paralogous proteins used to root the universal tree are so divergent that recovering a reliable phylogenetic signal to confirm or refute LBA seems impossible.

Subsequently, Fournier and Gogarten (2010) employed a different strategy to root the universal tree. They demonstrated, using the phylogeny of ribosomal proteins illustrated in Figure 3.2, that the branch leading from the trifurcation point to the bacteria was enriched with supposedly ancient amino acids[5], suggesting that LUCA should be positioned on this branch. The accumulation of molecular data

4 Long branch attraction is due to the conservation of a greater number of ancestral characters in organisms with short branches. It was first described by Felsenstein (1978).

5 Ancient amino acids would be the first to have been integrated into the genetic code. They would have had more time to become fixed in protein sequences before LUCA and would have been enriched in these sequences at the time of LUCA.

linking archaea more closely to eukaryotes has provided additional support for a "bacterial" root (Forterre 2015). Genome sequencing confirmed that nearly all archaeal proteins involved in "informational" molecular mechanisms, such as DNA replication, transcription and translation, were much more similar to those of eukaryotes than those of bacteria. In many cases, these proteins common to archaea and eukaryotes did not even have homologs in bacteria, only analogs[6].

The case of ribosomes is particularly significant (Lecompte et al. 2002; Forterre 2015; Bowman et al. 2020). Comparison of ribosomes from the three domains revealed that 34 ribosomal proteins are universal, and another 34 are common to archaea and eukaryotes but absent in bacteria (Figure 3.5). Conversely, there are no proteins common to archaea and bacteria that are absent in eukaryotes, nor any proteins common to bacteria and eukaryotes that are absent in archaea. The 33 ribosomal proteins common to archaea and eukaryotes are replaced in bacteria by 24 non-homologous proteins, which nevertheless interact with the same regions of the rRNAs. If the 33 proteins common to archaea and eukaryotes or the 24 bacterial-specific proteins were already present in LUCA (assuming a complex LUCA), it must be assumed that the 24 bacterial proteins were replaced in situ during evolution by the 33 proteins common to archaea and eukaryotes or vice versa (Figure 3.5). Such a scenario implies both losses and gains.

The most parsimonious hypothesis, which involves only gains, is that of a LUCA whose ribosomes included only the 34 universal proteins (simple LUCA) (Figure 3.5). The 33 proteins common to archaea and eukaryotes and the 24 bacterial-specific proteins were added later in the respective lineages.

Other scenarios can be imagined beyond the two presented in Figure 3.5, depending on whether we envision a simple or complex LUCA associated with different roots. The most parsimonious hypothesis remains that of a simple LUCA with the root located between bacteria and archaea. However, the hypothesis of a massive replacement of ribosomal proteins during evolution cannot be completely excluded. Such replacement has been observed in mitochondrial ribosomes (Desmond et al. 2011). Over the course of evolution, mitochondria have lost many proteins of bacterial origin and acquired new proteins of unknown origin. It is currently unclear what selection pressure could have led to the replacement of such a

6 Analogous proteins perform the same function. They may (or may not) be related. In contrast, homologous proteins are necessarily related because, by definition, they descend from a common ancestor. It is incorrect to speak of a percentage of homology. Two proteins are either homologous or they are not.

large batch of ancestral ribosomal proteins in proto-bacteria or in the lineage leading to archaea and eukaryotes.

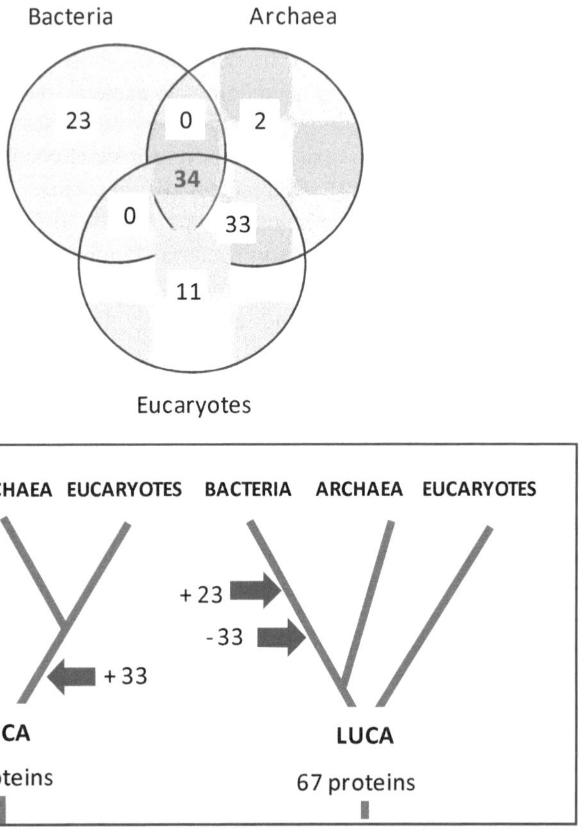

Figure 3.5. *Distribution of ribosomal proteins in the three domains. Scenario (a), with the root in the bacterial branch, is more parsimonious than scenario (b), with the root in the eukaryotic branch. For a color version of this figure, see www.iste.co.uk/clouet/archaea1.zip*

The hypothesis of a LUCA with ribosomes simpler than current ribosomes is also consistent with the fact that there is only one universal protein involved in ribosome biogenesis, whereas these proteins are numerous in each of the three domains of life, with several common to archaea and eukaryotes (Birikmen et al. 2021; Jüttner and Ferreira-Cerca 2022).

3.5. The nature of LUCA

3.5.1. *A simpler LUCA compared to the organisms of the three current domains*

Comparison of the mechanisms of replication, translation and transcription in bacteria, archaea and eukaryotes can provide important insights into the nature of LUCA. If we choose the most parsimonious scenario, with the root located between bacteria and archaea, we can conclude that not only were LUCA's ribosomes simpler than current ribosomes, but also that many other molecular mechanisms were less sophisticated than those we know today. For example, the fact that the transcription factors involved in promoter recognition are not homologous in bacteria compared to archaea and eukaryotes suggests that the transcription initiation mechanisms in LUCA were less specific than in modern organisms (Werner and Grohmann 2011). This scenario aligns with Carl Woese's initial hypothesis that the mechanisms of replication, transcription and translation improved independently in the three cell lineages.

Another important observation concerns ATP synthases, which are partly homologous but very different between bacteria on one side and archaea and eukaryotes on the other side. These complex enzymes consist of several homologous subunits, some of which are membrane-anchored, while others, particularly the two catalytic subunits, are cytoplasmic. The latter are connected to the membrane subunits by one or more subunits whose rotation transforms the whole into a molecular dynamo which provides the energy necessary for ATP synthesis in archaea and bacteria. However, unlike the other subunits, these subunits necessary for ATP synthesis are not homologous between bacteria and the other two domains. If the root of the universal tree is indeed located between bacteria and archaea, it is more parsimonious to think that these subunits were not present in LUCA. Some authors believe that the ancestor of these enzymes functioned in reverse in LUCA, consuming ATP to allow ion flows through the membrane (Mulkidjanian et al. 2007). In this case, LUCA would have obtained its ATP through fermentation.

3.5.2. *An RNA genome for LUCA?*

The most spectacular case distinguishing the molecular biology of bacteria from that of the other two domains is DNA replication. The three major bacterial proteins involved in this process – replicase, helicase and primase – have no homologues but only analogues in archaea and eukaryotes (Figure 3.6) (Olsen and Woese 1997). This observation led some authors to resurrect the hypothesis proposed by Woese and Fox in 1977 that LUCA still possessed an RNA genome (Mushegian and

Koonin 1996; Forterre 2006). According to this hypothesis, DNA and the mechanisms for its replication would have been introduced twice independently into the cell lineages that led to bacteria, on the one hand, and archaea and eukaryotes, on the other hand (Figure 3.6).

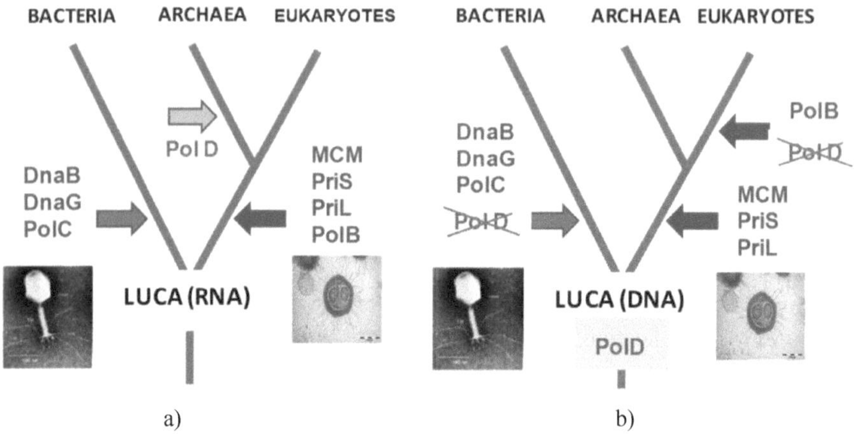

Figure 3.6. *Two evolutionary scenarios for DNA replication. For a color version of this figure, see www.iste.co.uk/clouet/archaea1.zip*

COMMENT ON FIGURE 3.6.– *From (a) an RNA LUCA (Forterre 2002) and (b) a DNA LUCA (Koonin et al. 2020a). Different color names indicate non-homologous proteins: DNA polymerases (Pol B, Pol C and Pol D)*[7]*, DNA primases (DNA G, PriS and PriL) and helicases (DnaB and MCM). Several other replication proteins – DNA ligase, processivity factors, SSB, etc. – are homologous but highly divergent between bacteria and archaea/eukaryotes. In the model on the left, these proteins were introduced by different viruses that also possessed them. In the model on the right, they could assist DNA polymerase D in LUCA.*

To explain this transition, I proposed in 2002, the hypothesis that bacteria, on the one hand, and archaea/eukaryotes, on the other hand, had inherited both DNA and two distinct sets of replication proteins from viruses[8] (Forterre 2002a). The idea that

7 Archaea possess a Pol B very divergent from eukaryotic Pol B, but except for *Crenarchaea* and *Marsarchaea*, they possess a Pol D, which they use as a replicase.

8 I even considered three independent transfers (Forterre 2006) because the replication mechanisms also show significant differences between archaea and eukaryotes, particularly the ancestral DNA replicases and type II DNA topoisomerases of archaea and eukaryotes are not homologous.

DNA and the proteins responsible for its replication have a viral origin is based on the observation that the various known viral lineages are characterized by a very high diversity of their DNA replication proteins. The replication proteins of archaea, bacteria and eukaryotes seem to correspond to a subset of viral replication proteins. By analogy, we can deduce their viral origin in the same way that the greater genetic diversity of African populations suggests the African origin of *Homo sapiens*. This is the "out of viruses" hypothesis for the origin of DNA and cellular replication proteins.

If LUCA still had an RNA genome, it is possible that the slowing of the rate of evolution between LUCA's time and the emergence of the three domains was a consequence of the transition from RNA genomes to DNA genomes. DNA is indeed more stable than RNA, and DNA polymerases are generally more accurate than RNA polymerases. Additionally, the deamination of cytosine to uracil, a chemical modification that frequently occurs spontaneously in nucleic acids, cannot be repaired in the case of RNA, since uracil is normally present in this molecule. In contrast, this spontaneous mutation from C to U will be recognized as an error in the case of DNA and can be repaired.

Some authors believe that RNA is not stable enough and that RNA replication is not accurate enough for this molecule to have served as a genome for LUCA (Takeuchi et al. 2011). However, it should be noted that there are RNA repair mechanisms and mechanisms that increase the fidelity of viral RNA replicases[9] (Poole and Logan 2005). Even if DNA repair mechanisms are more efficient, RNA could have been replicated faithfully enough to allow the functioning of a simpler LUCA than current cells.

Proponents of a DNA-based LUCA, Eugene Koonin and his collaborators, recently proposed a scenario in which LUCA's DNA was replicated by a DNA polymerase of the D family (Koonin et al. 2020a) (Figure 3.6). The DNA polymerases of this family are related to the RNA polymerases involved in transcription and currently serve as replicases for most archaea. According to these authors, all DNA polymerases are of viral origin except for DNA polymerase D, which uniquely appeared in a cellular lineage. This DNA polymerase would have been replaced by family B polymerases in eukaryotes and family C in bacteria[10] (Figure 3.6). If we assume that LUCA already had a DNA genome, it remains to be

9 This is the case with coronaviruses, which can increase or decrease their mutation rate depending on whether their environment is favorable or not, by modulating the activity of an enzyme capable of correcting replicase errors.

10 DNA polymerases of families A, B, C and D are not homologous. They probably evolved independently from different RNA polymerase families.

explained why the primases and helicases involved in replication are not homologous to those of archaea and eukaryotes. We must imagine that, in addition to the loss of Pol D, the primases and helicases of one of the two types (bacterium or archaea/eukaryote) present in LUCA were replaced by those of the other type in one of the two main branches of the universal tree, which is not very parsimonious. In my opinion, the current comparative biochemistry data suggest a LUCA much simpler than modern organisms (small ribosomes, imprecise transcription and absence of ATP synthase), which aligns well with an RNA-based LUCA. Thus, LUCA would have been very different from modern archaea and bacteria. The similarities observed between archaea and bacteria (small DNA genomes, genes grouped in operons) could correspond to a convergence phenomenon which led to the "prokaryote" phenotype.

Although comparative biochemistry data seem to support the concept of a progenote, they do not validate one of the predictions of Carl Woese and George Fox, namely, that the progenote could not produce proteins with high fidelity. Biochemists have identified several proteins involved in modifications of transfer RNAs (tRNAs) that must have already been present in LUCA (Forterre 2015; Van der Gulik and Hoff 2016)[11]. These modifications are crucial for allowing accurate translation of messenger RNA into proteins. Therefore, another hypothesis is needed to explain why the pace of evolution slowed between LUCA and the emergence of the three domains. The hypothesis that this slowdown was due to the transition from RNA genomes to DNA genomes currently seems the most probable to me.

3.5.3. *A presumably mesophilic LUCA*

The discovery of anaerobic hyperthermophilic archaea by Wolfram Zillig and Karl Stetter in the 1980s (see Chapters 1 and 2) revived the idea that life arose at high temperatures under the extreme environmental conditions of early Earth (Stetter 2013). In the 1990s, Karl Stetter redesigned Carl Woese's universal tree of life, highlighting in bold the lines leading to organisms whose optimal growth temperatures exceed 80°C. These clustered at the base of the tree, and Stetter highlighted in bold the lineages leading from the origins of life to the last common ancestors of archaea and bacteria, including LUCA (Figure 3.7) (Stetter 1996).

---

11 A particularly important universal modification for the reading of many codons is carried out by different complexes in bacteria and in archaea/eukaryotes. These complexes, however, include two universal proteins that are sufficient to perform this reaction in mitochondria. The most parsimonious scenario to explain the evolution of these complexes again involves a bacterial root (Forterre 2015).

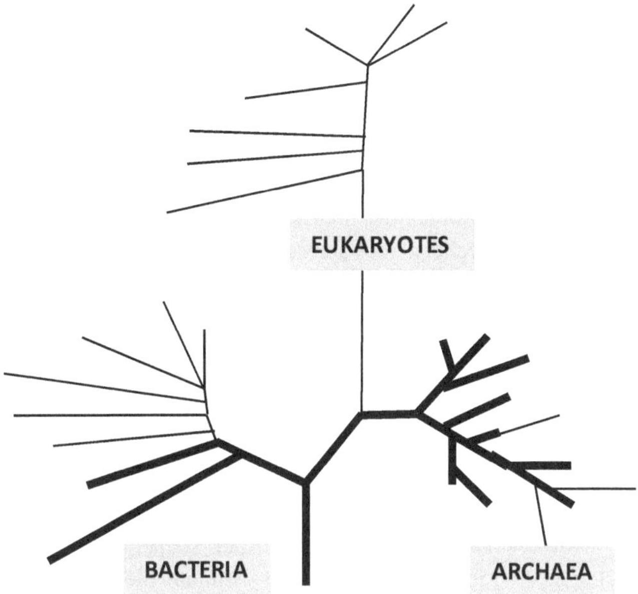

Figure 3.7. *The Woese tree based on 16S rRNAs redesigned by Karl Stetter to illustrate the hypothesis of a hyperthermophilic LUCA. The lines leading to the hyperthermophilic organisms known at the time are in bold*

I have criticized this hypothesis by pointing out that the clustering of hyperthermophiles around the root in the rRNA-based tree may be due to the GC-rich base pairs in these RNAs, whether in hyperthermophilic archaea or bacteria (Forterre 1995, 1996b). This GC richness contributes to their stability at high temperatures by facilitating the formation of hairpins with GC-rich stems at their bases. The maintenance of this GC richness, imposed by high temperatures, will limit the evolutionary possibilities of the corresponding sequences. Hyperthermophiles will have either convergently acquired G or C bases at the same positions and/or retained G or C bases present in their ancestors. In both cases, these mechanisms will slow the rate of evolution of their rRNAs, positioning them at the end of short branches in phylogenetic trees. Consequently, they will cluster due to a phenomenon which could be called the short branch attraction.

If the position of hyperthermophiles in RNA trees cannot be used to determine the temperature at which LUCA lived, how can we infer this? One approach is to examine whether LUCA possessed a reverse gyrase (see Chapter 2). This enzyme is deemed absolutely essential for life above 80°C, present without exception in all archaea and hyperthermophilic bacteria (and in many thermophilic organisms) and absent in all mesophilic organisms (Forterre 2002b; Catchpole and Forterre 2019)[12]. Phylogenetic analysis has shown that reverse gyrase was probably not present in LUCA (Forterre et al. 2000; Brochier-Armanet and Forterre 2007; Catchpole and Forterre 2019). Indeed, the reverse gyrases of archaea and bacteria are not separated by a long branch in the phylogenetic tree reconstructed from their sequences (Figure 3.8). On the contrary, the reverse gyrases of archaea and bacteria are intermixed within this tree, indicating that it was transferred from one domain to the other, presumably from archaea to bacteria. However, archaea and bacteria are almost always separated by long (or even very long) branches in the phylogenetic trees of universal proteins that were already present in LUCA, such as RNA polymerases or translation factors (see Figures 3.2 and 3.3, and also phylogenies in Da Cunha et al. (2017), and analyses by Berkemer and McGlynn, (2022) and Moody et al. (2022)). The absence of reverse gyrase in LUCA is thus a strong argument in favor of a mesophilic or moderately thermophilic LUCA.

In 1999, Manolo Gouy's team in Lyon attempted to determine the temperature at which LUCA lived by reconstructing the virtual sequence of its rRNAs using molecular phylogeny, which allowed them to estimate their GC base percentage. The GC pair percentage of an organism's rRNAs is a very good indicator of the temperature of its habitat. The result suggested that LUCA was probably not a hyperthermophilic organism, but rather a mesophile or a moderate thermophile (Galtier et al. 1999). Subsequently, the same team confirmed this result with a much larger number of organisms using rRNAs and also by determining the amino acid composition of LUCA's proteome[13] (Boussau et al. 2008). The results of these two independent analyses support a mesophilic LUCA, which aligns with the absence of reverse gyrase in LUCA.

Unexpectedly, this same study showed that the last common ancestors of archaea and bacteria (LACA and LBCA) lived at much higher temperatures than LUCA. The adaptation to extreme high temperatures thus appears to have occurred independently twice, once in the branch leading to archaea and once in the branch leading to bacteria (Figure 3.9).

12 The reverse gyrase gene formed from the fusion of two enzymes, a type I DNA topoisomerase and a DNA helicase. Organisms that lived before this fusion did not possess this enzyme and probably could not live above 80°C, unless the precise function of this enzyme – which remains unknown – was performed at the time by another enzymatic activity (Forterre 1996b).

13 The percentage of different amino acids differs in the proteomes of thermophilic and mesophilic organisms, allowing for the determination of a thermophilic index based on the proteome's amino acid composition (Zeldovich et al. 2007).

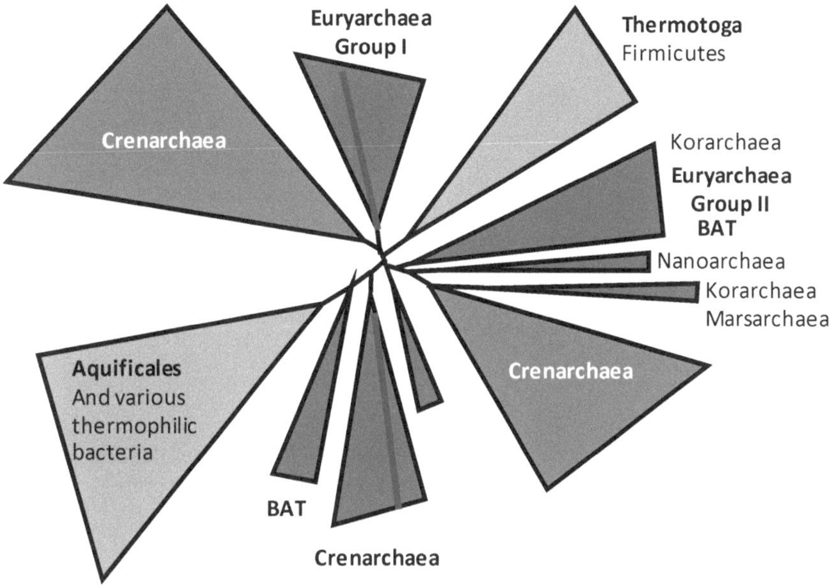

Figure 3.8. *Relationship between reverse gyrases of archaea and bacteria. For a color version of this figure, see www.iste.co.uk/clouet/archaea1.zip*

COMMENT ON FIGURE 3.8.– *The phylogenetic tree groups the 376 known sequences of reverse gyrases as of 2019 (Catchpole and Forterre 2019). Clades corresponding to reverse gyrases of archaea are shown in blue, and those corresponding to reverse gyrases of bacteria are in pink. The dark red lines indicate recent transfers from archaea to bacteria or between bacteria. The Crenarchaea (light blue) possess two reverse gyrases.*

To explain this observation, I proposed the hypothesis that adaptation to hyperthermophilic conditions favored the appearance of the prokaryotic phenotype: small cells allowing rapid renewal of macromolecules damaged by high temperature exposure and coupling of transcription and translation to protect messenger RNA from thermal degradation (Forterre 1995, 2013). This hypothesis might also explain why no eukaryotes are known to live at temperatures above 60°C (Chapter 2). Peter Gogarten and his collaborators proposed that the ancestors of archaea and bacteria were thermophilic because they descended from organisms that survived the intense meteorite bombardment about 3.9 billion years ago (known as the late heavy bombardment) (Gogarten et al. 1989).

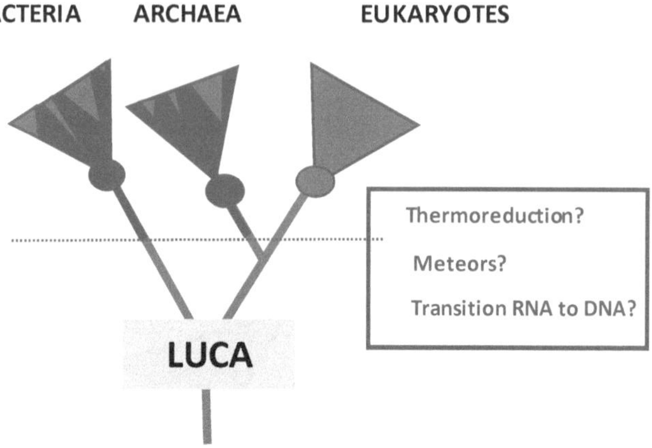

Figure 3.9. *Scenario of independent adaptation to hyperthermophilic life in proto-archaea and proto-bacteria. For a color version of this figure, see www.iste.co.uk/clouet/archaea1.zip*

COMMENT ON FIGURE 3.9.– *LACA and LBCA were thermophilic or hyperthermophilic (Boussau et al. 2009). Many groups of archaea and bacteria (in blue) subsequently adapted to life at lower temperatures. In a 2D tree where eukaryotes emerge from the archaea, it is suggested that eukaryotes also had a hyperthermophilic ancestor.*

Finally, if LUCA indeed retained an RNA genome, we could hypothesize that adaptation to very high temperatures was a key factor driving the selection of DNA as genetic material (Boussau *et al.* 2008). DNA is significantly more stable than RNA at high temperatures. It is conceivable that DNA first appeared in two lineages of hyperthermophilic organisms, which eventually gave rise to bacteria and archaea. Descendants of these early DNA-based organisms likely outcompeted RNA-based lineages over time. These hypotheses are not mutually exclusive and can be considered in combination.

The scenario of a mesophilic LUCA and the independent adaptation of proto-archaea and proto-bacteria to high-temperature environments may also explain why the lipids of hyperthermophilic archaea and bacteria have evolved to thrive under these conditions, while still retaining distinct characteristics typical of each domain. Hyperthermophilic bacteria exhibit fundamental lipid traits akin to those of mesophilic bacteria (especially their stereospecificity; see Chapter 2), albeit with partial resemblance to archaeal lipids. Some hyperthermophilic bacteria possess lipids featuring ether bonds between glycerol and fatty alcohols, while others have lipids consisting of glycerol tetraesters with fatty acids. If archaea and hyperthermophilic bacteria had diverged from a hyperthermophilic LUCA already equipped with thermally adapted lipids, we might expect these lipids not just to be similar (analogous) but homologous, originating from a common ancestral lipid type.

Moreover, the complexity of lipids in hyperthermophilic organisms underscores their evolution from mesophilic ancestors, driven by adaptive traits. Notably, the transition from simpler diether lipids to tetraethers and tetraesters likely involved the emergence of complex enzymes capable of fusing hydrophobic tails, as evidenced by the tetraether synthase recently identified in archaea, a sophisticated enzyme featuring an S-adenosylmethionine radical and four iron-sulfur centers (Lloyd et al. 2022; Zeng et al. 2022).

3.5.4. *LUCA's proteome*

The comprehensive identification of proteins across the three domains theoretically enables the reconstruction of LUCA's minimal proteome. Several research teams have conducted such studies, sometimes yielding conflicting results. Nevertheless, consensus exists among researchers that LUCA possessed foundational components of transcription and translation. Approximately a hundred proteins can be reasonably attributed to LUCA (Koonin 2003). It is important to note that many proteins present in LUCA may have been lost in various lineages during evolution and are no longer identifiable today as universal proteins. Nonetheless, the prevalence of RNA-interacting proteins among universal proteins suggests that RNA played a pivotal role during LUCA's era (Anantharaman et al. 2002; Delaye et al. 2005).

In 2005, Eugene Koonin and William Martin proposed the hypothesis that LUCA existed not as a cellular organism but rather as an association of molecules fixed on a mineral substrate within a submarine hydrothermal vent (Koonin and Martin 2014). However, the presence of membrane proteins among the universal proteins (Sec-type secretion proteins, membrane ATPases, proteins from the ribosome-associated complex involved in synthesizing membrane proteins) indicates that LUCA was indeed a cellular organism with a cytoplasmic membrane (Peretó et al. 2004; Lombard et al. 2012; Forterre 2015). Comparative analysis of lipid biosynthesis pathways across the three domains also supports the notion of LUCA possessing a membrane (Peretó et al. 2004; Lombard et al. 2012). This membrane likely already featured a glycoprotein envelope (Lombard 2016).

Recent studies have concluded that LUCA utilized FtsZ proteins (analogous to tubulin), involved in forming and positioning the division septum in bacteria and archaea (Pende et al. 2021; Santana-Molina et al. 2023). However, the phylogenetic trees presented show very short branches separating FtsZ proteins from archaea and bacteria, with archaea branching within different groups of bacterial FtsZ in one case. Since at least one form of FtsZ is present in the last bacterial ancestor (LBCA), it is plausible that FtsZ proteins were not present in LUCA but were later transferred from proto-bacteria to proto-archaea. In contrast, the presence of other more divergent tubulin homologs in archaea and the homology between FtsZ and eukaryotic tubulin, suggests that LUCA already possessed a homolog of these proteins involved in cell division.

What was LUCA's metabolism? This question presents greater challenges due to the susceptibility of metabolic proteins to gene transfer events. In 2015, William Martin and colleagues published a study identifying 338 proteins believed to have been present in LUCA (Weiss et al. 2016). From these proteins, they reconstructed LUCA's metabolism and concluded that LUCA was an autotrophic and hyperthermophilic organism. Contrary to earlier discussions, they suggested LUCA possessed reverse gyrase[14]. However, this study faced substantial methodological criticism later on (Berkemer and McGlynn 2022).

For Martin and his colleagues, a protein was considered present in LUCA if archaea and eukaryotes formed two monophyletic groups in their phylogeny. However, in most of the phylogenetic trees generated by Martin and his team, archaea and bacteria were separated by only very short branches (Figures 3.10(b)–(d)). However, as previously mentioned, phylogenies of proteins presumed present in LUCA typically show long, sometimes very long, branches separating archaea and bacteria (Da Cunha et al. 2017; Berkemer et al. 2020; Moody et al. 2022).

Furthermore, many trees published by Martin and colleagues included few species that adequately represented either domain, particularly archaea (trees b, c and d in Figure 3.10). Proteins that were present in LUCA and still exist today across all three domains must have been present in their respective ancestors. Consequently, these proteins are expected to be widely represented within these domains today (tree a). When Berkemer and McGlynn expanded the number of species in Martin's datasets, they frequently observed a mixing of archaeal and bacterial sequences in the phylogenies produced, as seen with reverse gyrase.

14 In their tree, based on 97 reverse gyrase sequences, archaeal and bacterial reverse gyrases formed two monophyletic groups but were separated by a very short branch. The addition of new sequences subsequently eliminated this branch (Catchpole and Forterre 2019).

This mixture suggests that the protein under study was not present in LUCA but was transferred between domains at a much later time.

Considering the criterion of branch length and the necessity for enough species representing the diversity within each domain, it is apparent that nearly 80% of the 338 proteins analyzed by Martin and his colleagues do not meet these criteria (Catchpole and Forterre 2019, personal observation). Therefore, there is currently no comprehensive and accurate study that has determined LUCA's proteome. This type of analysis remains to be undertaken.

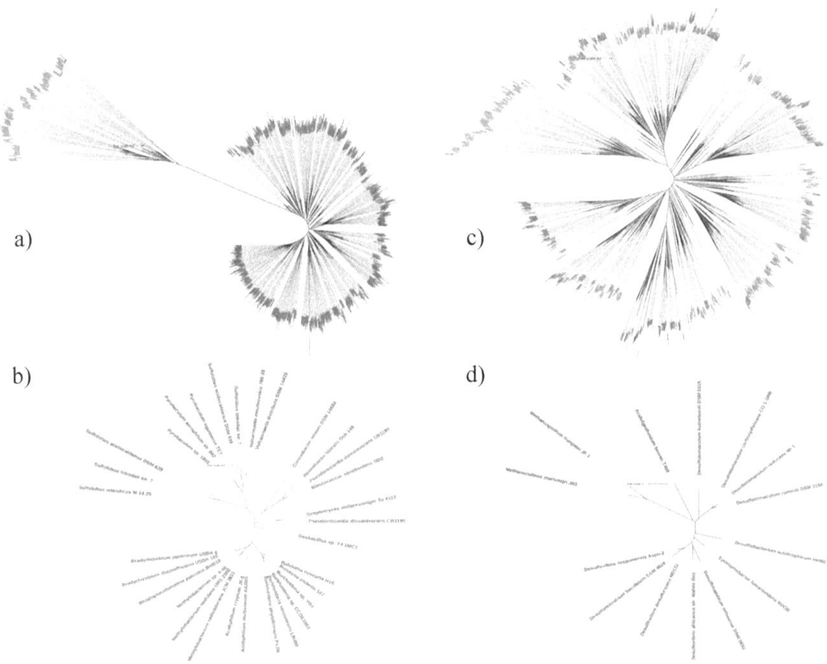

Figure 3.10. *Four trees corresponding to proteins assumed to be present in LUCA according to Martin and colleagues' publication*[15] *(Weiss et al. 2016). The 338 colored trees can be seen at https://osf.io/ypszh/. Fsor a color version of this figure, see www.iste.co.uk/clouet/archaea1.zip*

COMMENT ON FIGURE 3.10.– *Archaea are represented in blue and bacteria in red. Tree (a) illustrates a well-recognized universal protein. It features a long branch*

15 These trees are not presented in the initial publication. They were obtained by Ryan Catchpole from the datasets provided by the authors. They can be found in raw form in Weiss et al. (2018).

separating archaea and bacteria and includes a large number of represented species. This type of tree is highly uncommon in the published analysis. In trees (b), (c) and (d), however, the branch length is notably short, and the number of species is more limited. In tree (c), the representation of archaeal species is particularly sparse, while tree (d) exhibits an overall scarcity of species.

One aspect needing clarification pertains to the relationship between LUCA and the eukaryotic lineage. It is now recognized that the Woese tree lacks consensus among contemporary evolutionary biologists, with many favoring a tree in which archaea are the ancestors of eukaryotes (Figure 3.11). Under this scenario, proteins specific to eukaryotes (highlighted in red), absent in archaea and bacteria, could not have been present in LUCA.

Consequently, these proteins would inevitably have appeared in the branch of the universal tree specific to eukaryotes. If this scenario holds true, only proteins common to archaea and bacteria (in blue) should be considered for reconstructing LUCA's proteome, while eukaryotes can be excluded. This approach, for instance, was adopted by Martin and his colleagues (Weiss et al. 2016).

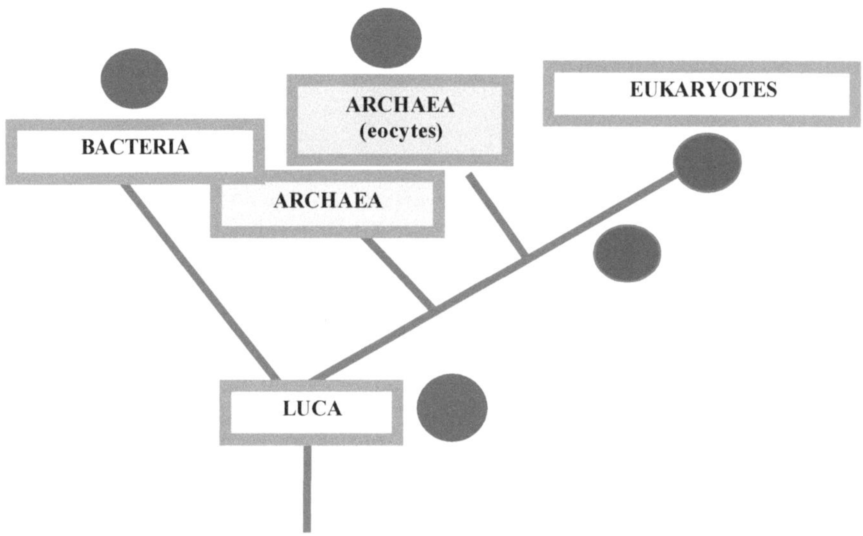

Figure 3.11. *Eukaryotic-specific proteins in the eocyte tree (2D). For a color version of this figure, see www.iste.co.uk/clouet/archaea1.zip*

COMMENT ON FIGURE 3.11.– *If eukaryotes emerged from archaea, eukaryotic-specific proteins (red circles) necessarily appeared in the specific branch of this domain.*

Only proteins common to archaea and bacteria (blue circles) should be considered to reconstruct LUCA's proteome, even if they were lost in eukaryotes, as shown in the example here.

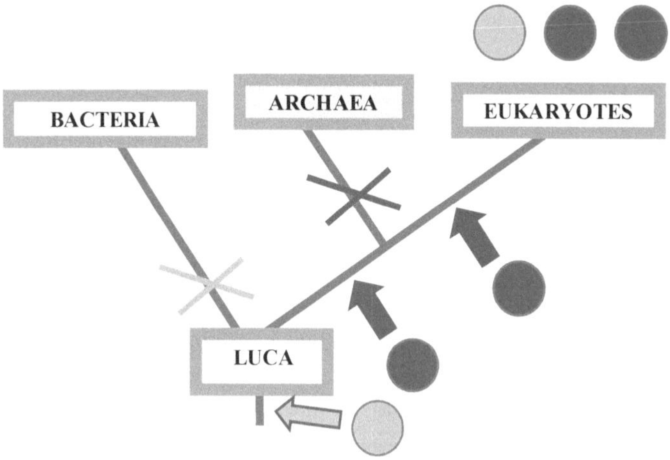

Figure 3.12. *The specific proteins of eukaryotes in the Woese's tree (3D). For a color version of this figure, see www.iste.co.uk/clouet/archaea1.zip*

COMMENT ON FIGURE 3.12.– *In the Woese's tree, some specific eukaryotic proteins may have already been present in LUCA (in red) before being lost in archaea and bacteria. Additionally, certain proteins found only in bacteria and eukaryotes could have also been present in LUCA (brown). It will be crucial to distinguish these from proteins of bacterial origin introduced into eukaryotes by mitochondria (blue).*

However, eukaryotes cannot be disregarded if the Woese's tree is indeed correct (Figure 3.12). In such a scenario, it is conceivable that certain eukaryotic-specific proteins were already present in LUCA (in red) and subsequently lost in the branches leading to bacteria and archaea. Other proteins (in brown) might have emerged in the branch common to archaea and eukaryotes before disappearing in archaea[16]. Therefore, determining the accurate topology of the universal tree is

16 To conclude that a protein that is absent in archaea but common in bacteria and eukaryotes was present in LUCA (in brown in Figure 3.12), this protein must be clearly different in eukaryotes and bacteria. In contrast, eukaryotic proteins derived from the bacterial ancestor of mitochondria (in blue), which can also be absent in archaea, generally remain very close to their bacterial ancestor and always branch in the middle of bacteria in many phylogenies.

pivotal for understanding LUCA's characteristics. Are we (eukaryotes) siblings of archaea, or are archaea our direct ancestors?

3.6. The topology of the universal tree under debate

3.6.1. *Early challenge to the Woese tree: the eocyte hypothesis*

The topology of the universal tree of life is currently the subject of vigorous debate among evolutionists, a debate that has a long history (for a review, see Gaia et al. (2018)). The division of the living world into three domains, proposed by Carl Woese, was initially questioned in 1984 by James Lake, an American researcher specializing in electron microscopy observation of ribosomes (Lake et al. 1984). Lake believed he had observed protrusions on the ribosomes of thermophilic archaea known at the *Thermoplasma*, *Thermococcus*, *Sulfolobus* and *Desulfurolobus* appeared absent in methanogenic and halophilic archaea but present in eukaryotes. He proposed dividing the living world not into three domains but into four: bacteria, eukaryotes, archaea (corresponding to methanogens and halophiles) and Eocytes, or dawn cells, corresponding to thermophilic archaea. This term again echoed the popular idea at the time of a hot origin of life. However, it soon became clear that the protrusions observed by James Lake on the ribosomes of thermophilic archaea were also present on the ribosomes of non-thermophilic methanogens (Stöffler-Meilicke et al. 1986).

Refuted on the basis of ribosomal evidence, James Lake did not consider himself defeated but turned to the study of molecular phylogeny and critique of the 16S rRNA tree. This tree is characterized by markedly different branch lengths between the trifurcation point and the positions of the last common ancestors of each of the three domains (Figure 3.1). Notably, the branch leading from the trifurcation point to the last common ancestor of archaea is very short. Observing this, James Lake hypothesized that this branch was an artifact caused by the attraction between the long branches of bacteria and eukaryotes.

The initial trees published by Woese and his colleagues were constructed using a relatively simple method – known as the distance method – which is susceptible to such artifacts. Returning to molecular phylogeny, James Lake developed a new method of phylogenetic reconstruction in the late 1980s, resulting in a 16S rRNA tree that grouped bacteria and archaea into the phylum *Euryarchaeota*, on the one hand, and eukaryotes and archaea into the phylum *Crenarchaeota*, which he termed Eocytes (Lake 1987, 1988) (Figure 3.11).

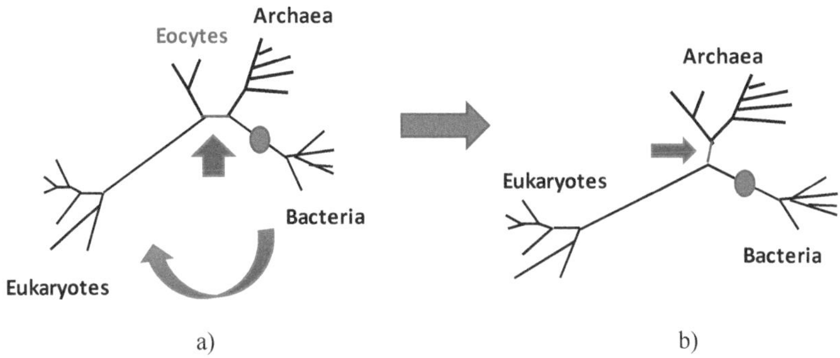

Figure 3.13. *James Lake's hypothesis explaining the Woese tree by long branch attraction. For a color version of this figure, see www.iste.co.uk/clouet/archaea1.zip*

COMMENT ON FIGURE 3.13.– *According to Lake, the Woese tree (b) resulted from the attraction of the long branch (blue arrow) of bacteria toward that of eukaryotes, artificially causing the small branch of archaea (red) to appear. The correct tree would be (a). The blue circle indicates the position of the bacterial root. Considering Eocytes as archaea, it is evident that eukaryotes emerge from archaea.*

In this new configuration, it is noteworthy that Eocytes no longer encompass all thermophilic archaea (with *Thermoplasma* and *Thermococcus* classified as *Euryarchaea*), but only the *Crenarchaea*. If the resulting tree is rooted in the bacterial branch, akin to Woese's tree, it places eukaryotes amidst archaea as currently defined (Figure 3.13). These findings sparked extensive debate over the following two decades among different research teams, paralleling the development of new phylogenetic analysis methods employing models to better accommodate protein sequence evolution (maximum likelihood methods and Bayesian methods) (Gaia et al. 2018). The results obtained were contradictory; however, by the late 1990s, due to notable similarities between *Crenarchaea* and other archaea, most researchers leaned toward Woese's tree, which grouped all archaea into the same clade, close to but distinct from eukaryotes.

3.6.2. *Searching for the archean ancestor of eukaryotes*

Starting in the 2000s, genome sequencing across the three domains of life enabled the construction of phylogenetic trees based on universally conserved proteins. This technique involves in silico assembly of sequences from all these proteins, known as concatenation, to build universal trees through comparative analysis. However, results from these new data failed to reconcile supporters of the Lake and Woese's trees. These trees are now commonly referred to as the two-domains (2D) tree and 3D tree, respectively. We adopt this nomenclature here, acknowledging concerns that archaea and eukaryotes may form a clade. Indeed, in such a scenario, Woese's tree could also be viewed as a 2D tree, grouping bacteria and *Arkarya* (Forterre 2015).

Throughout the 2000s, some research teams produced a 2D tree, while others favored a 3D tree. However, comparing these studies was challenging due to variations in species or protein datasets and slight differences in phylogenetic analysis methods. As described in a 2010 review by Simonetta Gribaldo and colleagues, the field appeared stuck in a "phylogenetic impasse" (Gribaldo et al. 2010). Nevertheless, the early 2010s witnessed rapid change. In 2011, Thijs Ettema and Lionel Guy introduced a 2D tree based on concatenating ribosomal proteins, placing eukaryotes as sister to a clade comprising *Thaumarchaeota*, *Aigarchaeota*, *Crenarchaeota* and *Korarchaeota* – the TACK superphylum (Guy and Ettema 2011). This nomenclature swiftly gained acceptance among archaea researchers, tipping the balance decisively toward 2D trees.

According to certain authors, such as Thijs Ettema and Martin Embley, these 2D trees were derived using more advanced phylogenetic analysis methods, particularly Bayesian methods, which better incorporate the mechanisms of protein sequence evolution and mitigate long-branch attraction phenomena. They argued that these methods facilitated the inclusion of species known to evolve rapidly, such as *Nanoarchaea* or *Methanopyrus kandleri* (see Chapter 2), into universal trees. Consequently, the new archaea with small genomes and rapid evolution from the DPANN superphylum (see Chapter 2) were progressively integrated into these universal trees.

On the other hand, other authors like Simonetta Gribaldo and Céline Brochier Armanet obtained a 2D tree that did not include DPANNs. In their tree, eukaryotes were placed as a sister group to *Candidatus Korarchaeum cryptophilum* (Raymann et al. 2015). These authors noted an unexpected finding: their 2D tree positioned archaea at the root within the *Euryarchaea*, thereby rendering them no longer monophyletic (see Chapter 2).

3.6.3. *Discovery of the Asgard archaea: validation of the 2D hypothesis?*

In 2015, Thijs Ettema's group in Sweden published the discovery and characterization of a new archaeal phylum, which they named *Lokiarchaeota* (named after the Germanic god Loki, the god of fire). This phylum encompassed three lineages identified as metagenome assembled genomes (MAGs) named Loki 1, Loki 2 and Loki 3. In the 2D phylogenies presented by these researchers, eukaryotes branched within the *Lokiarchaea*, specifically as the sister group to Loki 3 (Spang et al. 2015) (Figure 3.14). Subsequently, several other archaeal lineages related to *Lokiarchaea* were discovered and named after Germanic gods, each considered as new phyla (see Chapter 2).

In 2017, Thijs Ettema and collaborators proposed grouping all these newly discovered phyla into a "superphylum" called the Asgards, akin to Olympus for Greek gods (Zaremba-Niedzwiedzka et al. 2017). While Loki 1 remained within the *Lokiarchaeota*, Loki 2 and 3 were reclassified under the *Heimdallarchaeota* phylum. In the updated phylogeny by Ettema and collaborators, eukaryotes continued to emerge within the *Heimdallarchaea*. Formally, this finding indicated that eukaryotes themselves constitute a subgroup of *Heimdallarchaea*!

Analysis of their MAGs revealed eukaryotic-type proteins in the Asgard archaea[17], previously unknown among other archaea (Spang et al. 2015; Zaremba-Niedzwiedzka et al. 2017). Similar observations had been made previously in *Thaumarchaea* and *Aigarchaea* (Brochier-Armanet et al. 2008b; Nunoura et al. 2011), but the phenomenon was more pronounced in Asgard archaea.

Particularly noteworthy were their proteomes containing homologues closely related to eukaryotic cytoskeletal proteins such as actin and tubulin, along with a greater number of proteins involved in intracellular vesicle trafficking, akin to eukaryotes.

This discovery significantly contributed to convincing many evolutionary biologists of the validity of 2D trees and the hypothesis that Asgard archaea were our ancestors.

3.6.4. *Controversies over the position of Asgard archaea*

3.6.4.1. *Presence of rapidly evolving species in datasets favors 2D topologies*

The initial phylogeny published by Ettema and colleagues was based on concatenating 36 universal proteins. Unfortunately, the species dataset used included fast-evolving archaea: *M. kandleri*, DPANN, and *Korarchaea* (see Chapter 2).

17 Asgard archaea, originally considered a superphylum, are now regarded by some authors and in the GTDB (Genome Taxonomy Database) as a single phylum, with the different "phyla" of Asgard archaea being classes or orders (see Chapter 2). To avoid this debate, I will refer to lineages to designate the subgroups of Asgard archaea described so far.

Notably, *M. kandleri*, a *Euryarchaea* closely related to *Methanobacteriales*, did not group with *Euryarchaea* but instead rooted at the base of all archaea (Figure 3.14).

This observation highlighted that the a priori efficient Bayesian analysis method used by the authors was unable to mitigate the attraction of *M. kandleri*'s long branch toward that of bacteria.

In the meantime, Hervé Philippe and colleagues demonstrated that these methods failed to resolve artifacts caused by long-branch attraction when the outer group's branch is exceptionally long, as seen with the bacterial branch in the universal tree (Gouy et al. 2015).

By analyzing the individual phylogenies of the 36 universal proteins with and without fast-evolving species, Forterre and his colleagues observed that excluding these species significantly increased the number of 3D trees obtained, from 1 to 11 (Da Cunha et al. 2017). This outcome clearly demonstrated that the inclusion of fast-evolving species favored 2D trees. According to the authors, this result could be attributed to the attraction of fast-evolving archaea by the long branch of bacteria, which drew some archaea toward them. Consequently, the branch containing eukaryotes was situated within the archaea (Figure 3.15).

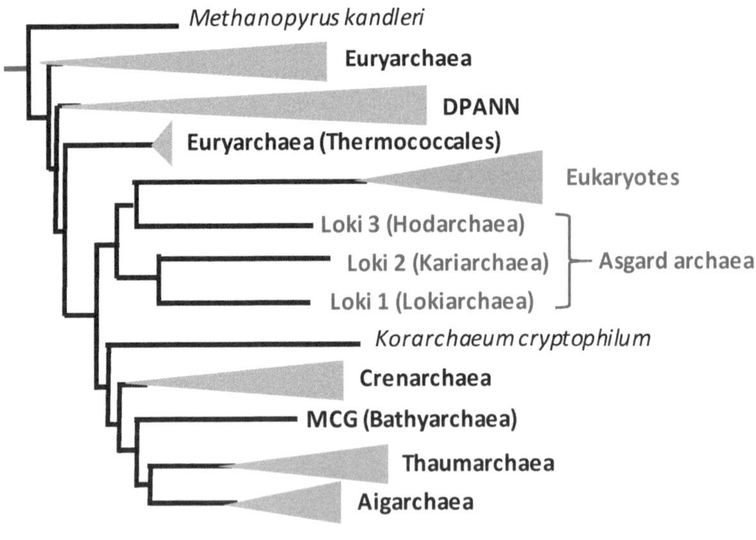

Figure 3.14. *The first universal tree suggesting the position of eukaryotes among Asgard archaea. For a color version of this figure, see www.iste.co.uk/clouet/archaea1.zip*

COMMENT ON FIGURE 3.14.– *In the initial tree published by Ettema's team (Spang et al. 2015), archaea are rooted in the branch of the fast-evolving species Methanopyrus kandleri. Eukaryotes are classified as Lokiarchaea. Subsequently, with Loki 2 and 3 becoming Kariarchaea and Hodarachaea, eukaryotes emerge from the latter. The long branch of bacteria is not depicted in this figure from the article. The positioning of Methanopyrus kandleri outside the Euryarchaea clearly results from its branch being attracted by bacteria.*

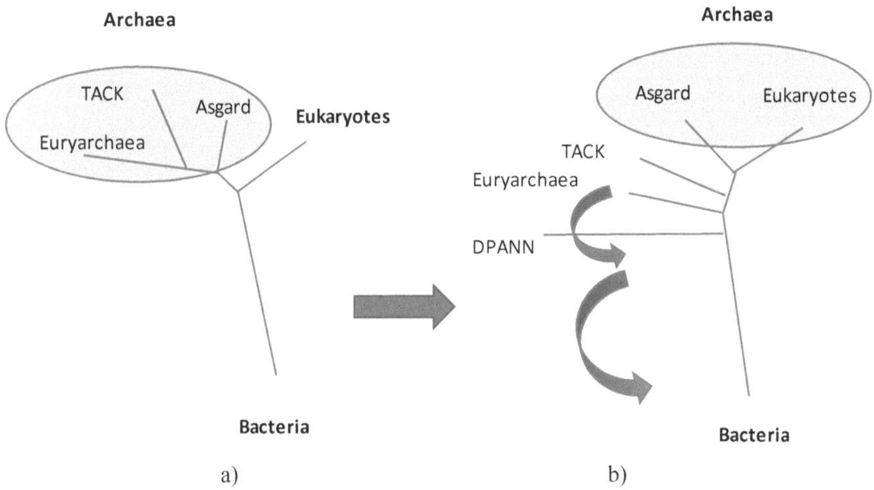

Figure 3.15. *Hypothesis explaining the transformation from a 3D tree to a 2D tree. The addition of DPANN and other rapidly evolving archaea favors the 2D (b) tree (Da Cunha et al. 2017). It is possible that the long branch of bacteria attracts that of DPANN, which in turn attracts those of other archaea. The correct tree (a) (3D) becomes a 2D tree. For a color version of this figure, see www.iste.co.uk/clouet/archaea1.zip*

3.6.4.2. *Importance of protein size in datasets*

By examining the 2D/3D distribution of the 36 trees obtained after removing fast-evolving species, Forterre and his colleagues observed a correlation between protein size and the topology of the trees obtained with these proteins (Figure 3.16) (Da Cunha et al. 2017). Phylogenies based on large proteins more frequently resulted in 3D trees, while those based on small proteins (mostly ribosomal) predominantly resulted in 2D trees. Independently concatenating sequences of proteins yielding robust 2D and 3D trees and using the same Bayesian methods as Ettema and colleagues for phylogenetic analyses, Forterre and his collaborators obtained highly robust 2D and 3D trees, respectively. This underscored that the obtained result depends more on the dataset than on the method used (Figure 3.17).

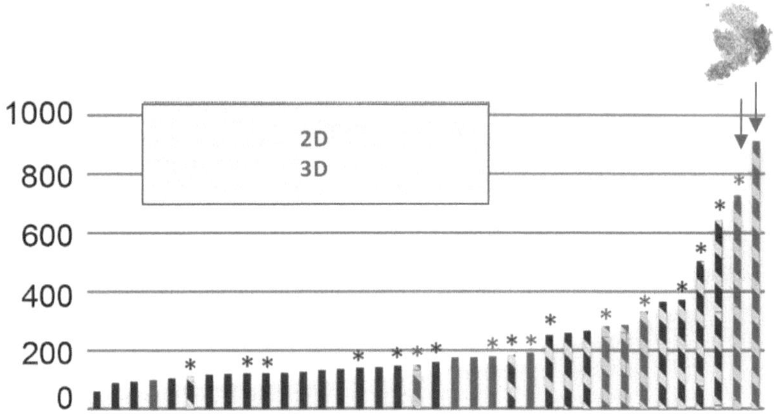

Figure 3.16. *Correlation between protein size and phylogeny (2D or 3D) obtained for these proteins. For a color version of this figure, see www.iste.co.uk/clouet/archaea1.zip*

COMMENT ON FIGURE 3.16.– *Each column represents one of the 36 individual proteins from Ettema and colleagues' publication (Spang et al. 2015) after excluding rapidly evolving species from the dataset (Da Cunha et al. 2017). Solid line columns correspond to ribosomal proteins. Stars indicate proteins yielding robust results that reject alternative topologies in statistical tests. Red arrows indicate the positions of the two RNA polymerase subunits (inset at the top right).*

The presence of conflicting signals in the dataset echoed the situation observed earlier with *Nanoarchaea* (see Chapter 2) and emphasized the critical need to assess the distribution of different topologies in individual trees.

The correlation between protein size and the resulting 2D or 3D topology could again be explained by the very short length of the specific branch of the archaea. If a protein sequence lacks a characteristic signal of this branch, obtaining a 3D tree is impossible. The larger the protein, the higher the likelihood of detecting this signal. This correlation between protein size and phylogenetic signal quality was later confirmed by Carolina Martinez-Gutierrez and Franck Aylward through their analysis of 41 proteins conserved between archaea and bacteria (Martinez-Gutierrez and Aylward 2021). They particularly showed that ribosomal proteins carry few phylogenetic signals.

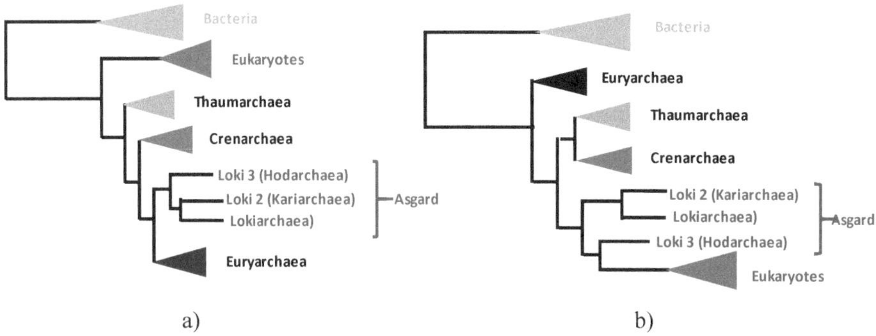

Figure 3.17. *Two opposing topologies obtained with the same method but two different sets of universal proteins. For a color version of this figure, see www.iste.co.uk/clouet/archaea1.zip*

COMMENT ON FIGURE 3.17.– *Concatenating two different sets of proteins using the same Bayesian method yielded two highly robust yet completely opposing phylogenies (according to Da Cunha et al. (2017)). (a) A 2D tree and (b) a 3D tree. The proteins used correspond to the 2D or 3D proteins marked with a star in Figure 3.16.*

The number of Asgard archaeal phyla proposed following the reconstruction of new MAGs from a wide variety of environments has surged in recent years, increasing from four in 2017 to 16 in 2022 (see Chapter 2; for review, see Da Cunha et al. (2022a)). To position Asgard archaea in universal topologies, researchers have continued to use datasets enriched with small proteins, such as ribosomal proteins, and often included rapidly evolving species (Liu et al. 2021; Xi et al. 2022). Consequently, they systematically obtained 2D trees with eukaryotes positioning themselves within the Asgard archaea or as a sister group to the Asgard archaea. Many other universal phylogenies resulting in 2D trees have been published in recent years (Imachi et al. 2020; Liu et al. 2021; Xie et al. 2021; Rodrigues-Oliveira et al. 2022). However, these studies are almost all based on datasets that include most small proteins and most also incorporate fast-evolving species[18]. Notably, Banfield and collaborators published a 2D tree based solely on ribosomal proteins,

18 The only exceptions I know of are the trees published by William Inskeep's group during the characterization of *Marsarchaea* (Jay et al. 2018). Their phylogenies are based on a concatenation of 55 universal proteins. These authors obtain a 2D tree in which eukaryotes do not branch from Asgard archaea but with the *Korarchaea*, *Candidatus Korarchaeum cryptofilum*. When they remove the DPANN, they get a tree where only this *Korarchaea* branches with the eukaryotes. If we (mentally) remove this rapidly evolving archaeon, we get a 3D tree.

in which many new and rapidly evolving mini-bacterial (CPR[19]) and mini-archaeal (DPANN) lineages are positioned at the base of their respective domains (Hug et al. 2016). Despite being clearly biased by the phenomenon of long-branch attraction, this tree, presented as a "new view of the tree of life", has been very successful as an illustration of the universal tree.

3.6.4.3. *Universal protein transfers from proto-eukaryotes to Asgard archaea leading to 2D phylogenies*

By analyzing the initial trees obtained with the datasets from Ettema's group, Forterre and his collaborators observed significant variability in the positions of the different groups of Asgard archaea and DPANN across various phylogenies, a variability not seen in other groups of archaea (Da Cunha et al. 2017). William Martin and his collaborators made a similar observation in the case of ribosomal proteins (Garg et al. 2021). Forterre and his collaborators noted that in some trees, certain Asgard archaea lineages clustered in the middle of the archaea, far from the eukaryotes, while other Asgard archaea formed a sister group to the eukaryotes. A particularly striking example was observed using data published by the Meng-Li group and collaborators in the case of the universal protein Kae1/TsaD, involved in a modification of tRNAs (Da Cunha et al. 2022a).

In the 2D tree obtained after removing rapidly evolving species from their dataset, only the Kae1/TsaD proteins of two Asgard archaea lineages, *Heimdallarchaea* and *Kariarchaea*, branched as sister groups to eukaryotes, while the Kae1/TsaD proteins of the other 10 Asgard archaea lineages branched between TACK and *Euryarchaea* (Figure 3.18). This observation strongly suggests that the ancestral archaeal-type proteins were replaced in an ancestor of *Heimdallarchaea* and *Kariarchaea* by homologous proteins from a proto-eukaryote. If the Kae1/TsaD proteins of these two Asgard archaea lineages are removed from the dataset, a 3D tree is obtained (Da Cunha et al. 2022a).

It is possible that other universal proteins have been affected by this type of transfer, which would result in a strong bias in favor of the 2D tree. Detailed analysis of individual trees remains to be done from this perspective, but a first example could be the elongation factor EF2. Forterre and his collaborators noted in

19 In this publication, CPR (Candidate Phylum Radiation), corresponds to a superphylum formed of a very large number of phyla that all branch at the base of bacteria. More recent work shows that all CPR may correspond to a single phylum, the *Patescibacteria*, which branches within bacteria as the sister group to the *Chloroflexi* (Coleman et al. 2020; Martinez-Guiterrez and Aylward 2021).

2017 the presence of several insertions common to the EF2 proteins of eukaryotes and Loki 3, which were absent in Loki 1, Loki 2 and all other archaea.

Forterre and his collaborators initially concluded that the MAG corresponding to Loki 3 was contaminated with sequences of eukaryotic origin (Forterre 2016; Da Cunha et al. 2017). This hypothesis became unlikely after the independent discovery of another MAG close to Loki 3 containing these insertions (Narrowe et al. 2018). The lineages corresponding to these two MAGs are now part of the *Hodarchaea*, and their EF2 elongation factors still branch as sister groups to eukaryotes, while other Asgard archaea EF2s branch far from them, between TACK and *Euryarchaea* (Narrowe et al. 2018). This result suggests that the ancestral EF2 protein of Asgard archaea was also replaced during their evolution by proteins from proto-eukaryotes (Da Cunha et al. 2022a)[20].

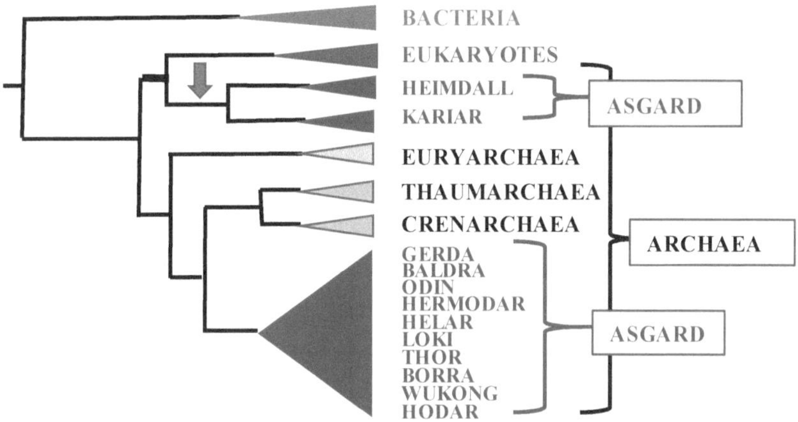

Figure 3.18. *Schematic phylogeny of the universal protein Kae1/TsaD indicating gene transfer between proto-eukaryote and two Asgard archaea lineages. For a color version of this figure, see www.iste.co.uk/clouet/archaea1.zip*

COMMENT ON FIGURE 3.18.– *This phylogeny was obtained from the dataset of Liu et al. (2021), after removing the rapidly evolving archaea (DPANN and Methanopyrus*

20 Removing the EF2 protein (in addition to the rapidly evolving species) from the concatenation of the 36 proteins in the dataset of Sang et al. (2015) transforms the 2D tree into a 3D tree (Da Cunha et al. 2017). It should also be noted that the removal of EF2 from their dataset probably explains why Inskeep and collaborators do not find the association between eukaryotes and Asgard Archaea (Jay et al. 2018a).

kandleri) that branched between bacteria and the remaining organisms (Da Cunha et al. 2022a). The blue arrow indicates a probable gene transfer between a proto-eukaryote and the ancestor of Heimdallarchaea and Kariarchaea, explaining why these two lineages do not cluster with the other Asgard archaea. If these two lineages are excluded, a 3D tree[21] *is obtained.*

3.6.4.4. *Overrepresentation of archaea in most universal trees*

Another issue that can affect the results of a phylogenetic analysis is an imbalance in the number of species represented in the study groups (Nasir et al. 2016; Rangel and Fournier 2019). The species sets used by most teams attempting to position the Asgard archaea in the tree of life are characterized by an imbalance between the number of species in each domain, with archaea generally being overrepresented. For instance, the first species set used by the Ettema group in 2015 included 10 bacteria, 10 eukaryotes and 87 archaea (Spang et al. 2015). More recently, the species set used by Meng-Li and collaborators included 448 archaea (including 162 Asgard archaea), 98 bacteria and 72 eukaryotes. Caetano-Anollés and collaborators demonstrated that this imbalance favors 2D trees by using the dataset from the first publication on Asgard archaea (Spang et al. 2015) and constructing different trees with 10 species randomly selected for each domain (Nasir et al. 2016). This phenomenon could again be due to the short branch that indicates the monophyly of archaea in 3D trees, as simulations have shown that adding species to a dataset reduces the length of internal branches in a phylogenetic tree (Rangel and Fournier 2023).

3.6.4.5. *Universal RNA polymerase tree in favor of 3D topology*

In 2017, Forterre and his collaborators obtained 3D trees by concatenating the sequences of the two longest universal proteins (the two large subunits of RNA polymerase involved in transcription). In their analyses, they used the same number of species (50) for each domain, avoiding the inclusion of fast-evolving archaea species (Da Cunha et al. 2017). In this tree, the Asgard archaea appeared at the base of the *Euryarchaea*, and the archaea were rooted in the branch leading to *Thaumarchaea*[22] (Figure 3.18). This topology was identical to that obtained by the same authors by concatenating the six universal proteins that robustly produced a

---

21 Several *Bathyarchaea* lineages that branch with Asgard archaea in the initial tree were not indicated. Their positions are likely due to transfers between *Bathyarchaea* and Asgard archaea.

22 By adding *Bathyarchaea* and *Aigarchaea* to their dataset, the archaeal tree was rooted in the branch, leading to the clade formed by these two lineages with *Thaumarchaea*, the BAT clade (Gaia et al. 2018).

3D tree in individual phylogenies (Figure 3.17), although one of the two large subunits of RNA polymerase was absent from this dataset.

As suggested previously by James Lake in the 1980s for the universal rRNA tree, the 3D protein trees obtained could be due to the attraction of the long branch of eukaryotes by the long branch of bacteria (Spang et al. 2017). To address this criticism, Forterre and his collaborators removed the bacteria from their species sets and demonstrated that the Asgard archaea still branched with the *Euryarchaea* (Da Cunha et al. 2017, 2018). If the 3D topology of these proteins were due to an attraction of the eukaryotes by the bacterial branch, and if the eukaryotes branched within the Asgard achaea, these should have branched with the eukaryotes after the removal of the bacteria.

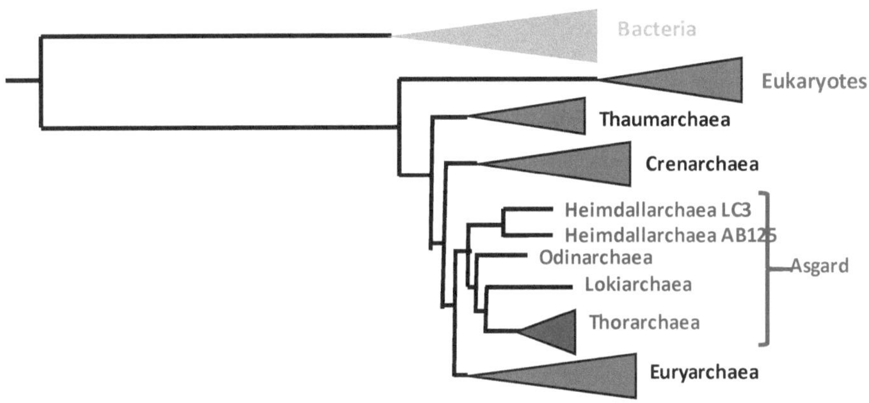

Figure 3.19. *Bayesian phylogenies obtained by concatenating the two major subunits of RNA polymerases (according to Da Cunha et al. (2017)). For a color version of this figure, see www.iste.co.uk/clouet/archaea1.zip*

In the case of RNA polymerases, Forterre and his collaborators employed a secondary approach to test the hypothesis that the 3D tree resulted from the attraction of eukaryotes by bacteria. In their initial phylogeny, they used eukaryotic RNA polymerase II[23] (Da Cunha et al. 2017). To shorten the eukaryotic branch, they then added the sequences of eukaryotic RNA polymerases I and III to their dataset, which did not alter the resulting tree topology (Figure 3.20) (Guglielmini et al. 2019; Da Cunha et al. 2022a). Thus, the 3D tree of RNA polymerases appears robust.

23 In eukaryotes, RNA polymerase II mainly transcribes protein-coding genes, RNA polymerase I transcribes ribosomal RNA genes, and RNA polymerase III transcribes transfer RNA genes.

The hypothesis that the long branch of eukaryotes is attracted to the bacterial branch to explain 3D trees was revisited by Martin Embley and his collaborators. They demonstrated that a 3D tree obtained by Forterre and colleagues, using 35 universal proteins and a *maximum likelihood estimation*[24] method, transformed into a 2D tree when Bayesian methods were applied (Williams et al. 2020). Furthermore, they used new parameters in their analysis, which were designed to better model the evolutionary mechanisms of protein sequences and reduce the long-branch attraction phenomenon. However, when Embley and his collaborators reanalyzed the datasets from Forterre and his colleagues on RNA polymerases, they also obtained a 3D tree using a Bayesian method, even with these new parameters (Figure 6 in the supplementary material; Williams et al. 2020).

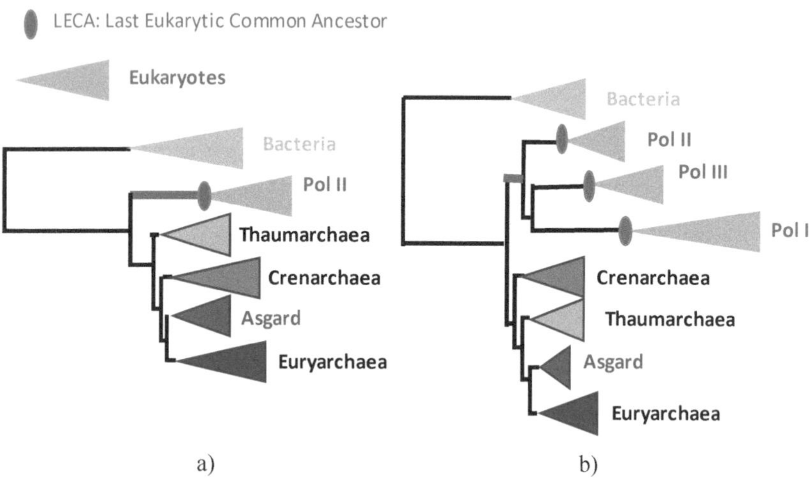

Figure 3.20. *Phylogenies compared: RNA polymerases with a single eukaryotic RNA polymerase (Pol II) and the three RNA polymerases (Pol I, II and III). For a color version of this figure, see www.iste.co.uk/clouet/archaea1.zip*

COMMENT ON FIGURE 3.20.– *(a) Simplified phylogeny of RNA polymerases from Figure 3.18. The eukaryotic RNA polymerase used is RNA polymerase II. (b) Eukaryotic RNA polymerases I and III were added to the dataset (Da Cunha et al.*

---

24 In this case, the 3D tree studied was not robust from the start because it included a majority of small ribosomal proteins (Da Cunha et al. 2017). The authors used a maximum likelihood method, supposed to be more efficient, to limit computation time because their goal was not to test the tree's topology but to assess the effect of removing one of the 36 proteins, the elongation factor EF2.

2022a). The branch leading to the eukaryotes is in bold and blue. All three eukaryotic RNA polymerases were already present in the LECA, which is thus represented three times in this phylogeny.

To finally obtain a 2D tree with RNA polymerases, Embley and his collaborators had to use a "recoding" approach, where the 20 letters corresponding to amino acids are replaced by four letters, each representing several biochemically similar amino acids (e.g. all hydrophobic amino acids) (*SR4 recoding*). However, the 2D tree obtained with this recoding shows numerous anomalies compared to the consensus topologies of archaea and eukaryotes. Specifically, the archaea are rooted within the paraphyletic *Crenarchaea* (with bacteria as the outgroup), the Asgard archaea form a sister group to the *Euryarchaea*, and the eukaryotes branch between the *Crenarchaea* and the rest of the archaea.

This aberrant phylogeny is likely due to the loss of phylogenetic signal caused by recoding, which leads to the loss of the signal corresponding to the short branch of archaea in the 3D tree. Hernandez and Ryan demonstrated through simulations that phylogenies obtained after recoding are generally of lower quality than those using the complete set of amino acids (Hernandez and Ryan 2021).

In conclusion, the RNA polymerase phylogenies obtained so far clearly support the 3D topology. Martinez-Gutierrez and Aylward recently showed that the two major subunits of RNA polymerase contain the most phylogenetic signal among universal proteins (Martinez-Gutierrez and Aylward 2021), which further strengthens the credibility of the phylogenies obtained with this enzyme.

3.6.4.6. *For more reliable universal trees*

The problem of signal and the interpretation of the long-branch attraction phenomenon are central to the 2D/3D debate. Proponents of the 2D tree suggest a possible attraction between eukaryotes and bacteria to explain the 3D trees, while proponents of the 3D tree suggest an attraction between *Euryarchaea* and bacteria to explain the 2D tree. The results obtained largely depend on the set of universal proteins used. Progress can certainly be made in this area. The datasets used by different authors so far only partially overlap (Gribaldo et al. 2010; Da Cunha et al. 2022a). Furthermore, these datasets are largely incomplete, comprising only a portion of the universal proteins that were likely present in LUCA. The most comprehensive datasets used so far include only about 50 proteins; however, the number of universal proteins available for this type of analysis is certainly higher, including large proteins such as subunits of ATP synthases. These have been used very little because they have often been transferred between archaea and bacteria. It is therefore necessary to distinguish the ATP synthases corresponding to their domain of origin, which can be used in concatenation to reconstruct a universal tree, from those resulting from gene transfers, which should not be included in the concatenation. This observation applies to all proteins that may have been transferred between the three domains at some point. Finally, some universal proteins exist in the form of several paralogs. These could have arisen by duplication or could correspond to proteins from now-extinct lineages that were transferred to one or more of the lineages leading to the three current domains (Fournier et al. 2015).

This appears to be particularly true for aminoacyl-tRNA synthetases, which are among the few large universal proteins present in the datasets used to construct universal trees. It is therefore essential to thoroughly analyze the topology of each universal protein before including it in a concatenation. This is challenging with the algorithms currently employed by most laboratories to automatically search for and classify universal proteins. These algorithms tend to reject problematic proteins, such as those from archaea and bacteria that seem mixed due to gene transfers, proteins that have duplicated (paralogous proteins), or genes that have split into two and are not present as single copies in each of the three domains. For instance, many datasets do not include the largest subunit of RNA polymerases because it is encoded by two genes in many archaea, complicating their identification and alignment with the complete sequences. However, this universal protein is undoubtedly one of those containing the most phylogenetic signals (Martinez-Gutierrez and Aylward 2021).

There is still much work to be done to analyze all universal proteins comprehensively and accurately. Additionally, significant efforts are needed to improve the datasets of species, especially for eukaryotes. The discovery of new lineages in the three domains will surely provide important new data to test the different models. However, in my opinion, the crucial point remains the analysis of individual trees and their variability to understand the origin of the different signals that favor one scenario over the other. For example, starting from the hypothesis that the 2D model is correct, Brochier-Armanet and his collaborators obtained an unrooted archaea tree with eukaryotes as the sister group to the Asgard archaea by concatenating 64 proteins conserved between archaea and eukaryotes (Aouad et al. 2022). In contrast, by concatenating 57 non-ribosomal proteins conserved between archaea and eukaryotes, Ettema and his collaborators found the eukaryotes within the Asgard archaea as a sister group to the *Hodarchaea*[25] (Eme et al. 2023). However, using the same strategy (absence of bacterial proteins) but focusing on the 11 proteins that provide a 3D topology in the individual trees, Forterre and his collaborators placed the eukaryotes as the sister group to the *Thaumarchae* (Da Cunha et al. 2018). It is important to identify which proteins in the datasets promote the association between eukaryotes and Asgard archaea, and to eliminate those that might have been transferred to Asgard archaea from proto eukaryotes, as seen in the cases of Kae1.TsaD and EF2 factor.

3.7. The origin of new eukaryotic-like proteins discovered in Asgard archaea

In the context of the 2D hypothesis, eukaryotic proteins whose distribution is limited to certain groups of archaea are called ESPs (eukaryotic signature proteins) because they suggest an evolutionary proximity between these archaea and eukaryotes. The distribution of ESPs within archaea is problematic in both the 2D and 3D models. All groups of archaea possess ESPs, although they are more abundant in TACK (especially BAT) and are concentrated in Asgard archaea.

However, it is striking that, except for actin, ESPs are never present in all Asgard archaea (Zaremba-Niedzwiedzka et al. 2017; Liu et al. 2021; Xie et al. 2021; Eme et al. 2023). In the 2D model, all ESPs were expected to be present in the eukaryotic ancestor Asgard archaea, but most ESPs are present in only one or two phyla; for instance, tubulin is only found in *Odinarchaea* (Xie et al. 2021). The erratic distribution of ESPs within Asgard archaea implies that many losses or acquisitions of ESPs by gene transfer have occurred during the evolution of Asgard archaea. In the 2D model, it must be imagined that such transfers eventually resulted in the accumulation of all ESPs in the Asgard archaea lineage leading to eukaryotes. For example, if the *Hodarchaea* are indeed the sister group of eukaryotes, the ancestor of eukaryotes would have had to acquire from other Asgard archaea the 21 ESPs missing from current *Hodarchaea* (Figure 3 in Eme et al. (2023)).

In the context of the 3D model, ESPs could have been present in the common ancestor of archaea and eukaryotes, and subsequently lost randomly during the evolution of archaea. This explanation might be valid for the ESPs abundantly present in the BAT, which are situated at the base of the archaea in the 3D trees

25 The *Hodarchaea* correspond to only one of the 16 known Asgard archaea lineages, which would mean that eukaryogenesis began long after the diversification of Asgard archaea. This hypothesis also contradicts the hypothesis that the actin conserved in all Asgard archaea comes from a proto-eukaryote (see section 3.7).

(Da Cunha et al. 2017). In this scenario, a single gene loss followed by gene transfers between BAT and Asgard archaea could explain the presence of these ESPs in these two groups.

This scenario is supported by phylogenetic analysis in the case of DNA *topoisomerases* of the IB family (Topo IB), an ESP first discovered in *Thaumarchaea* (Brochier-Armanet et al. 2008b) and later in *Aigarchaea* (Nunoura et al. 2011), *Bathyarchaea*, *Marsarchaea*, *Hadesarchaea*, *Thorarchaea* and some *Heimdallarchaea* (Zaremba-Niedzwiedzka et al. 2017; Jay et al. 2018). The Topo IB of archaea is much closer to that of eukaryotes than to that of bacteria, but it presents an archaic version, distinct from that of eukaryotes (Figure 3.21)[26]. The Topo IB tree, rooted with bacterial enzymes, is a 3D tree with archaea rooted within the BAT; the other lineages of archaea possessing Topo IB, including *Thorarchaea*, are positioned within the BAT (Figure 3.21). This topology suggests the presence of this enzyme in the common ancestor of archaea and eukaryotes. The Topo IB would then have been lost only once in the lineage leading to all archaea, except for the BAT, before being reacquired by a few other lineages of archaea through gene transfer.

The idea that all ESPs were present in LACA is, however, difficult to defend for several of these proteins, particularly those with a very limited distribution within archaea. We would have to imagine a series of losses that never affected the lineages leading to the few groups of archaea possessing these proteins, which is not very parsimonious. Another hypothesis to explain the existence of these ESPs could be gene transfers from ancient proto-eukaryotic lineages, as previously considered in the case of the universal proteins EF2 and Kae1/TsaD.

The hypothesis of gene transfer from proto-eukaryotes has been previously mentioned to explain the presence of actin and tubulin very similar to those of eukaryotes in some bacteria (Schlieper et al. 2005; Guljamow et al. 2007; Martin-Galiano et al. 2007). However, in the case of Asgard archaea, only proponents of the 3D model have advanced this hypothesis (Da Cunha et al. 2017, 2022a; Nasir et al. 2021).

The hypothesis of ESPs being transferred from proto-eukaryotes to Asgard archaea is strongly supported by phylogenetic analyses in the case of actin and tubulin (Zaremba-Niedzwiedzka et al. 2017; Da Cunha et al. 2022b; Santana-Molina et al. 2023).

---

26 Topo IB is never used in universal phylogenies because it is not sufficiently present in archaea.

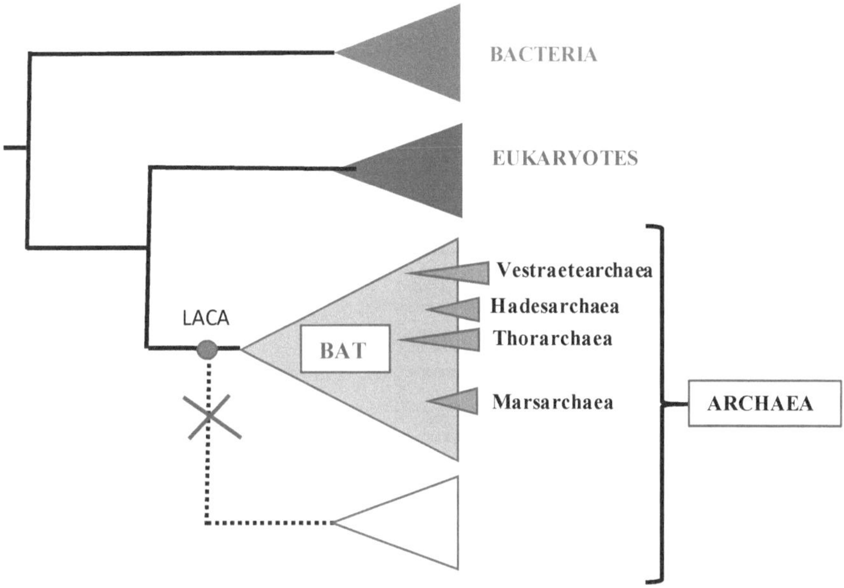

Figure 3.21. *Schematic phylogeny of DNA topoisomerases of the IB family (according to Brochier-Armanet et al. 2008b; Zaremba-Niedzwiedzka et al. 2017; Jay et al. 2018). For a color version of this figure, see www.iste.co.uk/clouet/archaea1.zip*

COMMENT ON FIGURE 3.21.– *If the BATs are indeed at the base of the archaea, this phylogeny suggests that Topo IB was present in the ancestor of archaea (LACA) and was subsequently lost (red cross) after the separation between the BAT and other archaea. It was then reacquired by some archaea through gene transfer.*

The various types of actin found in Asgard archaea (Asgardactins) are situated among the eukaryotic actin paralogs, known as ARPs (actin-related proteins).

Several groups of ARPs are present in nearly all eukaryotes, suggesting that these proteins were already present in LECA and diverged from each other during the evolution of proto-eukaryotes.

The position of Asgardactins among ARPs and eukaryotic actin suggests that Asgardactins originated from a transfer from proto eukaryotes to the Asgard archaea[27] (Da Cunha et al. 2022b) (Figure 3.22). If Asgardactins were the ancestors of ARPs and eukaryotic actin, as generally proposed (Stairs and Ettema 2020), we

27 This tree was rooted with archaeal Crenactins.

would expect to find them at the base – and quite distinct from – the ARPs and eukaryotic actin. Similarly, the tubulin of the Asgard archaea, which is only present in *Odinarchaea* (one of the 16 recognized lineages of Asgard archaea), branches within the eukaryotic tubulins and their paralogs (Santana-Molina et al. 2023)[28], a situation that can only be explained by a transfer from proto-eukaryotes to the *Odinarchaea*.

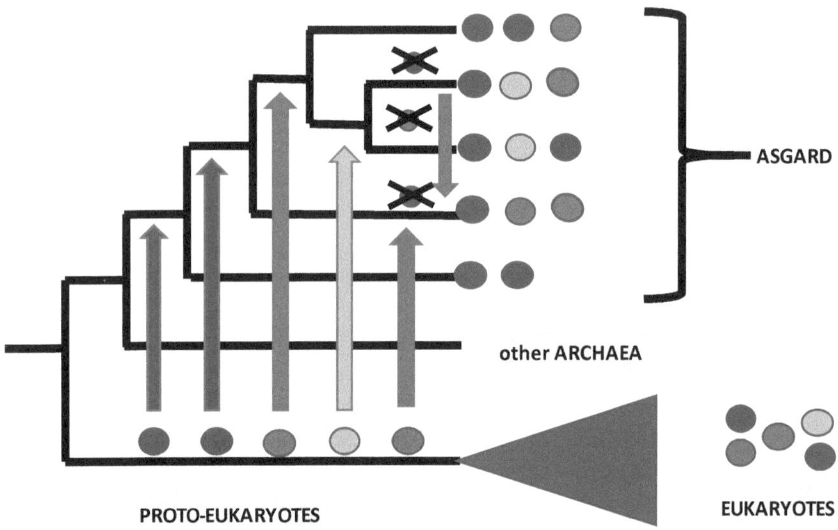

Figure 3.22. *Gene transfers coding for different ESPs (colored circles) at different times during the evolution of the Asgard archaea could explain the restricted distribution of ESPs within this group. For a color version of this figure, see www.iste.co.uk/clouet/archaea1.zip*

COMMENT ON FIGURE 3.22.– *The purple circle could symbolize the Asgardactin, which is present in all Asgard archaea today, and the blue circle, the tubulin, which is only present in Odinarchaea. Secondary losses and transfers may have complicated this picture.*

The hypothesis of gene transfer between proto-eukaryotes and Asgard archaea may explain why some ESPs are very similar (almost identical) to their eukaryotic counterparts, whereas the universal proteins of archaea, though closer to those of eukaryotes than bacteria, are distinctly different from those of eukaryotes.

28 *Odinarchaeal* tubulin corresponds to the isolated red branch in the middle of the green (eukaryotic) in Figure 1 of this article, where the authors do not mention this peculiarity.

Carl Woese early on noted sufficient differences among the universal proteins of each domain to distinguish three "versions": bacterial, archaeal and eukaryotic versions. In the case of Topo IB, as discussed earlier, the version found in archaea is notably different from that in eukaryotes. Conversely, we have observed that Asgard archaea's actin and tubulin integrate internally within eukaryotes. Interestingly, other groups of archaea exhibit homologues of actin and tubulin that diverge significantly more from the eukaryotic versions and could represent the "archaeal" forms of these proteins[29]. If the actin present in the most recent common ancestor of Asgard archaea did indeed originate from a proto-eukaryote, it suggests that proto-eukaryotes could not have emerged from a subgroup of Asgard archaea.

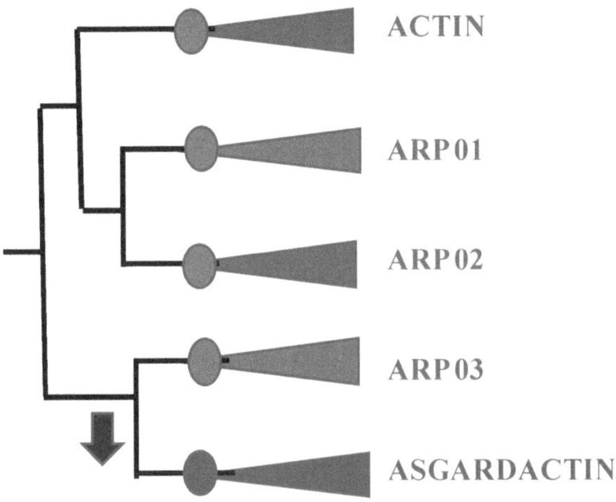

Figure 3.23. *Schematic representation of evolutionary relationships between asgardactin present in all Asgard archaea and their closest eukaryotic counterparts (according to Da Cunha et al. 2022b). For a color version of this figure, see www.iste.co.uk/clouet/archaea1.zip*

COMMENT ON FIGURE 3.23.– *The blue circles and the red circle indicate the respective positions of the last common ancestors to eukaryotes (LECA) and Asgard archaea (LAsCA). The black arrow signifies the likely transfer of this protein from a proto-eukaryote, the ancestor of the Asgard archaea. This segment of the complete tree emerges from other groups of ARPs and actin homologues present in Bathyarchaea and specific Asgard archaea lineages (Da Cunha et al. 2022b). Some*

29 In the case of actin, it is Crenactin, and in the case of tubulin, it is the CetZ and related proteins.

of these groups exhibit long branches, and their basal positions are likely due to an attraction with the long branch of Crenactin.

If the hypothesis of gene transfer is valid for certain ESPs, the question remains as to why ESPs are particularly abundant in Asgard archaea. We could speculate that these organisms live (and have always lived) in environments where proto eukaryotes once thrived. This hypothesis is interesting because it suggests that studying the physiology of Asgard archaea and their distribution in the environment could provide valuable insights into tracing the history of eukaryotic origins (eukaryogenesis), whether adopting a 2D or 3D scenario. It is also possible that transfers between proto-eukaryotes and archaea have involved other lineages of archaea. For instance, a "eukaryotic version" of actin has recently been identified in *Bathyarchaea* (Da Cunha et al. 2022b).

Note that in both scenarios, the study of ESPs in archaea provides significant information on eukaryogenesis, as these proteins either represent direct ancestors of eukaryotic proteins (2D scenario) or proteins that were present in the common ancestor of archaea and eukaryotes, or even proteins originating from proto-eukaryotes (3D scenario).

3.8. Asgard archaea and the origin of eukaryotes

3.8.1. *Asgard archaea at the origin of eukaryotes: a new paradigm*

The hypothesis that Asgard archaea are our ancestors has gained such popularity that it has motivated numerous teams to search within Asgard archaea for evidence that could support this hypothesis. For instance, the identification of condensed DNA nuclei in certain Asgard archaea, a phenomenon previously observed in bacteria (Floc'h et al. 2019), has been interpreted as an early separation between a nuclear and a cytoplasmic compartment (Avci et al. 2022). Similarly, the discovery of head–tail viruses (*Caudoviricetes*) containing genes homologous to those of eukaryotic viruses has been seen as an intermediate step in the prokaryote-to-eukaryote transition (Rambo et al. 2022), despite such genes also being present in head–tail viruses that infect bacteria (Kazlauskas et al. 2020; Guglielmini et al. 2022). ESPs are systematically investigated in Asgard archaea, but not in other groups of archaea. The serendipitous findings of actin (Da Cunha et al. 2022b) and tubulin (Forterre, personal observation) in Bathyarchaea suggest that this focus on Asgard archaea introduces a bias that distorts our broader understanding of these proteins' distribution.

3.8.2. *The inside-out model based on nanotubes discovered in Asgard archaea*

In the initial models of eukaryogenesis proposed after the discovery of Asgard archaea, these organisms were hypothesized to evolve into increasingly complex forms capable of engulfing the bacterium that gave rise to mitochondria (Zaremba-Niedzwiedzka et al. 2017). However, this model faced challenges due to the absence of proteins involved in phagocytosis found in eukaryotes within Asgard archaea (Burns et al. 2018). Furthermore, comparative genomic analyses revealed that the genomes of Asgard archaea contain only a minimal fraction of proteins that can be traced back to the LECA (Knopp et al. 2021). Additionally, the first two cultured Asgard archaea, *Candidatus Prometheoarchaeum syntrophicum* and *Lokiarchaeum ossiferum* (Imachi et al. 2020; Rodrigues-Oliveira et al. 2022), exhibit no visible complexity that would initially suggest they could be the missing link between archaea and eukaryotes. On the contrary, these are spherical cells approximately 500 nm in diameter, unusually small even for archaea.

These *Lokiarchaea*, which live in symbiosis with methanogenic archaea and bacteria, produce branched vesicles and membrane protrusions (nanotubes) containing Asgardactin[30] (Rodrigues-Oliveira et al. 2022). This observation led the authors to propose that the nanotubes of Asgard archaea may have served to encapsulate the aerobic bacteria that eventually became mitochondria (Imachi et al. 2020; Rodrigues-Oliveira et al. 2022). Subsequently, the protrusions of Asgard archaea could have expanded and fused to form the cytoplasm of the eukaryotic cell.

This scenario, wherein the host's cytoplasm extends outward to envelop the symbiont bacteria, was initially proposed before the discovery of Asgard archaea and is referred to as the inside-out model, in contrast to models where the bacterium enters the cytoplasm of the archaeal host (Baum and Baum 2014). It is important to note, however, that this type of expansion has not been observed in recently isolated *Lokiarchaea* or in other archaea or bacteria that produce nanotubes. Criticism of this model has arisen from calculations suggesting that approximately 50 nanotubes the size of those found in Asgard archaea would be necessary to completely enclose an

---

30 Several types of Asgardactins are present in Asgard archaea. The one found in the nanotubes of *Lokiarchaea* (Imachi et al. 2020; Rodrigues-Oliveira et al. 2022) is also the only common to all Asgard archaea (the one whose phylogenetic position is shown in Figure 3.22). It is therefore possible that all Asgard archaea produce nanotubes, which remains to be verified. It should be noted that other archaea, such as *Thermococcales*, which do not possess actin, are also capable of producing very long nanotubes (Marguet et al. 2013).

alpha-proteobacterium, considering the relative sizes of nanotubes and bacteria (Raval et al. 2022).

3.8.3. *The two-bacteria 2D model*

A 2D scenario involving two bacteria was first proposed by David Moreira and Purification Lopez-Garcia in 1998. According to this scenario, the alpha-proteobacterium, which gave rise to mitochondria, invaded a delta-proteobacterium containing an endosymbiont corresponding to a methanogenic archaeon (Moreira and Lopez-Garcia 1998). This archaeon was the precursor of the nucleus in eukaryotic cells. Recently, this scenario has been updated by substituting the methanogenic archaeon with an Asgard archaeon (López-García and Moreira 2020; Krupovic et al. 2023). This adaptation of the model offers the advantage of maintaining membrane continuity between the first bacterium and present-day eukaryotes (see section 3.7.4, Figure 3.24) and explains the absence of genetic similarity between viruses infecting Asgard archaea and those infecting eukaryotes (see section 3.9).

However, there is currently no known instance of intracellular archaea symbiotically associated with bacteria, and more broadly, there is no evidence of endosymbionts losing their membranes during intracellular existence. Furthermore, the relocation of Asgard archaeal ribosomes to the opposite side of this new membrane raises questions, necessitating the development of nuclear pores, while the ribosomes of the initial bacterium disappeared.

3.8.4. *The origin of eukaryotic cell complexity*

A central debate among proponents of the 2D model revolves around the timing of appearance of eukaryotic-specific features: which characteristics emerged prior to the association with the bacterium that gave rise to mitochondria, and which arose subsequently? These features encompass a wide array, including the nucleus, nucleolus, spliceosome, endoplasmic reticulum, Golgi apparatus, mRNA capping and polyadenylation, mitosis, meiosis and the advent of linear chromosomes organized with chromatin, telomeres and centromeres, among others. For some scholars, particularly William Martin and colleagues, mitochondrial endosymbiosis played a crucial role in providing the energy and selection pressures necessary for the emergence of these structures (Lane and Martin 2010; Tria et al. 2021; Raval et al. 2022). However, this viewpoint has faced critique from others, who point to the existence of modern eukaryotes lacking mitochondria yet possessing all the intricate attributes of this domain, suggesting that mitochondrial contributions were not essential to initiate eukaryogenesis (Hampl et al. 2019).

Proponents of the mitochondria's primary role in eukaryogenesis have argued that eukaryotes lacking mitochondria depend on other eukaryotes possessing these organelles (Raval et al. 2022). However, many unicellular eukaryotes do not primarily utilize their mitochondria for ATP production through oxidative phosphorylation, which constitutes the true innovation brought by these organelles, but instead use them to generate ATP through fermentation via glycolysis[31]. It is also notable that "petite" yeast mutants lacking mitochondrial DNA retain all eukaryotic characteristics, albeit with slowed growth (Slonimski et al. 1968). The timing of the endosymbiosis event that gave rise to mitochondria thus remains largely unresolved. Additionally, it should be noted that besides the bacterium responsible for mitochondria, other bacteria may have contributed genetically to proto-eukaryotes throughout eukaryogenesis.

3.9. The biological issues posed by the 2D model

We have observed that the 3D scenario leaves open the possibility that certain specific eukaryotic traits existed before the separation between archaea and eukaryotes, potentially being lost in archaea (and bacteria if these traits were already present in LUCA) (Figure 3.12). In contrast, the 2D model posits that all unique eukaryotic traits arose from the association between one or two bacteria and an Asgard archaeon (Figure 3.11). This requirement raises several issues regularly emphasized by advocates of the 3D model (Woese 2000; De Duve 2007; Forterre 2013, 2015; Nasir et al. 2021).

The one-bacteria 2D scenario involves replacing archaeal-type lipids (found in Asgard archaea) with bacterial-type lipids in eukaryotes (Figure 3.24). According to William Martin and colleagues, bacterial-type lipids would have gradually integrated into the membrane of Asgard archaea via membrane vesicles produced by the bacterium responsible for mitochondria (Gould et al. 2016). These vesicles, upon fusion with the Asgard archaeon membrane, would have progressively enriched it with bacterial-type lipids. This hypothesis aligns with the bacterial origin of enzymes involved in phospholipid biosynthesis in eukaryotes (Peretó et al. 2004; Lombard et al. 2012). Furthermore, studies demonstrating the expression of archaeal-type lipids in *Escherichia coli* suggest that the coexistence of both lipid types within the same organism is indeed feasible (Caforio et al. 2018).

31 Vladimir Hampl, personal communication.

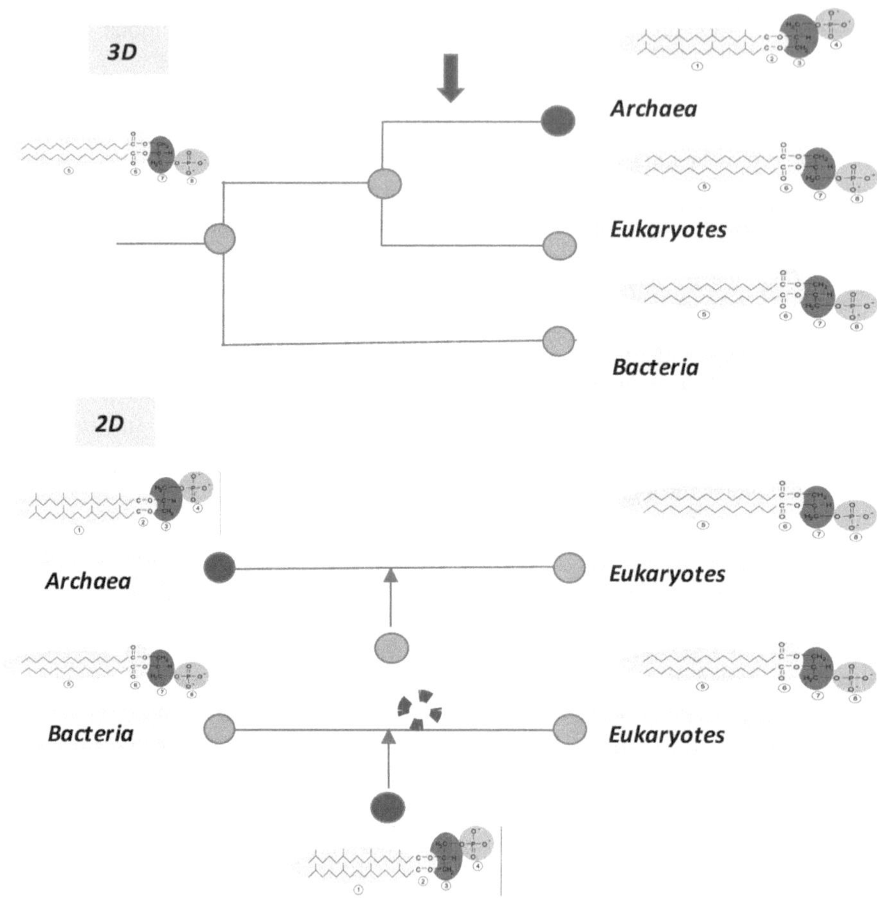

Figure 3.24. *Three possible scenarios of transitions between bacterial-type and archaeal-type lipids. For a color version of this figure, see www.iste.co.uk/clouet/archaea1.zip*

COMMENT ON FIGURE 3.24.– *In the 3D model, if LUCA possessed bacterial-type lipids, the transition to archaeal-type lipids could have occurred in the lineage leading to archaea to facilitate adaptation to high temperatures. In the one-bacterium 2D scenario, it is necessary to envision a replacement of archaeal lipids with the bacterial-type lipids of eukaryotes. In the two-bacteria 2D scenario, the archaeal lipids disappeared after the archaea entered the host bacterium.*

However, it should be noted that in this experiment, bacteria with membranes enriched in archaeal lipids are more resistant to heat shock than normal bacteria. It is therefore unclear why proto-eukaryotes would have replaced their archaeal lipids, which are more stress-resistant, with less efficient lipids from a bacterium (Da Cunha et al. 2019).

In the two-bacteria model, we could imagine continuity between the membrane of the first bacterium and that of current eukaryotes. This scenario, however, implies the disappearance of the archaeal type membrane that originally enclosed the nucleus and its replacement by a new membrane derived from the bacterial host, a phenomenon never observed in an endosymbiont.

The scenarios proposed to explain the distribution of different lipid types in the universal tree depend on the nature of the lipids present in LUCA. However, there is no consensus today on this point; it is unknown whether these lipids resembled those of archaea, bacteria or a mixture of the two. Authors who have conducted phylogenetic analyses of the enzymes involved in phospholipid biosynthesis have arrived at opposing conclusions (Lombard et al. 2012; Yokobori et al. 2016; Coleman et al. 2019). The history of these enzymes includes many gene transfers between the three domains, and some enzymes involved in lipid biosynthesis in archaea are present in many bacteria, where they seem to be involved in other mechanisms, and vice versa. Thus, the resulting phylogenies are difficult to interpret. In the 3D scenario, if LUCA had bacterial-type lipids, a single replacement in the archaeal branch would suffice to explain the current lipid distribution (Xu and Glansdorff 2002) (Figure 3.24). This replacement can be explained by the hyperthermophilic nature of LACA (Figure 3.24).

The 2D model also needs to explain why the ATP synthase present in the membrane of the Asgard archaeon ancestor of eucaryotes was transferred to the vacuole membranes of eukaryotic cells, where it now functions as an ATPase. Today, eukaryotes use ATP synthases of bacterial origin, present in large numbers in the mitochondrial membranes, to produce ATP. It could be imagined that eukaryotes would continue to use the ATP synthase from the archaea they emerged from, locating it in the membranes of the endoplasmic reticulum to increase their number. In the 3D model, this enzyme could have functioned as an ATPase in proto-eukaryotes, just as in LUCA. In this case, the transformation of this ATPase into ATP synthase in the proto-archaea would have provided a significant benefit to this lineage, while mitochondrial endosymbiosis would have allowed eukaryotes to utilize the bacterial ATPase.

In the 2D model, the association between one or two bacteria and an Asgard archaeon resulted in a radical change in all the major molecular mechanisms of the latter, particularly in the mechanisms of replication, transcription and translation. This is difficult to justity. It can be noted that the endosymbiosis of the cyanobacterium, which led to the emergence of green and red plants and algae, did not modify the fundamentals of the molecular biology of these eukaryotes, particularly the nature of their ribosomes or their replication mechanisms. In fact, all organisms forming close associations, whether endo- or ectosymbiotic, have retained the characteristics of their respective domains (Nasir et al. 2021). The case of mitochondria is exemplary, since their bacterial origin is still apparent today after more than two billion years of endosymbiotic existence. Why has the molecular biology of archaea been completely transformed to create all the characteristics of modern eukaryotes, while the remnants of bacterial molecular biology present in mitochondria are still typically bacterial? Why have mitochondrial ribosomal proteins remained very similar to those of bacteria, while the sequences of ribosomal proteins in archaea and eukaryotes now form two related but distinct groups[32]?

Finally, if the 2D model is correct, we may wonder why we do not find in nature today new examples of associations between archaea and bacteria giving rise to intermediate forms of proto-eukaryotes. In the 2D model, the emergence of eukaryotes appears as a unique, almost miraculous event, which could have occurred only once in the history of life on our planet.

3.10. Viruses and the universal tree of life

So far, we have left out viruses in our description of the universal tree of life. Viruses cannot be placed in this tree a priori because they have no ribosomes (therefore, no rRNA). In addition, there is no homologous protein common to all viruses. Finally, some authors reject in advance the very idea of locating viruses in the universal tree of life, because they consider that viruses are not alive (Moreira and Lope-Garcia 2009). Determining whether viruses are alive or not depends on the definition of life (Forterre 2016). However, it is indisputable that viruses are part of the living world (Raoult and Forterre 2008; Koonin et al. 2022). In fact, viruses seem primarily to co-evolve with their hosts, and we can imagine the viral evolutionary lines as lianas that wrap around the universal tree, from its trunk to the branches, and even the finest twigs.

32 These differences can be seen with the naked eye on a sequence alignment. The sequences of mitochondrial ribosomal proteins align perfectly with those of bacteria, while insertions and deletions are necessary to align the sequences of archaeal proteins with those of eukaryotes.

Like all organisms, archaea are infected by many highly diverse viruses. Their study has confirmed the existence of three distinct viral sets, corresponding to each of the three domains. The archaeal viruses (archaeal viruses) identified so far are all DNA viruses, almost all of which are double-stranded DNA (see Chapter 6) (Prangishvili 2013; Prangishvili et al. 2017). The study of archaeal viruses has revealed a dichotomy between two groups of viruses: one group represented by archaea-specific viruses and another group of so-called "cosmopolitan" viruses, which belong to viral lineages which also include viruses infecting bacteria (bacterioviruses) and eukaryotes (eukaryoviruses) (Prangishvili et al. 2017).

There are many families of archaea-specific viruses (Prangishvili 2013). The virions produced by these viruses are characterized by an extraordinary diversity of forms, some of which (e.g. lemon-shaped for *Fuselloviridae*, bottle-shaped for *Ampullaviridae*) are not found in viruses infecting bacteria or eukaryotes (see Chapter 6). This morphological diversity contrasts sharply with the great homogeneity of viruses infecting bacteria, 90% of which belong to the class *Caudoviricetes* (head–tail viruses). Archaea-specific viruses may have appeared in proto-archaea, especially if LUCA was still an RNA organism. However, it is also possible that these viruses were already present in LUCA and were subsequently lost in the lineages leading to bacteria and eukaryotes. The invention of peptidoglycan in protobacteria may have led to the elimination of many viral lineages in bacteria.

Cosmopolitan archaeal viruses belong to two major realms: *Duplodnaviria* (for double-stranded (*duplo*) *DNA* viruses) and *Varidnaviria* (for various (*vari*) *DNA* viruses)[33]. The existence of DNA viruses present in all three cellular domains is a strong argument in favor of a DNA LUCA (Krupovic et al. 2020). However, unlike what is observed for cellular organisms, the viruses of these spheres that infect archaea are not closer to those that infect eukaryotes than to those that infect bacteria. On the contrary, the *Duplodnaviria* and *Varidnaviria* that infect archaea are much more similar to those that infect bacteria than to those that infect eukaryotes, both morphologically and molecularly. If LUCA were still an RNA organism, these viruses could have appeared in one of the lineages leading to the three domains before being transferred between proto-archaea and proto-bacteria. This would explain the similarities between archaeal viruses and cosmopolitan bacterial viruses. We will now examine these similarities in greater detail.

---

33 All viruses in the same realm have homologous capsid proteins and DNA packaging ATPases. In contrast, these proteins are not homologous in viruses belonging to two different realms. For the ICTV (International Committee for Taxonomy of Viruses), realms correspond to the highest taxonomic level in virus classification (Koonin et al. 2020c).

The viruses within the *Duplodnaviria* sphere, whose capsid proteins are characterized by the presence of an "HK97" protein fold, include head–tail viruses that infect bacteria and archaea, *Herpesviruses* that infect animals, and a newly discovered family of viruses (identified as MAGs) that infect plankton organisms, known as *Mirusviruses* (Gaia et al. 2023). The head–tail viruses of archaea and bacteria are remarkably similar and have been classified into the same phylum, *Uroviricota*[34], and the same class, *Caudoviricetes*. They are very different from *Herpesviruses* and *Mirusviruses*, which have been grouped into different phyla, *Peploviricota*[35] and *Mirusviricota*[36].

Varidnaviria viruses are characterized by the presence of a fold in their capsid proteins called the double jelly[37] roll. Similarly, *Varidnaviria* that infect archaea and bacteria are very alike. In phylogenetic analysis based on the concatenation of capsid protein and ATPase sequences, they are intermixed and belong to the same class, *Tectiliviricetes* (Woo et al. 2021). This phylogeny distinctly separates them from *Varidnaviria* infecting eukaryotes, resulting in a universal tree of these viruses with an inverse topology to that of the universal tree of the three cellular domains (Figure 3.25). *Varidnaviria* infecting eukaryotes include small viruses producing icosahedral virions that superficially resemble those infecting archaea or bacteria, as well as much larger viruses belonging to the phylum *Nucleocytoviricota*, including the so-called "giant" viruses, which have no counterparts in archaea and bacteria. Some small eukaryotic viruses in this sphere infect these giant viruses; these are the *Lavidaviridae*, also known as virophages.

The study of archaeal viruses also raises several questions within the context of the debate between the 2D and 3D models. The *Duplodnaviria* and *Varidnaviria* discovered in Asgard archaea do not differ from the viruses in these spheres present in other archaea, which is difficult to explain within the 2D model involving a single bacterium (Medvedeva et al. 2022). This observation has been used as an argument in favor of the 2D model with two bacteria: the viruses infecting the intracellular Asgard archaea at the origin of the nucleus could not have reached it and, therefore, would not have contributed to the formation of eukaryotic viruses. In this scenario, eukaryotic viruses would have either descended from bacterial viruses or emerged de novo in the specific branch of eukaryotes (Krupovic et al. 2023).

34 From the Greek *ouros* for tail.

35 From the Greek *peplos* for garment.

36 From the Greek *mirus* for surprising, strange.

37 The name *jelly roll* corresponds to an English rolled cake.

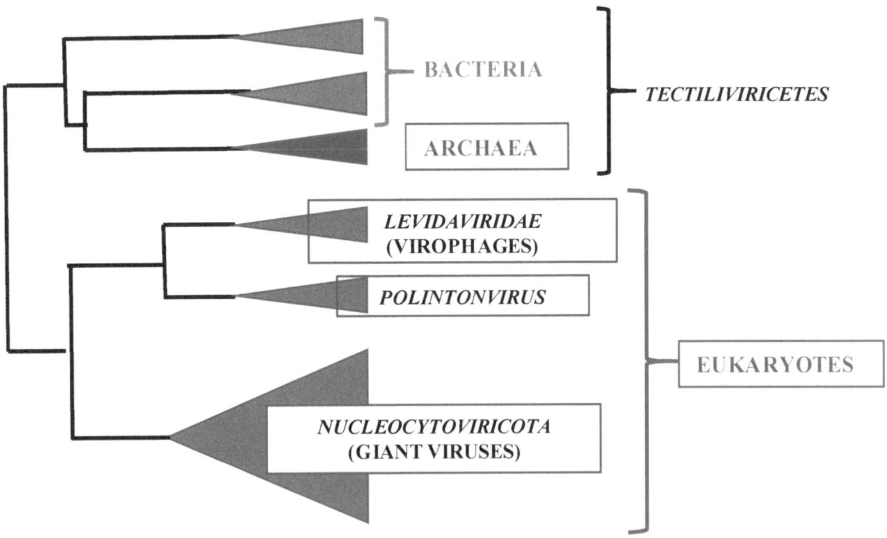

Figure 3.25. *Universal schematic phylogeny of Varidnaviria. This phylogeny was obtained by concatenating the sequences of their major capsid proteins and the ATPases involved in the packaging of their virion DNA (according to Woo et al. 2022)*[38]*. For a color version of this figure, see www.iste.co.uk/clouet/archaea1.zip*

However, this hypothesis is challenging to reconcile with the distribution of RNA viruses across the three domains. RNA viruses are extremely abundant and diverse in eukaryotes, rare in bacteria and completely absent in archaea. It is, therefore, difficult to imagine that all RNA eukaryotic viruses are derived from RNA bacterial viruses, especially since the latter seem to correspond to only a small subset of the overall diversity of RNA eukaryotic viruses (Zayed et al. 2022). It must then be assumed that RNA eukaryotic viruses appeared later, after the divergence of the eukaryotic lineage from those of other archaea.

In the 3D model, the RNA viruses of eukaryotes may be descendants of the RNA viruses that already existed during at the time of LUCA time. Their rarity in bacteria and their absence in archaea could then be explained by the thermophilic nature of their respective ancestors and by the fragility of RNA at high temperatures.

38 The *Varidnaviria* also include *Adenoviruses*. These eukaryotic viruses have been classified by the ICTV within the *Tectiliviricetes*, but their major capsid protein and ATPase are very divergent from those of other *Varidnaviria* and could not be included in the phylogeny described in Figure 3.25. In my opinion, their classification needs to be revised.

Specifically, we could imagine that RNA viruses disappeared in the lineage leading to archaea as their ancestors adapted to increasingly higher temperatures (Figure 3.9).

The distribution of DNA viruses also poses a problem. If *Duplodnaviria* and *Varidnaviria* were already present at the time of LUCA, it is necessary to explain why those infecting archaea and bacteria are so similar. In the 2D evolution scenario, we must imagine that these viruses remained virtually identical after the separation of the lineages leading from LUCA to bacteria and archaea, and then underwent dramatic transformation in the lineage leading from Asgard archaea to eukaryotes.

In the case of *Varidnaviria*, this transformation must have occurred very early during eukaryogenesis, as phylogenetic analysis indicates that the different families of *Nucleocytoviricota* (one of the three subgroups of eukaryotic viruses in the *Varidnaviria* realm) had already diverged by the time of LECA (Guglielmini et al. 2019). This explains why *Nucleocytoviricota* may have played a very important role in eukaryogenesis (Forterre and Gaia 2022). In particular, it has recently been shown that the nuclear RNA polymerases I and II of eukaryotes, as well as their DNA topoisomerase type II, originated from *Nucleocytoviricota* that infected proto-eukaryotes (Guglielmini et al. 2019, 2022). If the nucleus of the eukaryotic cell comes from an Asgard archaeon in the 2D model, it would rather be the result, in the 3D model, of one or more interactions between the proto-eukaryotes and the viral factories of the giant viruses that infected them (Gaia and Forterre 2023).

In the 3D scenario, the resemblance between the viruses of archaea and bacteria could be explained by virus transfers between proto-domains that occurred preferentially between archaea and bacteria. This hypothesis, compatible with an RNA or DNA LUCA, would generally explain the similarities observed between the mobilomes of archaea and bacteria. These mobilomes are indeed very similar to each other and radically different from that of eukaryotes (Forterre 2013). We can argue in favor of this hypothesis by noting that transfers of genetic material continued to occur preferentially between archaea and bacteria after the separation of the three domains. However, virus transfer between domains has never been observed in the current biosphere. It should be imagined that this type of transfer was easier at the time of the proto-domains or at the beginning of their diversification. The best-documented example to date of the transfer of a mobile element is that of conjugative plasmids present in archaea, which very clearly originate from bacterial conjugative plasmids, as shown by the phylogenetic analysis of the proteins of the conjugation complex (Guglielmini et al. 2013; Catchpole et al. 2023).

3.11. Conclusion

All models of eukaryogenesis today are based on the hypothesis of an association between an Asgard archaea and one or more bacteria (Dey et al. 2016; López-García and Moreira 2020; Mills et al. 2022; Raval et al. 2022). This model reinforces the traditional view that prokaryotes preceded eukaryotes, which, in my opinion, partly explains its success (Forterre 2022a, 2022b). While modern eukaryotes containing mitochondria necessarily appeared more recently than bacteria, things are less clear in my opinion in the case of proto-eukaryotes. The current 2D model is strikingly reminiscent of the one proposed by Lynn Margulis in the 1960s that Carl Woese strongly criticized (see Chapter 1; Forterre 2022b). Woese considered that modern cells were so complex, integrated, and individualized within each domain that we could not imagine they could evolve to produce cells of a new type corresponding to a new domain (Woese 2000). According to him, an archaeon could not turn into a eukaryote, even with the help of one or more bacteria. The so-called 3D topology is supported today by only a minority of biologists (Forterre 2015; Da Cunha et al. 2022a; Nasir et al. 2022). However, in my opinion, our phylogenetic analyses and the hypothesis that part of the ESPs originate from proto-eukaryotes strongly support the 3D model. The future discovery of new lineages belonging to the three domains, the development of new analytical methods allowing exhaustive identification of universal proteins, and their in-depth study may resolve the debate in the coming years, provided a new generation of evolutionists tackles this problem without the prejudices of previous generations.

3.12. References

Adam, P.S., Borrel, G., Brochier-Armanet, C., Gribaldo, S. (2017). The growing tree of Archaea: New perspectives on their diversity, evolution and ecology. *ISME J.*, 11, 2407–2425.

Anantharaman, V., Koonin, E.V., Aravind, L. (2002). Comparative genomics and evolution of proteins involved in RNA metabolism. *Nucleic. Acids. Res.*, 30, 1427–1464.

Aouad, M., Flandrois, J.P., Jauffrit, F., Gouy, M., Gribaldo, S., Brochier-Armanet, C. (2022). A divide-and-conquer phylogenomic approach based on character supermatrices resolves early steps in the evolution of the Archaea. *BMC Ecol. Evol.*, 22(1), 1. doi: 10.1186/s12862- 021-01952-0.

Avci, B., Brandt, J., Nachmias, D., Elia, N., Albertsen, M., Ettema, T.J.G., Schramm, A., Kjeldsen, K.U. (2022). Spatial separation of ribosomes and DNA in Asgard archaeal cells. *ISME J.*, 16, 606–610.

Baum, D.A. and Baum, B. (2014). An inside-out origin for the eukaryotic cell. *BMC Biol.*, 12(76). doi: 10.1186/s12915-014-0076-2.

Berkemer, S.J. and McGlynn, S.E. (2022). A new analysis of Archaea–Bacteria domain separation: Variable phylogenetic distance and the tempo of early evolution. *Mol. Biol. Evol.*, 37, 2332–2340.

Birikmen, M., Bohnsack, K.E., Tran, V., Somayaji, S., Bohnsack, M.T., Ebersberger, I. (2021). Tracing eukaryotic ribosome biogenesis factors into the archaeal domain sheds light on the evolution of functional complexity. *Front. Microbiol.*, 12, 739000. doi: 10.3389/fmicb.2021.739000.

Boussau, B., Blanquart, S., Necsulea, A., Lartillot, N., Gouy, M. (2008). Parallel adaptations to high temperatures in the Archaean eon. *Nature*, 456, 942–945.

Bowman, J.C., Petrov, A.S., Frenkel-Pinter, M., Penev, P.I., Williams, L.D. (2020). Root of the tree: The significance, evolution, and origins of the ribosome. *Chem. Rev.*, 120, 4848–4878.

Brochier-Armanet, C. and Forterre, P. (2007). Widespread distribution of archaeal reverse gyrase in thermophilic bacteria suggests a complex history of vertical inheritance and lateral gene transfers. *Archaea*, 2, 83–93.

Brochier-Armanet, C., Boussau, B., Forterre, P., Gribaldo, S. (2008a). Mesophilic *Crenarchaeota*: Proposal for a third archaeal phylum, the *Thaumarchaeota*. *Nat. Rev. Microbiol.*, 6, 245–252.

Brochier-Armanet, C., Gribaldo, S., Forterre, P.A. (2008b). DNA topoisomerase IB in *Thaumarchaeota* testifies for the presence of this enzyme in the last common ancestor of Archaea and Eucarya. *Biol. Direct*. doi: 10.1186/1745-6150-3-54.

Burns, J.A., Pittis, A.A., Kim, E. (2018). Gene-based predictive models of trophic modes suggest Asgard archaea are not phagocytotic. *Nat. Ecol. Evol.*, 2, 697–704.

Caforio, A., Siliakus, M.F., Exterkate, M., Jain, S., Jumde, V.R., Andringa, R.L.H., Kengen, S.W.M., Minnaard, A.J., Driessen, A.J.M., van der Oost, J. (2018). Converting *Escherichia coli* into an archaebacterium with a hybrid heterochiral membrane. *Proc. Natl. Acad. Sci. USA*, 115, 3704–3709.

Castelle, C.J. and Banfield, J.F. (2018). Major new microbial groups expand diversity and alter our understanding of the tree of life. *Cell*, 172, 1181–1197.

Catchpole, R.J. and Forterre, P. (2019). The evolution of reverse gyrase suggests a nonhy-perthermophilic last universal common ancestor. *Mol. Biol. Evol.*, 36, 2737–2747.

Catchpole, R.J., Barbe, V., Magdelenat G., Marguet, E., Terns, M., Oberto, J., Forterre, P., Da Cunha, V. (2023). A self-transmissible plasmid from a hyperthermophile that facilitates genetic modification of diverse Archaea. *Nat. Microbiol.*, 8, 1339–1347.

Coleman, G.A., Pancost, R.D., Williams, T.A. (2019). Investigating the origins of membrane phospholipid biosynthesis genes using outgroup-free rooting. *Genome Biol. Evol.*, 11, 883–898.

Coleman, G.A., Davín, A.A., Mahendrarajah, T.A., Szánthó, L.L., Spang, A., Hugenholtz, P., Szöllősi, G.J., Williams, T.A. (2021). A rooted phylogeny resolves early bacterial evolution. *Science*, 372(6542), eabe0511. doi: 10.1126/science.abe0511.

Da Cunha, V., Gaia, M., Gadelle, D., Nasir, A., Forterre, P. (2017). *Lokiarchaea* are close relatives of *Euryarchaeota*, not bridging the gap between prokaryotes and eukaryotes. *PLoS Genet.* doi: 10.1371/journal.pgen.1006810.

Da Cunha, V., Gaia, M., Nasir, A., Forterre, P. (2018). Asgard archaea do not close the debate about the universal tree of life topology. *PLoS Genet.* doi: 10.1371/journal.pgen.1007215.

Da Cunha, V., Gaia, M., Forterre, P. (2022a). The expending Asgard archaea and their elusive relationships with Eukarya. *mLife*, 1, 3–12. doi: 10.1002/mlf2.12012012.

Da Cunha, V., Gaia, M., Ogata, H., Jaillon, O., Delmont, T.O., Forterre, P. (2022b). Giant viruses encode novel types of actins possibly related to the origin of eukaryotic actin: The viractins. *Mol. Biol. Evol.*, 39(2), msac022. doi: 10.1093/molbev/msac022.

De Duve, C. (2007). The origin of eukaryotes: A reappraisal. *Nat. Rev. Genet.*, 8, 395–403.

Delaye, L., Becerra, A., Lazcano, A. (2005). The last common ancestor: What's in a name? *Orig. Life Evol. Biosph.*, 35, 537–554.

Desmond, E., Brochier-Armanet, C., Forterre, P., Gribaldo, S. (2011). On the last common ancestor and early evolution of eukaryotes: Reconstructing the history of mitochondrial ribosomes. *Res. Microbiol.*, 162, 53–70.

Dey, G., Thattai, M., Baum, B. (2016). On the archaeal origins of eukaryotes and the challenges of inferring phenotype from genotype. *Trends. Cell. Biol.*, 26, 476–485.

Eme, L., Tamarit, D., Caceres, E.F., Stairs, C.W., De Anda, V., Schön, M.E., Seitz, K.W., Dombrowski, N., Lewis, W.H., Homa, F. et al. (2023). Inference and reconstruction of the heimdallarchaeial ancestry of eukaryotes. *Nature*, 618, 992–999.

Felsenstein, J. (1978). Cases in which parsimony or compatibility methods will be positively misleading. *Systematic Biology*, 27, 401–410.

Floc'h, K., Lacroix, F., Servant, P., Wong, Y.S., Kleman, J.P., Bourgeois, D., Timmis, J. (2019). Cell morphology and nucleoid dynamics in dividing *Deinococcus radiodurans*. *Nat. Commun.*, 10(1), 3815. doi: 10.1038/s41467-019-11725-5.

Forterre, P. (1995). Thermoreduction, a hypothesis for the origin of prokaryotes. *C.R. Acad. Sci.*, 318, 415–422.

Forterre, P. (1996a). À la recherche de LUCA [Online]. Available at: https://web.archive.org/web/20060117033631/ and http://www-archbac.u-psud.fr/Meetings/LesTreilles/lesTreilles.html.

Forterre, P. (1996b). A hot topic: The origin of hyperthermophiles. *Cell*, 85, 789–792.

Forterre, P. (1999). Displacement of cellular proteins by functional analogues from plasmids or viruses could explain puzzling phylogenies of many DNA informational proteins. *Mol. Microbiol.*, 33, 457–465.

Forterre, P. (2002a). The origin of DNA genomes and DNA replication proteins. *Curr. Opin. Microbiol.*, 5, 525–532.

Forterre, P. (2002b). A hot story from comparative genomics: Reverse gyrase is the only hyperthermophile-specific protein. *Trends Genet.*, 18, 236–237.

Forterre, P. (2006). Three RNA cells for ribosomal lineages and three DNA viruses to replicate their genomes: A hypothesis for the origin of cellular domain. *Proc. Natl. Acad. Sci. USA*, 103, 3669–3674.

Forterre, P. (2013). The common ancestor of archaea and eukarya was not an archaeon. *Archaea*. doi: 10.1155/2013/372396.

Forterre, P. (2015). The universal tree of life: An update. *Front. Microbiol.*, 6, 717. doi: 10.3389/fmicb.2015.00717.

Forterre, P. (2016a). *Microbes from Hell*. Chicago University Press, Chicago.

Forterre, P. (2016b). To be or not to be alive: How recent discoveries challenge the traditional definitions of viruses and life. *Stud. Hist. Philos. Biol. Biomed. Sci.*, 9, 100–108.

Forterre, P. (2022a). Archaea: A goldmine for molecular biologists and evolutionists. *Methods. Mol. Biol.*, 2522, 1–21. doi: 10.1007/978-1-0716-2445-6_1.

Forterre, P. (2022b). Carl Woese: Still ahead of our time. *mLife*, 1, 359–367.

Forterre, P. and Gaia, M. (2022). Viruses and the evolution of modern eukaryotic cells. *Med. Sci.*, 38, 990–998.

Forterre, P. and Philippe, H. (1999). Where is the root of the universal tree of life? *Bioessays*, 21, 871–879.

Forterre, P., Bouthier De La Tour, C., Philippe, H., Duguet, M. (2000). Reverse gyrase from hyperthermophiles: Probable transfer of a thermoadaptation trait from archaea to bacteria. *Trends Genet.*, 16, 152–154.

Forterre, P., Gaia, M., Da Cunha, V. (2019). Engineered bacterium fuels evolution debate. *Nature*, 571, 326.

Fournier, G.P. and Gogarten, J.P. (2010). Rooting the ribosomal tree of life. *Mol. Biol. Evol.*, 27, 1792–1801.

Fournier, G.P., Andam, C.P., Gogarten, J.P. (2015). Ancient horizontal gene transfer and the last common ancestors. *BMC Evol. Biol.*, 15(70). doi: 10.1186/s12862-015-0350-0.

Gaia, M. and Forterre, P. (2023). From mimivirus to mirusvirus: The quest for hidden giants. *Viruses*, 15(8), 1758. doi: 10.3390/v15081758.

Gaia, M., Da Cunha, V., Forterre, P. (2018). The tree of life. In *Molecular Mechanisms of Microbial Evolution*, Rampelotto, P.H. (ed.). Springer, Heidelberg. doi: 10.1007/978-3-319-69078-0_3.

Gaia, M., Meng, L., Pelletier, E., Forterre, P., Vanni, C., Fernandez-Guerra, A., Jaillon, O., Wincker, P., Ogata, H., Krupovic, M. et al. (2023) Mirusviruses link herpesviruses to giant viruses. *Nature*, 616, 783–789.

Galtier, N., Tourasse, N., Gouy, M. (1999). A nonhyperthermophilic common ancestor to extant life forms. *Science*, 283, 220–221.

Garg, S.G., Kapust, N., Lin, W., Knopp, M., Tria, F.D.K., Nelson-Sathi, S., Gould, S.B., Fan, L., Zhu, R., Zhang, C. et al. (2021). Anomalous phylogenetic behavior of ribosomal proteins in metagenome-assembled asgard archaea. *Genome. Biol. Evol.*, 13(1). doi: 10.1093/gbe/evaa238.

Gogarten, J.P., Kibak, H., Dittrich, P., Taiz, L., Bowman, E.J., Bowman, B.J., Manolson, M.F., Poole, R.J., Date, T., Oshima, T. et al. (1989). Evolution of the vacuolar H+-ATPase: Implications for the origin of eukaryotes. *Proc. Natl. Acad. Sci. USA*, 86, 6661–6665.

Gogarten-Boekels, M., Hilario, E., Gogarten, J.P. (1995). The effects of heavy meteorite bombardment on the early evolution: The emergence of the three domains of life. *Orig. Life Evol. Biosph.*, 25, 251–264.

Gould, S.B., Garg, S.G., Martin, W.F. (2016). Bacterial vesicle secretion and the evolutionary origin of the eukaryotic endomembrane system. *Trends. Microbiol.*, 24, 525–534.

Gouy, R., Baurain, D., Philippe, H. (2015). Rooting the tree of life: The phylogenetic jury is still out. *Philos. Trans. R. Soc. Lond. B. Biol. Sci.*, 370, 20140329.

Gribaldo, S., Poole, A.M., Daubin, V., Forterre, P., Brochier-Armanet, C. (2010). The origin of eukaryotes and their relationship with the Archaea: Are we at a phylogenomic impasse? *Nat. Rev. Microbiol.*, 8, 743–752.

Guglielmini, J., de la Cruz, F., Rocha, E.P. (2013). Evolution of conjugation and type IV secretion systems. *Mol. Biol. Evol.*, 30, 315–331.

Guglielmini, J., Woo, A.C., Krupovic, M., Forterre, P., Gaia, M. (2019). Diversification of giant and large eukaryotic dsDNA viruses predated the origin of modern eukaryotes. *Proc. Natl. Acad. Sci. USA*, 116, 19535–19592.

Guglielmini, J., Gaia, M., Da Cunha, V., Criscuolo, A., Krupovic, M., Forterre, P. (2022). Viral origin of eukaryotic type IIA DNA topoisomerases. *Virus. Evol.*, 8(2), veac097. doi: 10.1093/ve/veac097.

Guljamow, A., Jenke-Kodama, H., Saumweber, H., Quillardet, P., Frangeul, L., Castets, M., Bouchier, C., Tandeau de Marsac, N., Dittmann E. (2007). Horizontal gene transfer of two cytoskeletal elements from a eukaryote to a cyanobacterium. *Curr. Biol.*, 17, R757–R759.

Guy, L. and Ettema, T.J. (2011). The archaeal 'TACK' superphylum and the origin of eukaryotes. *Trends. Microbiol.*, 19, 580–587.

Hampl, V., Čepička, I., Eliáš, M. (2019). Was the mitochondrion necessary to start eukaryogenesis? *Trends. Microbiol.*, 27, 96–104.

Hernandez, A.M. and Ryan, J.F. (2021). Six-state amino acid recoding is not an effective strategy to offset compositional heterogeneity and saturation phylogenetic analyses. *Syst. Biol.*, 70, 1200–1212.

Huet, J., Schnabel, R., Sentenac, A., Zillig, W. (1983). Archaebacteria and eukaryotes possess DNA-dependent RNA polymerases of a common type. *EMBO J.*, 2, 1291–1294.

Hug, L.A., Baker, B.J., Anantharaman, K., Brown, C.T., Probst, A.J., Castelle, C.J., Butterfield, C.N., Hernsdorf, A.W., Amano, Y., Ise, K. et al. (2016). A new view of the tree of life. *Nature Microbiol.* doi: 10.1038/nmicrobiol.2016.48.

Imachi, H., Nobu, M.K., Nakahara, N., Morono, Y., Ogawara, M., Takaki, Y., Takano, Y., Uematsu, K., Ikuta, T., Ito, M. et al. (2020). Isolation of an archaeon at the prokaryote–eukaryote interface. *Nature*, 577, 519–525.

Iwabe, N., Kuma, K., Hasegawa, M., Osawa, S., Miyata, T. (1989). Evolutionary relationship of archaebacteria, eubacteria, and eukaryotes inferred from phylogenetic trees of duplicated genes. *Proc. Natl. Acad. Sci. USA*, 86, 9355–9359.

Jay, Z.J., Beam, J.P., Dlakić, M., Rusch, D.B., Kozubal, M.A., Inskeep, W.P. (2018). *Marsarchaeota* are an aerobic archaeal lineage abundant in geothermal iron oxide microbial mats. *Nat. Microbiol.*, 3(6), 732–740. doi: 10.1038/s41564-018-0163-1.

Jüttner, M. and Ferreira-Cerca, S. (2022). Looking through the lens of the ribosome biogenesis evolutionary history: Possible implications for archaeal phylogeny and eukaryogenesis. *Mol. Biol. Evol.*, 39(4). doi: 10.1093/molbev/msac054.

Kazlauskas, D., Krupovic, M., Guglielmini, J., Forterre, P., Venclovas, C.S. (2020). Diversity and evolution of B-family DNA polymerases. *Nucleic. Acids. Res.*, 48, 10142–10156.

Knopp, M., Stockhorst, S., van der Giezen, M., Garg, S.G., Gould, S.B. (2021). The Asgard archaeal-unique contribution to protein families of the eukaryotic common ancestor was 0.3. *Genome Biol. Evol.*, 13(6), evab085. doi: 10.1093/gbe/evab085.

Koonin, E.V. (2003). Comparative genomics, minimal gene-sets and the last universal common ancestor. *Nat. Rev. Microbiol.*, 1, 127–136.

Koonin, E.V. and Martin, W. (2005). On the origin of genomes and cells within inorganic compartments. *Trends Genet.*, 21, 647–654.

Koonin, E.V., Krupovic, M., Ishino, S., Ishino, Y. (2020a). The replication machinery of LUCA: Common origin of DNA replication and transcription. *BMC Biol.*, 18(1), 61. doi: 10.1186/s12915-020-00800-9.

Koonin, E.V., Dolja, V.V., Krupovic, M. (2020b). The logic of virus evolution. *Cell. Host. Microbe*, 30, 917–929.

Koonin, E.V., Dolja, V.V., Krupovic, M., Varsani, A., Wolf, Y.I., Yutin, N., Zerbini, F.M., Kuhn, J.H. (2020c). Global organization and proposed megataxonomy of the virus world. *Microbiol. Mol. Biol. Rev.*, 84(2), e00061–19. doi: 10.1128/MMBR.00061-19.

Krupovic, M., Dolja, V.V., Koonin, E.V. (2020). The LUCA and its complex virome. *Nat. Rev. Microbiol.*, 18, 661–670.

Krupovic, M., Dolja, V.V., Koonin, E.V. (2023). The virome of the last eukaryotic common ancestor and eukaryogenesis. *Nat. Microbiol.*, 8, 1008–1017.

Lake, J.A. (1987). A rate-independent technique for analysis of nucleic acid sequences: Evolutionary parsimony. *Mol. Biol. Evol.*, 4, 167–191.

Lake, J.A. (1988). Origin of the eukaryotic nucleus determined by rate-invariant analysis of rRNA sequences. *Nature*, 331, 184–186.

Lake, J.A., Henderson, E., Oakes, M., Clark, M.W. (1984). Eocytes: A new ribosome structure indicates a kingdom with a close relationship to eukaryotes. *Proc. Natl. Acad. Sci. USA*, 81, 3786–3790.

Lane, N. and Martin, W. (2010). The energetics of genome complexity. *Nature*, 467, 929–934.

Lecompte, O., Ripp, R., Thierry, J.C., Moras, D., Poch, O. (2002). Comparative analysis of ribosomal proteins in complete genomes: An example of reductive evolution at the domain scale. *Nucleic. Acids. Res.*, 30, 5382–5390.

Liu, Y., Makarova, K.S., Huang, W.-C., Wolf, Y.I., Nikolskaya, A.N., Zhang, X., Cai, M., Zhang, C.J., Xu, W., Luo, Z. et al. (2021). Expanded diversity of Asgard archaea and their relationships with eukaryotes. *Nature*, 593, 553–557.

Lloyd, C.T., Iwig, D.F., Wang, B., Cossu, M., Metcalf, W.W., Boal, A.K., Booker, S.J. (2022). Discovery, structure and mechanism of a tetraether lipid synthase. *Nature*, 609, 197–203.

Lombard, J. (2016). Early evolution of polyisoprenol biosynthesis and the origin of cell walls. *PeerJ.*, 4, e2626. doi: 10.7717/peerj.2626.

Lombard, J., López-García, P., Moreira, D. (2012). The early evolution of lipid membranes and the three domains of life. *Nat. Rev. Microbiol.*, 10, 507–515.

López-García, P. and Moreira, D. (2020). The Syntrophy hypothesis for the origin of eukaryotes revisited. *Nat. Microbiol.*, 5, 655–667.

Marguet, E., Gaudin, M., Gauliard, E., Fourquaux, I., le Blond du Plouy, S., Matsui, I., Forterre, P. (2013). Membrane vesicles, nanopods and/or nanotubes produced by hyperthermophilic archaea of the genus *Thermococcus*. *Biochem. Soc. Trans.*, 41, 436–442.

Martin-Galiano, A.J., Oliva, M.A., Sanz, L., Bhatthacharyya, A., Serna, M., Yebenes, H., Valpuesta, J.M., Andreu, J.M. (2011). Bacterial tubulin distinct loop sequences and primitive assembly properties support its origin from a eukaryotic tubulin ancestor. *J. Biol. Chem.*, 286, 19789–19803.

Martinez-Gutierrez, C.A. and Aylward, F.O. (2021). Phylogenetic signal, congruence, and uncertainty across bacteria and archaea. *Mol. Biol. Evol.*, 38, 5514–5527.

Medvedeva, S., Sun, J., Yutin, N., Koonin, E.V., Nunoura, T., Rinke, C., Krupovic, M. (2022). Three families of Asgard archaeal viruses identified in metagenome-assembled genomes. *Nat. Microbiol.*, 7, 962–973.

Mills, D.B., Boyle, R.A., Daines, S.J., Sperling, E.A., Pisani, D., Donoghue, P.C.J., Lenton, T.M. (2022). Eukaryogenesis and oxygen in Earth history. *Nat. Ecol. Evol.*, 6, 520–532.

Moody, E.R.R., Mahendrarajah, T.A., Dombrowski, N., Clark, J.W., Petitjean, C., Offre, P., Szöllősi, G.J., Spang, A., Williams, T.A. (2022). An estimate of the deepest branches of the tree of life from ancient vertically evolving genes. *Elife*, 11, e66695. doi: 10.7554/eLife.66695.

Moreira, D. and Lopez-Garcia, P. (1998). Symbiosis between methanogenic archaea and delta-proteobacteria as the origin of eukaryotes: The syntrophic hypothesis. *J. Mol. Evol.*, 47, 517–530.

Moreira, D. and Lopez-Garcia, P. (2009). Ten reasons to exclude viruses from the tree of life. *Nat. Rev. Microbiol.*, 7, 306–311.

Mulkidjanian, A.Y., Makarova, K.S., Galperin, M.Y., Koonin E.V. (2007). Inventing the dynamo machine: The evolution of the F-type and V-type ATPases. *Nat. Rev. Microbiol.*, 5, 892–899.

Mushegian, A.R. and Koonin, E.V. (1996). A minimal gene set for cellular life derived by comparison of complete bacterial genomes. *Proc. Natl. Acad. Sci. USA*, 93, 10268–10273.

Narrowe, A.B., Spang, A., Stairs, C.W., Caceres, E.F., Baker, B.J., Miller, C.S., Ettema, T.J.G. (2018). Complex evolutionary history of translation elongation 2 and diphthamide biosynthesis in archaea and parabasalids. *Genome Biol. Evol.*, 10, 2380–2393.

Nasir, A., Kim, K.M., Da Cunha, V., Caetano-Anollés, G. (2016). Arguments reinforcing the three-domain view of diversified cellular life. *Archaea*. doi: 10.1155/2016/1851865.

Nasir, A., Mughal, F., Caetano-Anollés, G. (2021). The tree of life describes a tripartite cellular world. *Bioessays*, 43, e2000343.

Nunoura, T., Takaki, Y., Kakuta, J., Nishi, S., Sugahara, J., Kazama, H., Chee, G.J., Hattori, M., Kanai, A., Atomi, H. et al. (2011). Insights into the evolution of Archaea and eukaryotic protein modifier systems revealed by the genome of a novel archaeal group. *Nucleic. Acids. Res.*, 39, 3204–3223.

Olsen, G.J. and Woese, C.R. (1997). Archaeal genomics: An overview. *Cell*, 89, 991–994.

Pace, N.R., Olsen, G.J., Woese, C.R. (1986). Ribosomal RNA phylogeny and the primary lines of evolutionary descent. *Cell*, 45, 325–326.

Pende, N., Sogues, A., Megrian, D., Sartori-Rupp, A., England, P., Palabikyan, H., Ritt-mann, S.K.R., Graña, M., Wehenkel, A.M., Alzari, P.M. et al. (2021). SepF is the FtsZ anchor in archaea, with features of an ancestral cell division system. *Nat. Commun.*, 12(1), 3214. doi: 10.1038/s41467-021-23099-8.

Peretó, J., López-García, P., Moreira, D. (2004). Ancestral lipid biosynthesis and early membrane evolution. *Trends. Biochem. Sci.*, 29, 469–477.

Philippe, H. and Forterre, P. (1999). The rooting of the universal tree of life is not reliable. *J. Mol. Evol.*, 49, 509–523.

Poole, A.M. and Logan, D.T. (2005). Modern mRNA proofreading and repair: Clues that the last universal common ancestor possessed an RNA genome? *Mol. Biol. Evol.*, 22, 1444–1455.

Prangishvili, D. (2013). The wonderful world of archaeal viruses. *Annu. Rev. Microbiol.*, 67, 565–585.

Prangishvili, D., Bamford, D.H., Forterre, P., Iranzo, J., Koonin, E.V., Krupovic, M. (2017). The enigmatic archaeal virosphere. *Nat. Rev. Microbiol.*, 15, 724–739.

Rambo, I.M., Langwig, M.V., Leão, P., De Anda, V., Baker, B.J. (2022). Genomes of six viruses that infect Asgard archaea from deep-sea sediments. *Nat. Microbiol.*, 7, 953–961.

Rangel, L.T. and Fournier, G. (2023). Fast-evolving sites are highly informative for reconstructions of deep tree of life phylogenies. *Microorganisms,* 11. doi: 10.3390/microorganisms11102499.

Raoult, D. and Forterre, P. (2008). Redefining viruses: Lessons from mimivirus. *Nat. Rev. Microbiol.*, 6, 315–319.

Raval, P.K., Garg, S.G., Gould, S.B. (2022). Endosymbiotic selective pressure at the origin of eukaryotic cell biology. *Elife*. doi: 10.7554/eLife.81033.

Raymann, K., Brochier-Armanet, C., Gribaldo, S. (2015). The two-domain tree of life is linked to a new root for the Archaea. *Proc. Natl. Acad. Sci. USA*, 112, 6670–6675.

Rodrigues-Oliveira, T., Wollweber, F., Ponce-Toledo, R.I., Xu, J., Rittmann, S.K., Klingl, A., Pilhofer, M., Schleper, C. (2022). Actin cytoskeleton and complex cell architecture in an Asgard archaeon. *Nature*. doi: 10.1038/s41586-022-05550-y.

Salcher, M.M., Andrei, A.Ş., Bulzu, P.A., Keresztes, Z.G., Banciu, H.L., Ghai, R. (2020). Visualization of *Lokiarchaeia* and *Heimdallarchaeia* (*Asgardarchaeota*) by fluorescence *in situ* hybridization and catalyzed reporter deposition (CARD-FISH). *mSphere*, 5(4), e00686-20. doi: 10.1128/mSphere.00686-20.

Santana-Molina, C., Del Saz-Navarro, D., Devos, D.P. (2023). Early origin and evolution of the FtsZ/tubulin protein family. *Front. Microbiol.*, 13, 1100249. doi: 10.3389/fmicb.2022.1100249.

Schlieper, D., Oliva, M.A., Andreu, J.M., Löwe, J. (2005). Structure of bacterial tubulin BtubA/B: Evidence for horizontal gene transfer. *Proc. Natl. Acad. Sci. USA*, 102, 9170–9175.

Slonimski, P.P., Perrodon, G., Croft, J.H. (1968). Ethidium bromide induced mutation: Complete transformation of cells into respiratory deficient non-chromosomal petites. *Biochem. Bioph. Res. C.*, 30, 232–239.

Spang, A., Saw., J.H., Jørgensen, S.L., Zaremba-Niedzwiedzka, K., Martijn, J., Lind, A.E., van Eijk, R., Schleper, C., Guy, L., Ettema, T.J.G. (2015). Complex archaea that bridge the gap between prokaryotes and eukaryotes. *Nature*, 521, 173–179.

Spang, A., Eme, L., Saw, J.H., Caceres, E.F., Zaremba-Niedzwiedzka, K., Lombard, J., Guy, L., Ettema, T.J.L. (2018). Asgard archaea are the closest prokaryotic relatives of eukaryotes. *PLoS Genet.* doi: 10.1371/journal.pgen.100708.

Stairs, C.W. and Ettema, T.J.G. (2020). The archaeal roots of the eukaryotic dynamic actin cytoskeleton. *Curr. Biol.*, 30(10), R521–R526. doi: 10.1016/j.cub.2020.02.074.

Stetter, K.O. (1996). Hyperthermophiles in the history of life. *Ciba Found. Symp.*, 202, 1–10.

Stetter, K.O. (2013). A brief history of the discovery of hyperthermophilic life. *Biochem. Soc. Trans.*, 41, 416–420.

Stöffler-Meilicke, M., Böhme, C., Strobel, O., Böck, A., Stöffler, G. (1986). Structure of ribosomal subunits of *M. vannielii*: Ribosomal morphology as a phylogenetic marker. *Science*, 231, 1306–1308.

Takeuchi, N., Hogeweg, P., Koonin, E.V. (2011). On the origin of DNA genomes: Evolution of the division of labor between template and catalyst in model replicator systems. *PLoS Comput Biol.*, 7(3), e1002024. doi: 10.1371/journal.pcbi.1002024.

Tria, F.D.K., Brueckner, J., Skejo, J., Xavier, J.C., Kapust, N., Knopp, M., Wimmer, J.L.E., Nagies, F.S.P., Zimorski, V., Gould, S.B. et al. (2021). Gene duplications trace mitochondria to the onset of eukaryote complexity. *Genome Biol. Evol.*, 13(5), evab055. doi: 10.1093/gbe/evab055.

Van der Gulik, P.T. and Hoff, W.D. (2016). Anticodon modifications in the tRNA set of LUCA and the fundamental regularity in the standard genetic code. *PLoS One*, 11(7), e0158342. doi: 10.1371/journal.pone.0158342.

Weiss, M.C., Sousa, F.L., Mrnjavac, N., Neukirchen, S., Roettger, M., Nelson-Sathi, S., Martin, W.F. (2016). The physiology and habitat of the last universal common ancestor. *Nat. Microbiol.*, 1(9), 16116. doi: 10.1038/nmicrobiol.2016.116.

Weiss, M.C., Preiner, M., Xavier, J.C., Zimorski, V., Martin, W.F. (2018). The last universal common ancestor between ancient Earth chemistry and the onset of genetics. *PLoS Genet.* doi: 10.1371/journal.pgen.1007518.

Werner, F. and Grohmann, D. (2011). Evolution of multisubunit RNA polymerases in the three domains of life. *Nat. Rev. Microbiol.*, 9, 85–98.

Williams, T.A., Szöllősi, G.J., Spang, A., Foster, P.G., Heaps, S.E., Boussau, B., Ettema, T.J.G., Embley, T.M. (2017). Integrative modeling of gene and genome evolution roots the archaeal tree of life. *Proc. Natl. Acad. Sci. USA*, 114, 4602–4611.

Williams, T.A., Cox, C.J., Foster, P.G., Szöllősi, G.J., Embley, T.M. (2020). Phylogenomics provides robust support for a two-domains tree of life. *Nat. Ecol. Evol.*, 4, 138–147.

Woese, C.R. (1981). Archaebacteria. *Scientific American*, 244, 98–125.

Woese, C.R. (2000). Interpreting the universal phylogenetic tree. *Proc. Natl. Acad. Sci. USA*, 97, 8392–8396.

Woese, C.R. and Fox, G.E. (1977a). Phylogenetic structure of the prokaryotic domain: The primary kingdoms. *Proc. Natl. Acad. Sci. USA*, 74, 5088–5090.

Woese, C.R. and Fox, G.E. (1977b). The concept of cellular evolution. *J. Mol. Evol.*, 10, 1–6.

Woese, C.R., Kandler, O., Wheelis, M.L. (1990). Towards a natural system of organisms: Proposal for the domains Archaea, Bacteria, and Eucarya. *Proc. Natl. Acad. Sci. USA*, 87, 4576–4579.

Woo, A.C., Gaia, M., Guglielmini, J., Da Cunha, V., Forterre, P. (2021). Phylogeny of the *Varidnaviria* morphogenesis module: Congruence and incongruence with the tree of life and viral taxonomy. *Front. Microbiol.*, 12, 704052. doi: 10.3389/fmicb.2021.704052.

Xie, R., Wang, Y., Huang, D., Hou, J., Li, L., Hu, H., Zhao, X., Wang, F. (2021). Expanding Asgard members in the domain of Archaea sheds new light on the origin of eukaryotes. *Sci. China Life Sci.* doi: 10.1007/s11427-021-1969-6.

Xu, Y. and Glansdorff, N. (2002). Was our ancestor a hyperthermophilic procaryote? *Comp. Biochem. Physiol. A. Mol. Integr. Physiol.*, 133, 677–688.

Yokobori, S.I., Nakajima, Y., Akanuma, S., Yamagishi, A. (2016). Birth of archaeal cells: Molecular phylogenetic analyses of G1P dehydrogenase, G3P dehydrogenases, and glycerol kinase suggest derived features of archaeal membranes having G1P polar lipids. *Archaea*, Sep 28(2016), 1802675. doi: 10.1155/2016/1802675.

Zaremba-Niedzwiedzka, K., Caceres, E.F., Saw, J.H., Bäckström, D., Juzokaite, L., Vancaester, E., Seitz, K.W., Anantharaman, K., Starnawski, P., Kjeldsen K.U. et al. (2017). Asgard archaea illuminate the origin of eukaryotic cellular complexity. *Nature*, 41, 353–358.

Zayed, A.A., Wainaina, J.M., Dominguez-Huerta, G., Pelletier, E., Guo, J., Mohssen, M., Tian, F., Pratama, A.A., Bolduc, B., Zablocki, O. et al. (2022). Cryptic and abundant marine viruses at the evolutionary origins of Earth's RNA virome. *Science*, 376, 156–162.

Zeldovich, K.B., Berezovsky, I.N., Shakhnovich, E.I. (2007). Protein and DNA sequence determinants of thermophilic adaptation. *PLoS Comput. Biol.*, 3, 62–72.

Zeng, Z., Chen, H., Yang, H., Chen, Y., Yang, W., Feng, X., Pei, H., Welander, P.V. (2022). Identification of a protein responsible for the synthesis of archaeal membrane-spanning GDGT lipids. *Nat. Commun.*, 13(1), 1545. doi: 10.1038/s41467-022-29264-x.

Zillig, W., Stetter, K.O., Janeković, D. (1979). DNA-dependent RNA polymerase from the archaebacterium *Sulfolobus acidocaldarius*. *Eur. J. Biochem.*, 96, 597–604.

4

Archaea: Habitats and Associated Physiologies

Karine ALAIN, Marc COZANNET, Maxime ALLIOUX, Sarah THIROUX and Jordan HARTUNIANS
BEEP-BIOMEX, CNRS, Ifremer, Université de Bretagne Occidentale, Plouzané, France

4.1. Introduction

Archaea are widespread within planetary ecosystems and are present in all biotopes. They inhabit a wide range of aquatic, terrestrial and atmospheric habitats, where they sometimes constitute a significant fraction of microbial communities, occasionally accounting for up to 20% of the biomass of total microorganisms (DeLong and Pace 2001).

Detected as early as 1776 in swamps by the physicist and chemist Alessandro Volta, archaea were not identified as such until 1977, following the work of Carl Woese and George Fox on ribosomal RNAs (Woese and Fox 1977). In the late 1970s, known archaea consisted mainly of extremophilic species, that is, organisms fulfilling their lifecycle at the extreme limits of life, in environmental conditions fatal to most living beings. These archaea included methanogenic archaea (producing methane (CH_4) under anaerobic conditions), halophilic archaea (thriving

For a color version of all figures in this chapter, see www.iste.co.uk/clouet/archaea1.zip.

Biology of Archaea 1,
coordinated by Béatrice CLOUET-D'ORVAL, Bruno FRANZETTI and Philippe OGER.

in hypersaline environments), and thermo-acidophilic archaea (growing at high temperatures and under acidic conditions).

For about 15 years, archaea were therefore associated, in the collective mind, with extreme environments (Figures 4.1 and 4.3). For years, research on archaea focused on the most hostile environments on the planet. Archaea were successively isolated from salt lakes, deep ocean hydrothermal vents, terrestrial geothermal springs, solfatares and soda lakes. Later, in the early 1990s, molecular methods that bypassed the culturing step revealed that these microorganisms were more widely distributed than previously assumed and were not strictly confined to extreme environments.

Archaea thriving under more ordinary conditions were thus discovered in habitats such as soil, the ocean and freshwater lakes. Today, it is known that they are ubiquitous. They are also present in the human microbiome (e.g. intestine, skin, oral and respiratory systems) and are, in some cases, associated with infections or allergies (Bang and Schmitz 2015).

The last decade has witnessed significant advancements in our understanding of archaea diversity, phylogeny, metabolism and potential physiological capacities, driven by the remarkable progress in genomic reconstruction methods from metagenomes or single-cell genomes, and the exploration of previously unstudied environments. By 2016, 26 *phyla* of archaea had been identified, but only 13 had representatives isolated in clonal culture (Hug et al. 2016). Nevertheless, for many of these phyla, physiological hypotheses have been proposed by analyzing the coding sequences of reconstructed genomes within their community and ecosystem contexts. Today, it is clear that due to their extensive environmental distribution, metabolic activities and/or abundance, archaea play key roles in natural environments.

This chapter aims to describe the various habitats inhabited by archaea, beginning with the most extreme environments. Subsequently, we will discuss their more common habitats. For each habitat, the main *taxa* and associated metabolic types will be identified. Where possible, these will be correlated with environmental parameters. The third and final part of the chapter will introduce archaeal lineages that currently have no cultured members and will present physiological hypotheses regarding them.

Figure 4.1. *Photographs of extreme environments harboring archaea*

COMMENT ON FIGURE 4.1.– *(a) Sea ice: an ecosystem populated by psychrophilic archaea (Iceland, A. Postec). (b) Permafrost: frozen soil inhabited by several lineages of methanogenic archaea (A. Postec). (c) Frozen pond, favorable to the development of psychrophilic archaea (Wanuskewin Heritage Park, Canada, M. Allioux). (d) Solfatare, hosting thermophilic and acidophilic archaea (Iceland, S. Thiroux). (e) Fumaroles, characterized by lower emissions of hydrogen sulfide, hydrogen fluoride and sulfur dioxide than solfatares, and sheltering thermophilic archaea (Iceland, S. Thiroux). (f) Hot spring favorable to the development of thermophilic and thermotolerant archaea (China, K. Alain).*

4.2. Archaea of extreme habitats: extremophiles

Archaea are ubiquitous in the planet's most extreme and hostile environments, such as habitats with temperatures close to 0°C or even negative, and those with very high temperatures, sometimes exceeding 100°C. They also inhabit sites characterized by very acidic or very alkaline pH levels, high salinities, or exposure to radiation or high hydrostatic pressures (Table 4.1) (Figures 4.1 and 4.2). While some archaea merely tolerate the harsh conditions of their habitat, many require these conditions for their development. This section introduces extremophilic archaea within the context of their environments, focusing on their metabolic types in relation to the sources of carbon, energy and electron acceptors available in their natural habitats.

Environment	Temperature (°C)	pH	Pressure (MPa)	Salinity (% NaCl)	Remarks
Soda lake	–0.5 to 83	6.9–11.2	Nr	0.64–37.1	A record pH of 11.2 was recorded for Lake Magadi (Kenya)
Terrestrial hot springs/geothermal waters	15 to 270	0.02–9.8	0.1–7.2	0.0002–saturation	7.2 MPa measured for a geothermal water source in the Moxi Fault (China)
Deep-sea hydrothermal vents	<1 to 464	4–11	2.1–50.7	0.1–8	A record low temperature (<1°C) measured for hydrothermal fluid in a volcanic arc in the Mariana Trench
Deep ocean floor and trenches	–1.9 to 13.8	7.3–8.1	2.1–112	3.4–3.9	13.8°C measured for deep marine waters in the Mediterranean Sea. Geothermally influenced sediments not considered

Environment	Temperature (°C)	pH	Pressure (MPa)	Salinity (% NaCl)	Remarks
Deep anoxic hypersaline basins	10 to 65	5.4–8.6	2.1–40.5	4–50	Includes thalassohaline (NaCl-dominant) and athalassohaline ($MgCl_2$-dominant) basins
Polar environments	–98.6 to 24.3	4.6–9.8	0.1–35.5	0–40.2	35.3 MPa measured for Lake Vostok in Antarctica
Subsurface ecosystems	3.25 – <400	~1–12.8	<800	0.05–saturation	400°C: theoretical maximum temperature at the crust/mantle boundary 800 MPa: highest pressure estimated at the collision of the Indian and Eurasian plates
Deserts	–19.4 to 70	6.8–10	Nr	0.02–30.8	Does not include polar deserts
Mine drainage	1–47	~3.6–13.3	6–14	0.008–7.6	A pressure of 14 MPa measured at the Pyhäsalmi mine (Finland)
Serpentinites	10–365	2.8–12.6	<900	0.03–49.68	Includes both marine and terrestrial serpentinization sites

Table 4.1. *Environmental characteristics of extreme microbial habitats. Translated from Merino et al. (2019a, 2019b). Nr = not reported*

4.2.1. *Psychrophiles*

The majority of Earth's biosphere is cold, subjected to temperatures below 5°C (Feller 2017). This biosphere is populated by microorganisms called psychrophiles, which are adapted to life at (very) low temperatures (Cavicchioli 2016). In general, psychrophiles can develop at temperatures ranging from negative values to +20°C (Banerjee et al. 2016). Psychrophiles can be categorized based on their tolerance to cold: stenopsychrophiles and eurypsychrophiles (Bakermans 2012; Cavicchioli 2016; Raymond-Bouchard et al. 2018). Eurypsychrophiles have a broad tolerance and can develop from sub-zero temperatures up to +20°C. Stenopsychrophiles, on the other hand, develop within a narrow temperature range (Δ 10°C) of low temperatures. While psychrophilic bacteria have been extensively studied, little information is available on archaea from cold or glacial environments. These environments are numerous, ranging from deep seas to the Arctic and Antarctic poles and their oceans to permafrost areas, and even high mountain environments.

Ocean floor temperatures are often below 4°C (see section 4.2.6). Psychrophilic and piezophilic archaea, taxonomically affiliated with *Thaumarchaeota* (previously named *Crenarchaeota* of Marine Group I (MG-I)) and *Haloarchaea* of Group IV, are found in these deep environments (Bakermans 2012).

The Earth also has extensive glacial environments at its poles: the Arctic and Antarctica. Exploration of their oceans has led to the isolation of archaea belonging to *Thaumarchaeota* and *Candidatus Thalassoarchaea* (*Euryarchaeota*) (Bakermans 2012). The water temperature in these oceans ranges from −1.9°C to +5°C. Lakes found beneath glaciers in Antarctica are formed due to the geothermal warming of the underlying rocks. The thickness of the glacier influences the in situ pressure, which can be high enough to lower the melting point of the ice, allowing the water to remain liquid, even at sub-zero temperatures. Antarctica contains more than a hundred subglacial lakes, the most well-known being Lake Vostok, with an area of 15,690 km^2 and an average depth of 344 m. Anaerobic haloarchaea and the earliest known psychrophilic methanogenic archaea, *Methanococcoides burtonii* (eurypsychrophile) and *Methanogenium frigidum* (stenopsychrophile), have been isolated from this polar region, from samples taken from Lake Ace (Siddiqui et al. 2013; Bowman 2017). Archaea belonging to the *Thaumarchaeota* (formerly MG-I) and *Candidatus Izemarchaea* (formerly Marine Benthic Group D) are also found in polar marine sediments (Bakermans 2012).

The pack ice is a distinctive cold environment for psychrophiles, characterized by a semi-solid state and gradients of temperature and salinity (Figure 4.1(a)). During its formation, dissolved salts are expelled from the solid matrix.

"Veins", where liquid water flows, also form within it depending on the fluctuating temperature. The temperature can drop to −35°C, and salinity can reach 14.5% at −10°C, while it is only 3.4% at −1.9°C. These veins in the pack ice contain high concentrations of dissolved organic matter, allowing the development of many psychrophilic and halotolerant microorganisms. Few archaea have been isolated from this environment since they represent only 0–3% of the total number of cells. It is mainly bacteria that have been cultured from pack ice (Cayol et al. 2015).

Glaciers, found mostly in Greenland and Antarctica, contain about 80% of Earth's fresh water. Their thickness varies from a few meters to 4 km. The ice composing them is formed by the compression of snow and contains liquid veins. Older glaciers may retain ancestral species of microorganisms (Cayol et al. 2015) (Figure 4.1(a)).

Permafrost zones are soils where the temperature has been below 0°C for at least two consecutive years; they are also considered reservoirs of ancient microbial strains (Figure 4.1(b)). They occupy about 20% of Earth's surface. Active microbial life has been detected at temperatures between −20°C and 0°C, with observations of oxidation activity of organic compounds and CO_2 fixation. Methanogenic archaea, including some halophiles (see section 4.2.5), have been isolated from these unique environments. These include *Methanomicrobiaceae*, *Methanosarcinaceae* and *Methanosaetaceae* (Margesin and Miteva 2011; Cayol et al. 2015). Notably, at these low temperatures, methanogenesis activity increases with soil depth. Genetic sequences affiliated with *Thaumarchaeota* have also been detected in Siberian permafrost.

Snow and sediment found in high mountain environments also harbor psychrophiles, such as *Methanolobus psychrophilus* (*Methanosarcinales*) (Bowman 2017).

It is important to emphasize that methanogenic archaea are present in cold environments, particularly in the cryosphere (Figure 4.1(c)). Melting ice releases a significant amount of methane (CH_4) into the atmosphere. Estimates indicate that 0.2–2.5 Gt of the carbon contained in permafrost has been released as methane and carbon dioxide over the past 60 years (Anthony et al. 2016). Methane is a greenhouse gas 25 times more potent than carbon dioxide in terms of global warming potential and is responsible for about 20% of current global warming. Methane emissions are about 500–600 Tg per year, with the majority of these emissions resulting from archaeal methanogenesis (about 70%) (Lyu et al. 2018). Melting ice will potentially release microorganisms, including methanogens, further contributing to global warming.

Figure 4.2. *Photographs of extreme environments hosting microorganisms*

COMMENT ON FIGURE 4.2.– *(a) Geyser, including thermophiles (A. Grosche). (b) Underwater hydrothermal chimney where thermophilic and piezophilic archaea live (Ifremer, Bicose). (c) Organic composters where high temperatures are conducive to the growth of thermophilic archaea (M. Cozannet). (d) Mud volcano, populated by methanotrophic and methanogenic archaea (mud volcano Paclele Mici, Romania, K. Alain). (e) Salt marshes: ecosystems favorable to the development of halophilic archaea (Guérande, France, E. Vince). (f) Food cans sterilized by irradiation, from which radiotolerant microorganisms have been isolated (M. Cozannet).*

4.2.2. *Thermophiles/hyperthermophiles*

Microorganisms that grow at high temperatures have been known since the 1960s, thanks to the work of Thomas D. Brock and his team. As early as 1967, they isolated microorganisms with growth *optima* greater than 75°C from hot springs in Yellowstone National Park. The study of these microorganisms has subsequently advanced, leading to the description of strains such as *Pyrolobus fumarii* (*Thermoproteota*) and *Methanopyrus kandleri* (*Methanopyrales*) with optimal growth at 106°C and 105°C, respectively (Blöchl et al. 1997; Takai et al. 2008). It should be noted that the vast majority of hyperthermophilic organisms are archaea.

(Hyper)thermophiles are therefore capable of proliferating and colonizing habitats where temperatures exceed 45°C. These environments, characterized by high temperatures, include both natural hot environments and environments associated with human activities. These environments are widely distributed on the planet and can be marine or terrestrial.

These environments are most often associated with active tectonic zones that generate terrestrial or underwater volcanic activities (Chaban et al. 2006; Mehta and Satyanarayana 2013). Among continental hot sites, land-based hot springs of various types allow high-temperature life: fumaroles, solfatares and geysers (Figures 4.1(d)–(f) and 4.2(a)). Their formation results from movements of the Earth's crust through relatively similar processes (Nordstrom et al. 2000; Colman et al. 2018): initially, water infiltrates the Earth's crust, heats up near a heat source (e.g. magmatic rocks), then leaches surrounding rocks and becomes enriched with minerals and gases before resurfacing under pressure.

Solfatares (Figure 4.1(d)) differ from fumaroles by their high concentrations of hydrogen sulfide (H_2S), hydrogen fluoride (HF) and sulfur dioxide (SO_2), and by the accumulation of metals in the surrounding soil (Nordstrom et al. 2000). The emitted vapor condenses to form small acidic pools that support the development of acidophilic microorganisms. Solfatares are often grouped into fields including soils, surface water and heated sludge. Notable examples are found in Italy (Solfatare, Vulcano), Papua New Guinea (Lihir), Japan (Tsurumi) and Iceland. These environments host sulfur-dependent archaea, primarily from the genera *Sulfolobus* and *Acidianus* (*Thermoproteota*).

The formation of geysers (Figure 4.2(a)) or terrestrial hot springs results from the percolation of groundwater to the Earth's surface. In geysers, water is ejected at regular intervals due to high pressure. While geysers are rare and localized in volcanic areas, terrestrial hot springs are widespread (Mehta and Satyanarayana 2013). Examples include Yellowstone National Park, the Dallol site in the Danakil Depression (Ethiopia/Eritrea), Japan, China, Malaysia, the Philippines and Iceland. The primary lineages of archaea inhabiting these ecosystems are affiliated with *Sulfolobales* (*Thermoproteota*) and *Thermoplasmatales* (*Thermoplasmatota*). Molecular signatures of *Korarchaeota*, Arman (Archaeal Richmond Mine Acidophilic Nanoorganisms) and sequences affiliated with the candidate phylum *Ca. Geoarchaeota* have also been detected in some Yellowstone hot springs (Reigstad et al. 2010; Kozubal et al. 2013; Colman et al. 2018). Available data on these new lineages suggest they are sometimes acidophilic (Colman et al. 2018).

Mud volcanoes form due to an accumulation of gases (methane, carbon dioxide, nitrogen and water vapor) at depth, followed by their emission as semi-liquid clays and gas-enriched mud matrices, when the cohesion of the sediments/rocks above the reservoir is insufficient (Figure 4.2(d)). Mud volcanoes are found in both terrestrial and marine environments, particularly in areas of compression/subduction or in river deltas with high sedimentation rates. These geological structures are characterized by the effusion of mud, often laden with hydrocarbons or salts, sometimes acidic and sometimes hot. In the latter case, they are favorable for the development of thermophilic archaea.

Marine hot environments are mainly represented by deep ocean hydrothermal vents (Figure 4.2(b)). These are among the most extreme biotopes on Earth (Dick 2019). Found between 700 and 5,000 m deep, they are distributed along mid-ocean ridges and back-arc basins at over 200 (confirmed or presumed) oceanic sites in the Pacific, Atlantic and Indian Oceans (Jebbar 2015).

The formation of hydrothermal vents results from the activity of subduction zones where seawater infiltrates. This water heats up on contact with the magma chamber, becoming enriched with dissolved gases, minerals and various elements as it rises to the ocean floor. Under the effects of heat and pressure, the hot and enriched fluid is discharged via hydrothermal vents, which are formed through the precipitation of elements from the hot fluid (350°C) upon contact with cold seawater (2°C). The archaeal communities inhabiting these ecosystems include methanogens from the genera *Methanococcus*, *Methanocaldococcus* (*Methanococcales*), *Methanopyrus* (*Methanopyrales*) and *Methanosarcinales*, performing methanogenesis, as well as chemo-organotrophic *Thermococcales* from the genera *Thermococcus* and *Pyrococcus* (Dick 2019).

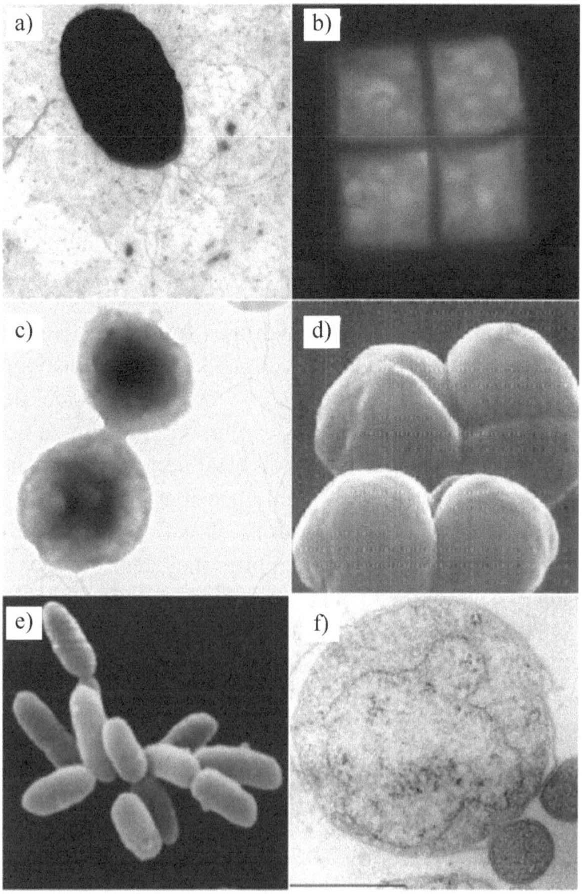

Figure 4.3. *Microphotographs of archaea living in extreme or non-extreme environments*

COMMENTS ON FIGURE 4.3.– *(a) Pyrococcus yayanosii, an obligate hyperthermophilic and piezophilic archaea of hydrothermal origin (UMR 6197, BEEP). (b) Haloquadratum walsbyi, a phototrophic halophilic archaea from a brine lake. (c) Thermococcus piezophilus, a hyperthermophilic and piezophilic archaea from the deepest hydrothermal sources of the planet (Cayman Trench) (UMR 6197, BEEP). (d) Methanosarcina barkeri fusaro, a methanogenic archaea isolated from lacustrine sediments. (e) Halobacterium sp., an aerobic halophilic archaea frequently encountered in hypersaline lakes. (f) Hydrothermal thermophilic archaea Ignicoccus hospitalis with its two symbionts Nanoarchaeum equitans.*

Human activities create non-natural microbial habitats hosting thermophilic organisms, such as mine drainage sites, organic waste composters, biological waste repositories and waste treatment systems (Mehta and Satyanarayana 2013). Thermophilic archaea identified in these unnatural environments belong to the genera *Ferroplasma* (*Thermoplasmatota*), *Sulfolobus*, *Acidianus*, *Metallosphaera* and *Stygiolobus* (*Thermoproteota*).

4.2.3. *Acidophiles*

Microorganisms that grow at acidic hydrogen (pH) potentials are referred to as acidophiles. Two categories of acidophilic organisms can be distinguished (Johnson 1998, 2007): acid-tolerants (more rarely called moderate acidophiles) and obligate acidophiles. Acid-tolerant microorganisms exhibit optimal growth between pH 3 and 5.5 but can grow at slightly lower pH levels (Hallberg and Johnson 2001; Johnson 2007). Obligate acidophiles exhibit optimal growth at pH less than 3. According to Johnson (1998, 2007), acidophiles can also be classified based on their temperature tolerance. Thus, the group of thermo-acidophiles includes both archaea and bacteria, whereas the group of acidophilic hyperthermophiles includes only archaea.

Two main mechanisms lead to the acidification of aqueous ecosystems where acidophiles develop: the oxidation of metal sulfides and the degassing of magmatic rocks (Nordstrom et al. 2000). The first case can be illustrated by acid mine drainage (AMD), which are highly acidic continuous mineral flows resulting from a slow process of oxidation by water and air of metal sulfides present in exposed rocks (Baker and Banfield 2003). While AMDs can form spontaneously, the majority derive from anthropogenic activities (excavations, pumping in mines, quarries, etc.) and appear as acidic runoff loaded with metals (silver, copper, iron, nickel, gold, lead, zinc) (Akcil and Koldas 2006). Their pH generally ranges from 2 to 8, with a record low of pH – 3.6 in a California mine (Iron Mountain) (Nordstrom et al. 2000; Skousen et al. 2019) Pyrite (FeS_2) is the predominant metal sulfide in mines. The high concentration of metals and the acidity of AMDs constitute a major source of pollution for nearby rivers, lakes and wetlands (Skousen et al. 2019).

It is important to emphasize that the oxidations of pyrite and ferrous iron (Fe^{2+}) can be catalyzed, or even regulated, by various metabolic activities of iron-oxidizing prokaryotes (Hallberg and Johnson 2001; Baker and Banfield 2003). These microbial activities can lead to acidification of the environment to below 2 pH units (Baker and Banfield 2003). *Sulfolobales*, *Thermoplasmatales* and Arman (archaeal Richmond Mine Acidophilic Nanoorganisms) (*Candidatus Micrarchaeota* and *Ca. Parvarchaeota*) are the main lineages of moderately thermophilic archaea developing in AMDs, sometimes as biofilms (Baker et al. 2006; Chen et al. 2018).

Among the *Sulfolobales*, *Acidianus brierleyi* has been shown to actively participate in the oxidation of Fe^{2+} and pyrite (Larsson et al. 1990). Other genera of *Sulfolobales* have also been identified in AMDs, notably the genera *Sulfolobus* and *Metallosphaera* (e.g. *Metallosphaera prunae*) (Baker and Banfield 2003). Among the *Thermoplasmatales* present in these unique environments, *Ferroplasma acidarmanus* is notable for lacking a cell wall and for developing at pH 0 in arsenopyrite (FeAsS) microcavities (<0.5 μm) formed by the biotic oxidation of the mineral (Baker and Banfield 2003; Pikuta et al. 2007).

Geological activities such as volcanism and/or plate tectonics lead to the formation of several types of naturally hot environments, such as hot springs, solfatare fields and oceanic hydrothermal vents, which are, in the vast majority of cases, acidic (Nordstrom et al. 2000; Chaban et al. 2006).

Oceanic hydrothermal vents (Figure 4.2(b); see section 4.2.2) are characterized by the emission of very hot fluids, with pH values varying between 1 and 12, depending on the lithological context (Dick 2019). The main archaeal lineages found in hydrothermal ecosystems are mostly neutrophilic or acid-tolerant, and more rarely acidophilic (Reysenbach et al. 2006). The species *Aciduliprofundum boonei* (*Thermoplasmatota*) is the first thermo-acidophilic archaeon to have been isolated from this type of habitat (Reysenbach et al. 2006). This species grows at pH values between 3.3 and 5.8, with optimal growth at pH 4.2–4.8.

Solfatares and some terrestrial hot springs are conducive to the development of acidophilic archaea (see section 4.2.2). Several *taxa* growing at acidic pH have been detected or cultivated from these environments, including archaea affiliated with *Thermoproteota* (*Acidilobales*, *Sulfolobales* and *Thermoproteales*) and *Thermoplasmatota* (*Thermoplasmata*) (Pikuta et al. 2007; Plumb et al. 2007; Sato et al. 2013). The *Thermoplasmatota* species *Picrophilus oshimae* and *Picrophilus torridus* grow at pH values between –0.2 and +3.5 (Schleper et al. 1995; Pikuta et al. 2007). These species, isolated in Japan, currently hold the record for life at low pH with optimal growth at pH 0.7 (Schleper et al. 1995).

In addition to the ecosystems described above, other acidic environments host acidophilic and/or acid-tolerant archaea. This is particularly true for peatlands. The formation of these complex sites and the associated archaeal communities are described in section 4.3.1.3. Different types of peatlands can be distinguished based on plant composition, water chemistry (including dissolved nutrient concentration) and hydrological characteristics (Andersen et al. 2013). Some archaeal lineages developing in these environments are methanogenic and acid tolerant. For example, *Methanoregula boonei* (*Halobacteriota*) is a *Methanomicrobiales* with optimal growth at pH 5.1. It was isolated from a bog with an in situ pH of 4.0–4.5 (Bräuer et al. 2011).

4.2.4. *Alkalophiles*

Microorganisms growing at alkaline pH values are referred to as alkalophiles (more rarely as alkaliphiles or alkalinophiles). Three categories of alkalophiles can be distinguished: obligatory alkalophiles, facultative alkalophiles and alkalotolerants (or alkalitolerants) (Bowers and Wiegel 2011). Obligatory alkalophiles show optimal growth around pH 9 (or even between 10 and 12). Their tolerance to neutral pH ($\geq$7.5) is very limited. Facultative alkalophiles preferentially grow at neutral or even slightly acidic pH values (pH 6.5), with optimal growth around pH 8.5. Finally, alkalotolerants exhibit optimal growth between pH 6 and 8.5 but can withstand pH values greater than 9. It is important to note that alkaline environments are often highly concentrated in salts. Microorganisms developing in these environments are thus qualified as haloalkaliphiles. All known alkalophilic archaea are also haloalkaliphilic (see section 4.2.5) (Grant 2004).

Two main types of alkaline environments can be distinguished based on their temporal stability and their concentration of Ca^{2+} and Mg^{+} ions.

The first type corresponds to soda lakes and some deserts. These environments are stable and have a low concentration of Ca^{2+} and Mg^{2+} cations. These deserts and soda lakes primarily result from the evaporation of closed bodies of water (Kulkarni et al. 2016). Evaporation induces a new equilibrium between CO_2, bicarbonate (HCO_3^-) and dissolved carbonates (CO_3^{2-}), leading to rapid saturation and subsequent precipitation of Ca^{2+} and Mg^{2+} ions in the form of calcium carbonate ($CaCO_3$), magnesium carbonate ($MgCO_3$) and dolomite ($(CO_3)_2$) (Grant et al. 1990). This phenomenon increases the pH, sometimes exceeding 11.5 (Jones et al. 1998). The precipitation of carbonates is often concomitant with that of Na^+ and Cl^- ions, resulting in a significant increase in salinity in soda lakes (>33 $g{\cdot}L^{-1}$ NaCl). The main soda lakes studied to date are located in the Great Rift Valley (Lake Magadi, Kenya), the Natron Valley (Libyan Desert and Egypt), China (Lakes Aiding, Baer, Bange, Chahannor and Zabuye), the United States (Lake Mono), India (Lakes Lonar and Sambhar), Mexico (Lake Texcoco), Mongolia (Lake Hulubeir) and Russia (Kulunda Steppe) (Grant and Jones 2016; Kulkarni et al. 2016). Two main clades of haloalkaliphilic *Halobacteriota* are found in soda lakes: halophiles of the class *Halobacteria* (=*Haloarchaea*) and certain methanogens (various classes of *Methanobacteriota*). About twenty *Halobacteria* species have been described so far (Grant and Jones 2016). Among the described species, *Natronobacterium gregoryi* (Table 2.2) holds the record for tolerance to high pH, with optimal growth at pH 9.5. The methanogenic archaea developing in soda lakes are mainly affiliated with the genera *Methanobacterium*, *Methanocalculus* and *Methanosalsum*. These archaea are all methylotrophic, using methanol, methylamines and/or dimethyl sulfide as substrates (Jones et al. 1998).

The second major type of alkaline environment corresponds to stable sites concentrated in Ca^{2+} ions and associated with serpentinization reactions. Serpentinization is the process of hydration and alteration of olivine ($MgFeSiO_4$) and pyroxene ($MgCaFeSiO_3$) present in ultramafic rocks of the Earth's mantle. This phenomenon induces the release of small organic compounds, dihydrogen (H_2), methane (CH_4) and alkaline fluids (pH ≥ 10) loaded with OH^-, Ca^{2+} and Mg^{2+} ions. Serpentinization is frequently observed in groundwater, aquifers and continental or submarine hydrothermal systems. The main continental sites studied are located in Portugal (Cabeço de Vide aquifer), Norway (Leka complex), the United States (The Cedars sources), Canada (Tablelands), Oman, Jordan and Cyprus (Grant et al. 1990; Postec 2015). Some of these terrestrial alkaline hot springs are conducive to the development of alkalophilic and/or alkalotolerant archaea. This is the case for *Sulfophobococcus zilligii* (*Thermoproteota*), an alkalotolerant hyperthermophile isolated from an Icelandic hot spring (Hensel et al. 1997). This species has optimal growth at pH 7.5 but can grow at pH 9.5. The submarine hydrothermal systems of the Bay of Prony (New Caledonia, France) and Lost City (mid-Atlantic oceanic ridge) are among the few serpentinized sites whose microbial diversity has been studied. In these sites, the main groups of archaea found are affiliated with *Methanosarcinales* (*Halobacteriota*) and *Thaumarchaeota* (Postec et al. 2015).

In addition to the two main types of alkaline environments described above, other types of alkaline ecosystems are identified: alkaline micro-niches and anthropogenic ecosystems. Alkaline micro-niches are most often ephemeral. The pH of these microenvironments is directly influenced by various metabolic activities such as photosynthesis, ammonification or sulfate reduction (Grant et al. 1990). Anthropogenic activities generating alkaline biotopes are primarily cement and paper production plants, certain mines, electroplating and leaching of condiments (Grant et al. 1990). However, little data are available to date on the microbial diversity associated with these ecosystems.

4.2.5. *Halophiles*

Halotolerant and halophilic organisms can accommodate or require high concentrations of salts for development. In addition to these capabilities, halophilic archaea (or *Haloarchaea*) are notable for their ability to survive under extreme conditions of desiccation, oligotrophy and radiation, sometimes over geological time scales. Their globally varied habitats approach salt saturation (concentrations above 300 $g \cdot L^{-1}$) (Purdy et al. 2004).

The literature describes various natural geological environments inhabited by *Haloarchaea*, such as terrestrial or underwater brines (e.g. Kryos Athalassic Lake, Mediterranean Sea), salt marshes, mangroves, salt lakes (e.g. Lake Meyghan, Iran), the Dead Sea and hot oil reservoirs (Oren 2010; Bhattacharyya et al. 2015; Pontefract et al. 2017; Pannekens et al. 2019). Hypersaline environments can also be of direct anthropogenic origin, including soils adjacent to salt manufacturing or processing plants, salt mines (frequently hosting species of the genus *Halolamina*), commercial salts or traditional Asian fermented/salted foods (e.g. jeotgal) (Roh et al. 2010; Kambourova et al. 2016; Bachran et al. 2018). Other non-natural habitats include living organisms, such as the plumage of certain captive pink flamingos, where archaea of the genera *Halococcus* and *Halogeometricum* impart a red color, or certain sea sponges (Lee et al. 2011; Yim et al. 2015).

Although *Haloarchaea* were long thought to be exclusively aerobic and organoheterotrophic, recent discoveries have revealed the existence of a lithoheterotrophic anaerobic lifestyle among these microorganisms (*Halanaeroarchaeum*). Hypersaline anoxic habitats host numerous uncultivated archaea, whose metabolic and ecological roles remain uncertain, but anaerobic halophiles are likely ubiquitous in these environments (Sorokin et al. 2017).

More brackish environments are also populated by halophilic archaea. Certain orders of *Haloarchaea*, such as *Halobacteriales* (*Halobacteriota*), are found in both saline and non-saline environments (e.g. the Zodletone spring in Oklahoma, a freshwater source rich in sulfur and sulfides, with a high presence of *Halobacteriales*) (Elshahed et al. 2004). In less saline habitats, these *Haloarchaea* coexist with more diverse prokaryotic *taxa*. Some isolates from hypersaline environments have also demonstrated the ability to grow at lower salinities (25 $g \cdot L^{-1}$ NaCl) (Purdy et al. 2004).

These extremophiles inhabit a wide variety of sites, with some species resisting UV and gamma irradiation or even vacuum conditions. There are psychrotolerant *Haloarchaea* species; some are phototrophic and others are anaerobic. These characteristics lead the scientific community to consider the possibility of haloarchaean life on other planets or moons, particularly Mars, Enceladus or Europa (Oren 2014; Pontefract et al. 2017; Merino et al. 2019a).

4.2.6. *Piezophiles*

Although most terrestrial biotopes are subject to high pressures (hydrostatic and/or lithospheric), and it is estimated that the deep biosphere could contain up to 70% of all cells on the planet (Oger and Jebbar 2010), few archaea cultivated and characterized as piezophiles are known. These strains, growing at pressures exceeding 10 MPa (Jannasch and Taylor 1984), in specialized high hydrostatic pressure incubators, are almost all (hyper)thermophiles belonging to the orders *Thermococcales* (e.g. *Thermococcus piezophilus* (Dalmasso et al. 2016), *Thermococcus barophilus* (Marteinsson et al. 1999), *Pyrococcus yayanosii* (Zeng et al. 2009; Birrien et al. 2011), *Desulfurococcales* (e.g. *Pyrolobus fumarii*; Blöchl et al. 1997), or *Methanococcales* (e.g. *Methanococcus jannaschii*; Miller et al. 1988). These archaea originate from deep ocean hydrothermal vents (Figure 4.2(b)), chemosynthetic, anoxic and hot ecosystems (see section 4.2.2). The fluids have varied geochemical compositions, resulting from the percolation of seawater through faults in the subsurface oceanic crust. Their plumes, rich in potential energy sources (e.g. H_2S, H_2, Fe^{2+}, Mn^{2+}, CH_4), can reach hundreds of meters in height and disperse over hundreds of kilometers (Dick 2019).

However, piezophilic microorganisms are extremely prone to sampling and enrichment biases because technologies for maintaining in situ pressure until strain isolation remain limited (Sievert and Vetriani 2012). It is also reasonable to assume that strains collected from great depths, even if not characterized as piezophiles, would show better growth at high pressure (Oger 2018). Thus, although archaea have not necessarily been isolated and characterized as piezophiles, their potential habitats are varied: soils, water columns and ocean floors of the deep ocean, invertebrates and fish from great depths, deep brines, lakes (e.g. Baikal), deep marine sediments, deep karst aquifers, deep basaltic or granitic rock and oil reservoirs.

Additionally, a handful of mesophilic archaea have also been cultivated at pressures representative of the deep biosphere, even though they were not isolated from hyperbaric environments. Communities of methane-oxidizing anaerobic archaea, anaerobic methanotrophic (ANME) archaea, from marine lake sediments at 45 m depth, have been cultivated at 20 and 40 MPa (Cassarini et al. 2019). Similarly, a ferric iron-reducing acidophilic strain, *Ferroplasma acidiphilum*, isolated from a bioleaching plant, has been cultivated at 25 MPa (Zhang et al. 2018).

Piezophilic archaea, which can (or prefer to) develop at high pressures, could therefore populate the majority of the biotopes on our planet.

4.2.7. *Radiotolerant archaea*

Radiotolerant archaea have been isolated from deep ocean hydrothermal vents, shallow marine sediments with geothermal activity, salt marshes, salt lakes and solfatares (Ranawat and Rawat 2017). Among the species that tolerate high doses of gamma and/or ultraviolet radiation are *Thermococcus gammatolerans* (30 kGy), *Pyrococcus furiosus* (3 kGy), *Pyrococcus abyssi* (3 kGy), *Sulfolobus solfataricus* (200–280 nm UV) and *Halobacterium salinarium* (8 kGy) (Jolivet et al. 2003; Jung et al. 2017; Ranawat and Rawat 2017). Among these *taxa*, only species from deep hydrothermal environments rich in heavy metals are exposed in their native habitat to doses of natural radioactivity hundreds of times higher than those in terrestrial environments (e.g. East-Pacific Ridge). This may explain their radioresistant nature. For others, it seems that the ability to resist radiation may derive from their (hyper)thermophilic or halophilic nature, acquired indirectly. Indeed, the mechanisms of resistance to high temperatures, desiccation and radioresistance are largely similar (Ranawat and Rawat 2017). These resistance strategies involve protecting the genome and proteome from oxidation (via enzymatic defense mechanisms and the synthesis of powerful antioxidant agents, condensing the nucleoid), as well as efficiently repairing damaged DNA (Ranawat and Rawat 2017).

4.2.8. *Poly-extremophilic archaea*

Many extreme environments are subject to two or more non-standard parameters and host poly-extremophilic organisms. For example, active deep ocean hydrothermal vents are often characterized by high hydrostatic pressure, high temperatures, acidic pH and high concentrations of toxic metals and gases. Solfatares, on the other hand, are characterized by very acidic pH and high temperatures.

While the individual impacts of extreme parameters on archaea have been extensively studied for a wide range of factors (temperature, pH, salinity, radiation, desiccation, pressure, etc.), studies aiming to define their collective influences on living organisms are rare (Seckbach and Rampelotto 2015). The molecular mechanisms of evolution that underlie adaptation to multiple extreme conditions are just beginning to be studied (Seckbach and Rampelotto 2015). However, in the current state of knowledge, poly-extremophilic microorganisms seem to possess less extreme developmental capacities than mono-extremophilic microorganisms (Bowers et al. 2009). For example, halothermophilic archaea isolated from brines are less thermophilic than some archaea isolated from other environments, where the only extreme parameter is temperature, perhaps due to the high bioenergetic demand inherent in adapting to multiple extreme factors.

The study of poly-extremophilic archaea is crucial to define the habitable environments of the planet and the limits of terrestrial life. The habitability of a niche is governed by a set of physical, chemical and environmental constraints such as temperature, pressure, salinity, pH and the availability of water, energy and nutrients, which act individually and synergistically to make an area habitable or not. However, many environments located at the frontiers of life are characterized by multi-extreme conditions and are poorly accessible, resulting in under-studied poly-extremophilic organisms. To date, the limits of habitability on Earth remain unknown.

4.2.9. *Record-holding archaea*

Records of survival in extreme conditions are not exclusive to archaea or organisms from extreme environments. For example, the bacterium *Deinococcus radiodurans*, isolated from a can of minced meat (Figure 4.2), and water bears, invertebrates of the *Tardigrada* phylum, found in various "normal" habitats, also exhibit exceptional survival abilities under a variety of extraordinary conditions, such as radiation, space vacuum, extreme temperatures, desiccation and nutritional deficiencies (Cox 2010; Hashimoto et al. 2016).

However, in the current state of knowledge, archaea seem to be present, sometimes predominantly, as we approach the biotic fringe. This poorly understood boundary marks the transition from the realm of the living to the inert, mineral world. Several records of life in improbable conditions, close to the theoretical physicochemical limits of life, are held by archaea (Merino et al. 2019a). Table 4.2 lists these records.

Record Held	Archaea	Habitat	Growth conditions	References
Life at high temperature	*Methanopyrus kandleri* Most hyperthermophilic organism	Hydrothermal sediments of Guaymas Basin, Gulf of California	80–110°C, optimum 98°C (at atmospheric pressure) Growth at 122°C at 20 MPa	(Kurr et al. 1991) (Takai et al. 2008)
Life at high temperature and pressure	*Pyrococcus yayanosii* Unique obligate hyperthermophilic piezophile	Hydrothermal vent at Ashadze site (–4,100 m), Mid-Atlantic Ridge	80–108°C, optimum 98°C 20–120 MPa, optimum 52 MPa	(Zeng et al. 2009; Birrien et al. 2011)
	Thermococcus piezophilus Broadest range of growth pressure	Hydrothermal vent at Beebe site (–4,964 m), Cayman Trench	60–90°C, optimum 75°C 0.1–125 MPa, optimum 50 MPa	(Dalmasso et al. 2016)

Record Held	Archaea	Habitat	Growth conditions	References
Life at acidic pH	*Picrophilus oshimae* Most acidophilic organism	Solfatara in Japan	Optimum 60°C pH 0–3.5, optimum pH 0.7	(Schleper et al. 1995)
Life at alkaline pH	*Natronobacterium gregoryi* Most alkaliphilic organism	Lake Magadi, Kenya	pH 8.5–11.0, optimum pH 9.5	(Tindall et al. 1984) (Xu et al. 1999)
Radiotolerance	*Thermococcus gammatolerans* Most radioresistant organism	Hydrothermal vent in Guaymas Basin, Gulf of California	Optimum 88°C Withstands 30 kGy of gamma radiation without lethality	(Jolivet et al. 2003)
Life in hypersaline environments	Various genera and species of *haloarchaea*	Hypersaline lakes or seas, salt marshes, Dead Sea	Environments with near-saturation salt levels	–
Life with very little water	Haloarchaeal strains GN-2 and GN-5 Lowest water availability (a_W = 0.635)	Salt marshes in Mexico	–	(Javor 1984)

Table 4.2. *Archaea breaking all records for life in extreme conditions*

4.3. Archaea populating ordinary, non-extreme environments

Over the past two decades, the use of molecular techniques based on PCR amplification of genes encoding 16S rRNAs has enabled an assessment of microbial diversity independent of culturing. Remarkably, these techniques have revealed a wide distribution of mostly uncultivated archaea in normal habitats, such as ocean waters, lake waters and soil (Jarrell et al. 2011).

The majority of archaea, like bacteria, are thought to reside in natural habitats in the form of biofilms (Flemming and Wuertz 2019).

According to some hypotheses, archaea have adapted to mesophilic conditions through genetic imports from bacteria (Flemming and Wuertz 2019).

Section 4.3 introduces non-extremophilic archaea living in these "ordinary", "common" environments and focuses on their physiology in relation to the nutrients available in their habitat (Figure 4.4).

Figure 4.4. *Photographs of non-extreme environments populated by archaea*

COMMENT ON FIGURE 4.4.– *(a) The rhizosphere is an environment strongly influenced by plants and favorable to the development of nitrifying and methanogenic archaea (M. Cozannet). (b) Rice fields host a significant taxonomic diversity of methanogenic and nitrifying archaea (Thai rice field, K. Alain). (c) Peatlands are reservoirs of organic matter and host methanogenic archaea (Mougau peat bog, France, M. Cozannet). (d) Many lakes host methanogenic and methanotrophic archaea (Candle Lake, Canada, M. Allioux). (e) Seas and oceans are large spaces favorable to the development of a significant taxonomic diversity of archaea (Iroise Sea, France, M. Cozannet). (f) Corals and sponges live in symbiosis with archaea participating in the recycling of nitrogenous matter (Lophelia sp. and Hexadella sp., Ifremer Bobeco, Gouillou). (g) The rumen of cattle is populated by methanogenic archaea participating in the digestion process (Bos taurus, M. Cozannet). (h) In humans, some archaea participate in the oxidation of ammonia present in perspiration (Homo sapiens, M. Cozannet).*

4.3.1. *Phytobiomes, rice paddies and peatlands*

4.3.1.1. *Phytobiomes*

Plants live in close interaction with a significant diversity of microorganisms; the different microbiomes of a plant form the plant microbiome or phytobiome. This occupies all plant tissues (*endophytic microbiome*: roots, stems, leaves, seeds, fruits and pollens), aerial parts (*phyllospheric microbiome*: on and under the leaves) and underground surfaces (*rhizospheric microbiome*: near the roots) (Gopal and Gupta 2016; Moissl-Eichinger et al. 2018). Its microbial composition is specific to each species, and sometimes even to each plant genotype. The phytobiome directly affects the health, growth, productivity and potentially the environmental adaptation of the plant (Moissl-Eichinger et al. 2018).

The rhizosphere can be defined as the part of the soil extending between the root surface (rhizoplane) and the surrounding raw soil (Chaban et al. 2006; Buée et al. 2009) (Figure 4.4(a)). This space is colonized by the root microbiota (rhizobiome). The rhizospheric microbial community differs from that of the surrounding soil. Indeed, the different types of metabolites and exudates released by the roots act as chemical *stimuli* promoting the preferential colonization of the rhizosphere by certain *taxa* through chemotaxis. The exudates produced by the plant also serve as a source of nutrients and carbon for the rhizobiome (Hartmann et al. 2009). Similarly, the exchange of different types of metabolites and exudates produced by the plant and microorganisms directly influences communication and activity within the rhizobiome. The soil type, physiology, ecology and age of the plant host also structure rhizospheric microbial communities (Buée et al. 2009; Neori and Agami

2017). The rhizobiome plays a key role in nutrient assimilation by the roots by releasing fragmented or modified molecules (e.g. organic acids). Archaea are generally sparsely abundant in the rhizosphere.

Thaumarchaeota and *Methanobacteriota* are the two most frequently encountered phyla. *Thermoprotei* (*Thermoproteota*) have also been found in this habitat, though more anecdotally.

Some slightly anoxic rhizospheres are populated by ammonium-oxidizing archaea (AOA) affiliated with the genus *Nitrososphaera* (*Thermoproteota*), which play a key role in nitrate (NO_3^-) input for plants via nitrification (Alegria et al. 2016; Moissl-Eichinger et al. 2018). Ammonium (NH_4^+) is converted to nitrate in three major steps: oxidation of ammonium to hydroxylamine (NH_2OH), followed by oxidation of hydroxylamine to nitrite (NO_2^-) by AOA, and finally, oxidation of nitrite to nitrate by autotrophic nitrite-oxidizing bacteria. Under anaerobic conditions, AOA can convert nitrate to nitrous oxide (N_2O), a potent greenhouse gas. Plants can regulate and inhibit these reactions through the excretion of exudates (Alegria et al. 2016; Moissl-Eichinger et al. 2018).

Many lineages in the rhizosphere are methanogenic. At least six orders of methanogenic archaea described to date have been detected in the rhizosphere of rice (*Oryza sativa*) (see section 4.3.1.2; Figure 4.4(b)). The methane (CH_4) produced by these archaea is released through the plant's aerenchyma. Interestingly, some methanogens in the rhizosphere synthesize antioxidant enzymes, making them aerotolerant (Buée et al. 2009). Methanogenesis can be carried out from simple molecules such as dihydrogen (H_2) and carbon dioxide (CO_2), formate (HCO_2^-) (hydrogenotrophic methanogenesis), acetate ($CH_3CO_2^-$) (acetoclastic methanogenesis), or methylated compounds ($R\text{-}CH_3$) (methylotrophic methanogenesis), all resulting from the degradation of organic matter.

The endosphere includes all microorganisms developing in the tissues of a host plant. The composition of the endophytic microbiome varies according to the geographical location and growth stage of the host. For example, a study on the endosphere of olive branches and leaves (*Olea europea*) showed that archaea can represent between 5.3% and 67.3% of the microorganisms in the plant's endophytic microbiome (Müller et al. 2015). Phylogenetically, the endophytic archaea identified in this study were mainly affiliated with *Nitrosophaerales* (*Thermoproteota*), *Halobacteriales* (*Halobacteriota*), *Methanomicrobia* (*Halobacteriota*), *Marine Benthic Group A* and, more marginally, *Parvarchaea* (*Parvarchaeota*) and *Thermoplasmatales* (*Thermoplasmatota*). *Thaumarchaeota* present in plant tissues could influence the health of the endophytic microbiome (Müller et al. 2015).

The aerial part of the plant constitutes the phyllosphere (Vorholt 2012). This surface is mainly colonized by commensal bacteria and microfungi, sometimes pathogenic. Archaea appear to represent only a very small fraction of the phyllospheric microbiome (≈1%) (Vorholt 2012; Moissl-Eichinger et al. 2018).

4.3.1.2. *Rice fields*

With 155 million hectares of cultivated arable land, rice (*Oryza sativa*) is one of the most extensively grown crops worldwide (Ma et al. 2010) (Figure 4.4(b)). Ninety percent of rice production occurs in Asia, particularly in China, India, Indonesia, Japan, the Philippines, Thailand and Vietnam. Rice fields play a crucial role in global warming, being responsible for an estimated 33–110 Tg (33–110 million tonnes) of atmospheric methane (CH_4) emissions annually, which is substantial (Liesack et al. 2000; Kirschke et al. 2013). Rice is typically cultivated in lightly flooded soils. The soil in rice fields can be divided into three compartments based on depth: an oxic surface horizon, an anoxic root-free horizon and the rhizosphere (Liesack et al. 2000). Significant phylogenetic diversity of *Methanobacteriota* and *Thaumarchaeota* has been described or detected in these environments.

Historically, *Methanobrevibacter arboriphilus* (*Methanobacteriales*) and *Methanosarcina mazei* (*Methanosarcinales*) were the first archaea isolated from rice field (Asakawa et al. 1993, 1995). In the initial molecular diversity inventories of rice paddies, environmental DNA sequences were classified under the name rice clusters (RC), numbered from RC-I to RC-VI (Großkopf et al. 1998; Chin et al. 1999).

The distribution and physiology of methanogenic archaea in rice fields vary across compartments. Hydrogenotrophic methanogens such as *Methanomicrobiales*, *Methanobacteriales* and *Methanocellales* (RC-I) are primarily found in the rhizosphere of mature plants. The methane produced by these archaea is released into the atmosphere through plant channels (Buée et al. 2009). Notably, *Methanocellales* can tolerate low partial pressures of dihydrogen (H_2) and relatively oxic conditions by synthesizing various antioxidant enzymes (Buée et al. 2009; Alpana et al. 2017). This order is responsible for significant methane production in such ecosystems. Acetoclastic methanogens (both obligate and non-obligate) affiliated with *Methanosarcinaceae* and *Methanosaetaceae* (*Methanosarcinales*) are primarily found in the anoxic soil horizon (Alpana et al. 2017). Methanogenic archaeal sequences from candidate family *Candidatus Methanoflorentaceae* (RC-II) (hydrogenotrophic methanogens), *Methanomassiliicoccales* (RC-III) (hydrogen-dependent methylotrophs) and the *Euryarchaeota* cluster RC-V (= PlSA1) have also been identified in rice paddies. As with most microorganisms, factors influencing the distribution and activity of methanogenic archaea include soil composition, variety and age of surrounding plants, fertilizer amendments, crop rotation, pollution and seasons (Alpana et al. 2017).

In recent publications, the RC-IV (also called "Group C2 freshwater *Crenarchaeota*") and RC-VI (also called "Group I.1.b terrestrial *Crenarchaeota*") clusters have been reclassified within the *Thaumarchaeota* (Itoh et al. 2013; Adam et al. 2017). The RC-VI cluster now constitutes the order *Nitrososphaerales*. Current knowledge establishes that *Thaumarchaeota* populate all compartments of rice fields and coexist with methanogenic archaea. Various exudates (including oxygen and CO_2) produced by the roots seem to influence the activity of *Thaumarchaeota* (Buée et al. 2009; Ke et al. 2016).

4.3.1.3. *Peatlands*

Peatlands are complex wetland terrestrial ecosystems characterized by soil that is consistently saturated with water (Figure 4.4(c)). These sites are predominantly found in the Northern Hemisphere (Northern Europe, Russia, Canada, United States), as well as in South America (Chile), Southeast Asia (Indonesia), Africa (Republic of Congo) and Oceania (Australia). Despite covering approximately 3% of terrestrial surfaces, peatlands sequester nearly one-third of terrestrial organic matter in the form of peat (Andersen et al. 2013; Dargie et al. 2017). This accumulation results from an imbalanced ratio between primary production and organic matter decomposition. Peatlands are significant sources of atmospheric methane (CH_4), contributing between 3% and 5% of global methane emissions (Chaban et al. 2006). Two primary types of peatlands can be differentiated based on hydrology, dissolved element concentrations and the plant communities they support: ombrotrophic peatlands and minerotrophic peatlands (Andersen et al. 2013). Ombrotrophic peatlands are oligotrophic and acidic ($pH \approx 4$), receiving water and nutrients primarily from precipitation or snowmelt. In contrast, minerotrophic peatlands are fed by runoff and groundwater, and can be oligotrophic with acidic pH or mesotrophic with acidic to basic pH levels.

Peat forms through the gradual accumulation and partial (or complete) decomposition of organic matter, predominantly of plant origin. This decomposition occurs across three defined horizons based on depth: the acrotelm, mesotelm and catotelm (Andersen et al. 2013). The acrotelm is the oxic upper layer dominated by aerobic bacteria and saprotrophic fungi initiating organic matter breakdown. The mesotelm serves as a moderately oxic intermediate horizon conducive to the growth of methanotrophic bacteria. The catotelm constitutes the lower, constantly anoxic, water-saturated horizon of peatlands, where methanogenic archaea thrive, representing the predominant archaeal lineages in these ecosystems.

The composition of methanogenic communities in peatlands varies depending on site, peatland type, plant species, pH and depth (Yavitt et al. 2012; Andersen et al. 2013). Hydrogenotrophic methanogenic archaea (*Methanobacteriaceae*, *Methanocellales*, *Methanomicrobiales*) often seem to dominate the upper part of the catotelm, while acetoclastic methanogenic archaea (some *Methanosarcinales*) appear to be primarily found in the lower part of the horizon. Genomic sequences associated with the candidate family *Candidatus Methanoflorentaceae* and the order *Methanomassiliicoccales* have also been identified in peatlands (Andersen et al. 2013; Söllinger et al. 2016). Additionally, sequences affiliated with *Thaumarchaeota* and candidate divisions *Ca. Bathyarchaeota*, *Ca. Izemarchaea* (*Marine Benthic Group D*) and *Ca. Micrarchaeota* have been found in peatlands in the USA to a lesser extent (Hawkins et al. 2014; Adam et al. 2017). The roles and physiology of these archaeal *taxa* within these ecosystems remain poorly characterized to date.

4.3.2. *Aquatic habitats: lakes, oceans and estuaries*

Aquatic habitats dominate terrestrial landscapes and exhibit highly diverse physicochemical conditions in terms of temperature, light availability, salinity, depth and nutrient and oxygen concentrations. These parameters vary significantly, depending on the nature and location of the habitats. Due to their vast expanse and the wide range of environmental parameters they encompass, aquatic environments harbor considerable biological richness and extensive microbial diversity. Assuming there are potentially 1.2×10^{30} bacteria and archaea cells on Earth (Flemming and Wuertz 2019), aquatic habitats represent substantial reservoirs of microbial biomass. Estimates suggest there are approximately 1×10^{29} prokaryotic cells in the ocean and 5×10^{28} prokaryotic cells in marine sediments (Flemming and Wuertz 2019), although the specific proportions of archaea versus bacteria in each habitat remain inadequately documented. Among aquatic habitats, hydrothermal vents and freshwater environments are the primary reservoirs for discovering new species and archaeal lineages, rendering them potentially intriguing environments for studying novel archaeal clades. The discovery of certain archaeal lineages has reshaped our understanding of aquatic ecosystem ecology. Indeed, some archaeal groups play pivotal ecological roles within their natural habitats, roles that have long been underappreciated, such as the roles of *Thaumarchaeota* in the nitrogen cycle of marine environments (see section 4.3.2.2) (Santoro et al. 2019). Consequently, certain understudied aquatic archaeal *taxa* warrant detailed analysis of their distribution patterns and ecological impacts (Auguet et al. 2010).

4.3.2.1. *Freshwater environments*

Archaea populate numerous freshwater sites ranging from oligotrophic water bodies to eutrophic environments, including low- to high-altitude lakes, boreal lakes, wetlands (marshes, peatlands, fens, rivers, etc.) and various types of lake sediments (Chaban et al. 2006) (Figure 4.4(d)). Environmental conditions can vary significantly within these freshwater habitats. Depending on their geographical location, different parameters such as temperature, altitude, salinity (fresh or brackish), oxygenation level, oxygen stratification, water flow dynamics, geographical isolation, or anthropogenic impacts such as pollution may differ. Generally, aquatic environments are studied both at the surface and at depth (where sediments are located); and in the case of stratified lakes, based on their stratification into, for example, epilimnion, metalimnion and hypolimnion. Taxonomic and functional diversity, as well as the abundance of archaea, vary greatly depending on the study site and the analyzed horizon. Similar to bacterial communities, archaeal communities differ significantly between the water column and sediments in these environments (Yang et al. 2016).

Archaeal communities in freshwater environments generally stratify according to light exposure and dissolved oxygen concentration. Those in the anoxic zones of freshwater habitats notably include a significant proportion of methanogenic archaea producing methane as a metabolic byproduct (see section 4.3.1.1), a potent greenhouse gas. Estimates suggest that wetlands contribute approximately 20-39% of annual atmospheric methane production, while natural lake environments contribute 6-16% (Laskar et al. 2018). These emissions are predominantly biogenic, arising from methanogenesis by archaea inhabiting the sediments of these environments (Laskar et al. 2018). Methanogenic *taxa* frequently detected in freshwater sediments, often among the most abundant, include genera such as *Methanosarcina*, *Methanosaeta* (or *Methanothrix*), *Methanobacterium*, *Methanomicrobium*, *Methanoculleus* and *Methanomethylovorans* (Chaban et al. 2006; Laskar et al. 2018). Wetlands also harbor anaerobic methanotrophic archaea capable of methane utilization, affiliated with the ANME-2 group, alongside methanogenic archaea (both confirmed and putative) belonging to *Methanobacteriota* and *Candidatus Bathyarchaeota* (Narrowe et al. 2017).

Numerous new species of methanogenic archaea have been recently discovered in lake sediments; although they remain relatively sparse within total microbial communities, they are crucial for the carbon cycle of these ecosystems (Narrowe et al. 2017).

Non-methanogenic archaea in freshwater habitats are less well documented. Although their study is limited, it is known that they participate in various nutrient cycles and exhibit considerable metabolic diversity. For instance, in the epilimnion of lakes, archaea affiliated with *Euryarchaeota*, *Crenarchaeota* and *Thaumarchaeota* have been identified (Madigan et al. 2018). In wetland sediments, archaea affiliated with *Ca. Woesearchaeota* and *Ca. Bathyarchaeota* have been found, potentially involved in symbioses or possessing fermentative metabolisms (Narrowe et al. 2017). In nitrogen-rich freshwater habitats, specialized archaeal *taxa* involved in ammonia (NH_3) oxidation have been detected. These archaea may play a pivotal role in the nitrogen cycle but remain poorly characterized to date. Lastly, some freshwater anoxic sediments contain significantly more archaea than bacteria. For instance, estimates suggest that the total benthic prokaryotic community in Qinghai Lake (China) comprises between 51% and 86% archaea (Jiang 2008). Archaeal communities in the water column also exhibit seasonal variations (Yang et al. 2016).

4.3.2.2. *Oceanic environments*

Seas and oceans harbor a diversity of oceanic provinces and ecological niches inhabited by microorganisms, including archaea (Figure 4.4(e)). These provinces range from coastal environments to open seas, from the photic to the aphotic zones, encompassing both pelagic and benthic regions of the water column, as well as unique environments such as deep coral zones, cold seep areas, hydrothermal vents, submarine mud volcanoes, deep sediments and hypersaline anoxic basins (see section 4.2.5). Factors such as light, temperature, geographic location, oxygen and nutrient concentrations, pressure, seasonality, salinity and the presence of fauna and flora significantly influence marine microbial communities and thereby archaea.

Similar to bacteria and microalgae, there are pronounced spatiotemporal variations in the abundance of archaea in marine and oceanic environments (Chaban et al. 2006). A study in the Pacific Ocean indicated that between 1 and 5 km deep, bacterial and archaeal abundances appear to be comparable (Madigan et al. 2018). Analyses based on the 16S rRNA gene across various seas and oceans have shown that archaea constitute between 14% and 60% of the total prokaryotic cells (Chaban et al. 2006). Moreover, investigations into deep marine sediments off the coast of Peru suggest that archaea can comprise up to 98% of the total microbial community at certain deep sedimentary layers (Biddle et al. 2006).

Marine pelagic archaea are more extensively studied compared to archaea in freshwater environments. Apart from methanogenic archaea, planktonic marine archaea can be categorized into four distinct groups known as MG, based on their 16S rRNA gene sequence (note: this classification is under revision) (Chaban et al. 2006; Adam et al. 2017; Santoro et al. 2019). *Thaumarchaeota* (MG-I) are chemoautotrophic and are prevalent in most known marine environments.

They significantly contribute to carbon and nitrogen cycles through ammonia (NH_3) oxidation (Santoro et al. 2019). Archaea belonging to *Ca. Thalassoarchaea* (MG-II) are typically heterotrophic and photoheterotrophic, specializing in high molecular weight molecule utilization within the water column. Their abundance correlates closely with phytoplankton and associated organic matter, varying significantly with seasons and diel cycles (Santoro et al. 2019). Archaea in *Ca. Pontarchaeota* (MG-III) are heterotrophic and are found in the water column, particularly at depth; this group remains poorly characterized to date (Santoro et al. 2019). The MG-IV archaeal group currently includes only a few strains from polar ocean floors, North Atlantic and Mediterranean regions, phylogenetically close to archaea in the *Halobacteria* order (Santoro et al. 2019).

In the water column, a significant fraction of *Thaumarchaeota* (MG-I) oxidizes ammonia for energy, a metabolic feature observed in certain *Nitrosopumilales* (Chaban et al. 2006; Madigan et al. 2018; Santoro et al. 2019). These archaea, highly adapted to ammonia, dominate quantitatively in the pelagic zone, particularly in the mesopelagic aphotic regions. Some locations globally also host *Pontarchaea* (MG-III) and *Ca. Thalassoarchaea* members in the water column (Adam et al. 2017). Photoheterotrophic archaea are also prevalent in oceanic waters, particularly among some *Euryarchaeota* of Marine Group II, utilizing light via proteorhodopsins (Tully 2019). Nonetheless, archaea in the pelagic zone remain partially unexplored due to their lower abundance compared to bacteria (Madigan et al. 2018). Knowledge of pelagic archaea is mainly confined to *Euryarchaeota*, *Crenarchaeota* and *Thaumarchaeota.*

Marine sediments differ significantly in environmental parameters compared to the water column, notably due to the absence of light. They harbor diverse communities of archaea, both taxonomically and functionally, particularly in the deeper sedimentary horizons. Temperature is lower than in surface ocean waters, often reaching 3–4°C in the upper meters of sediments, increasing with depth thereafter. Pressure also increases with depth (due to the weight of water and sediments/rocks), while oxygen and other strong oxidants progressively decrease. The structure and composition of sedimentary microbial communities are primarily influenced by available carbon sources and electron acceptors across various lithological layers. Marine sediments are notably populated by methanogenic archaea of the *Methanomicrobia* class and genera *Methanococcus*, *Methanogenium* and *Methanoculleus* (Chaban et al. 2006; Orcutt et al. 2011). Similar to other ecosystems, marine sediments exhibit highly diverse environmental characteristics, including lithological sequences, and physical and chemical parameters, influenced by the study site's location, geological context and sediment depth. It is therefore not surprising to find rich and diverse communities of archaea there (Parkes et al.

2014). Generally, there is a low abundance of archaea affiliated with *Thaumarchaeota* (MG-I) and *Crenarchaeota* in oxic surface sediments (Orcutt et al. 2011). At the methane–sulfate transition zone (SMTZ), syntrophic associations are often observed in the form of consortia or microbial mats between methane-oxidizing methanogenic archaea (ANME) and sulfate-reducing bacteria. In addition to consuming methane, archaea belonging to *Euryarchaeota* ANME-1 are likely involved in its production (Beulig et al. 2019).

Additionally, heterotrophic archaea inhabit marine sediments, including *Candidatus Izemarchaea* (Marine Benthic Group D) and *Candidatus Bathyarchaeota* (Adam et al. 2017).

4.3.2.3. *Deltas and estuaries*

At the interface between freshwater and saltwater lie unique environments: deltas and estuaries. These environments are characterized by an intermediate salinity between seawater and rivers, as well as a high concentration of nutrients. Archaea appear to be ubiquitous and diverse in these unique environments (Chaban et al. 2006; Liu et al. 2018).

The composition of archaeal communities seems to vary significantly, depending on the geographical location of the estuary. Indeed, the latitude of the estuary is the most influential parameter on the composition of archaeal communities (Liu et al. 2018). Studies of microbial communities in 24 deltas worldwide have shown that certain lineages of archaea are frequently encountered and predominate in these habitats, namely, *Thaumarchaeota*, *Bathyarchaeota* and *Euryarchaeota* (Liu et al. 2018). Other studies have highlighted the presence of archaea affiliated with *Theionarchaea*, *Altiarchaeales* and *Hadesarchaea* in some estuaries (Adam et al. 2017).

In conclusion, based on the current state of knowledge, the archaeal communities of aquatic environments are extremely diverse, both taxonomically and physiologically. Archaea are abundant in specific aquatic niches. Some lineages of archaea are endemic to particular habitats, while others have a more ubiquitous distribution (Auguet et al. 2010). For instance, archaea from the lineages of *Lokiarchaeota, Thorarchaeota* and *Ca. Izemarchaea* are found in both freshwater and marine sediments, with the latter also present in estuaries (Adam et al. 2017). Methanogenic archaea of the *Methanomassiliicoccales* order have been detected in oceans, wetlands and a wide variety of non-aquatic environments, including terrestrial and digestive environments (Evans et al. 2019). Some archaea can develop in association with other organisms, such as metazoans, plants, fungi, protozoa and bacteria, forming symbiotic associations (Chaban et al. 2006;

Wrede et al. 2012). In all likelihood, aquatic environments harbor a diversity of archaeal lineages and microbial functions, which are still unknown. For example, archaea inhabiting sediments are only partially known due to the concentration of studies on methanogenic archaea alone. A vast diversity of archaea, though seemingly less abundant, remains to be discovered within sediments, along with their functions and impact on the ecosystem. Moreover, many environments related to aquatic habitats and populated by archaea, such as the oceanic crust or certain underground aquifers, have been understudied (Flemming and Wuertz 2019). While numerous investigations have focused on the specific and functional diversity of archaea in aquatic environments, the role of many archaeal *taxa* in the trophic chain of their habitats has not been addressed, and their symbiotic and predatory relationships remain largely unknown and will be a major focus in the coming years (Santoro et al. 2019).

4.3.3. *Environments linked to human activities: example of waste treatment and anaerobic digesters (methanizers)*

Human activities such as metal extraction, acid mine drainage, oil extraction, anaerobic digesters, solid household waste, composting, water distribution networks and wastewater treatment create numerous habitats favorable to the development of archaea. Waste and wastewater treatment, for example, is a major societal issue, as waste represents a source of pollution that, if poorly managed, could have a severely detrimental impact on human health and the environment.

Wastewater treatment plants provide various habitats for microorganisms, including archaea, which are actively involved in waste treatment. Depending on the wastewater treatment applied, waste can be left in anaerobic or aerobic conditions. If the conditions are anoxic, anaerobic digestion will occur, involving specific microbial communities, particularly methanogenic archaea. Wastewater can also be treated under oxic conditions, primarily involving communities of heterotrophic microorganisms that degrade organic compounds.

Anaerobic digestion relies on a syntrophic chain of metabolic reactions within a complex ecosystem. It is schematically divided into four successive stages: hydrolysis, acidogenesis, acetogenesis and methanogenesis.

Within ecosystems for the methanization of organic waste, archaea are generally affiliated with the orders *Methanosarcinales* (which perform acetoclastic, hydrogenotrophic or methylotrophic methanogenesis), *Methanobacteriales* (hydrogenotrophic methanogenesis) and/or *Methanomicrobiales* (hydrogenotrophic methanogenesis) (Nguyen et al. 2018). The temperature, concentration of organic matter, incubation time and nature of the substrate for anaerobic digestion significantly impact the total microbial communities, including archaea (Nguyen et al. 2018). Depending on the nature of the waste and the composition of the microbial community, different degradation products will be generated and then used as substrates in various pathways of methanogenesis. Thus, the primary degradation of organic waste into CO_2, formate, acetate and/or methylated compounds will specifically structure the communities of methanogenic archaea involved in the terminal degradation of organic matter into methane. This process modifies the abundances and proportions of archaea present and selects the archaeal *taxa* most suited to these ecological niches.

Methanogenic archaea play a major role in anaerobic digestion by intervening at the end of the syntrophic chain to degrade small carbonaceous molecules into methane and/or CO_2 (Chaban et al. 2006). Additionally, hydrogenotrophic methanogenic archaea degrade the carbon dioxide and dihydrogen generated by anaerobic digestion, which further reduces the system pressure and limits the inhibitory effect of dihydrogen on the growth of certain microorganisms (Nguyen et al. 2018).

Besides the aforementioned *taxa*, archaea belonging to the lineages of *Ca. Methanofastidiosa* and *Ca. Verstraetearchaeota*, sometimes also detected in freshwater sediments, are found in anaerobic digesters (or methanizers) (Adam et al. 2017). In wastewater treatment plants, it is common to find representatives of the genera *Methanosaeta*, *Methanobrevibacter*, *Methanosarcina* and *Methanobacterium* (Chaban et al. 2006).

Non-methanogenic archaea are also present in methanizers. They are affiliated with *Crenarchaeota*, *Ca. Bathyarchaeota* and *Thaumarchaeota*. These archaea, likely a very small minority compared to methanogens, are involved in the oxidation of ammonia, degradation of proteins and/or degradation of aromatic compounds. Their presence appears to be closely related to that of acetoclastic methanogenic archaea.

Water networks also host methanogenic archaea, and these archaea could, in some cases, be associated with corrosion. Specifically, the presence of hydrogenotrophic methanogens affiliated with the genera *Methanogenium*, *Methanoplanus* and *Methanocalculus* is potentially associated with direct corrosion by cathodic depolarization; these *taxa* use H_2 from the metal as an electron donor. Additionally, some *Methanosaeta*-affiliated acetoclastic methanogenic archaea could indirectly cause corrosion by growing in association with sulfate-reducing bacteria, whose interaction with materials causes direct microbiological corrosion or accelerates ongoing corrosion processes (Zhang et al. 2003; Chaban et al. 2006).

4.3.4. *Animal microbiomes*

The microbiota can be defined as the entire community of microorganisms developing in a specific environment within a plant or animal host. The microbial composition of microbiota varies according to the species of the host, its health status and the specific organ being colonized (or microbiome). The collective group of archaea within the microbiome of a given host is referred to as the archaeome.

4.3.4.1. *Microbiome of corals and sponges*

The sponge microbiome is composed of microscopic fungi, protists, microalgae, (cyano)bacteria and archaea growing in the mesohyl (or extracellular matrix) of the host (Hentschel et al. 2012) (Figure 4.4(f)). Archaea associated with sponges are distinct from those found in the natural environment of sponges and can sometimes co-evolve with their host (Hallam et al. 2006; Taylor et al. 2007). The main archaeal lineages associated with sponges belong to the *Thaumarchaeota* and *Euryarchaeota* (Hentschel et al. 2012). For example, the *Thaumarchaeota Cenarchaeum symbiosum* and *Nitrosopumilus maritimus* participate in the elimination of nitrogenous waste produced by their host (*Axinella spp.*) through the oxidation and nitrification of urea and ammonia (Hallam et al. 2006).

Several lineages of archaea develop in the mucus produced by coral polyps. *Thaumarchaeota*, and mainly the species *Nitrosopumilus maritimus*, constitute the main line of archaea populating the coral mucus (Nachshon 2008). As with sponges, *Thaumarchaeota* could potentially play a role in the elimination of nitrogenous waste from the host. The archaeal cohort associated with corals also contains several clades of *Euryarchaeota* (*Thalassoarchaeota*, *Pontarchaeota*, *Thermoplasmatales* and *Halobacteria*).

4.3.4.2. *Intestinal microbiota of termites and ruminants*

The digestive tracts of various organisms are populated by methanogenic archaea. Methanogens have been demonstrated in the intestinal microbiota of arthropods, birds, reptiles and mammals. Methanogenesis allows for the indirect elimination of excess dihydrogen from microbial fermentation, which is potentially toxic to the rest of the microbial community. To date, all methanogenic archaea identified in human or animal digestive tracts are affiliated with *Euryarchaeota*.

Subtropical termites release between 2 and 22 Tg (2,000,000 and 22,000,000 tonnes) of methane into the atmosphere annually (Kirschke et al. 2013). Depending on their diet and the presence or absence of protozoan symbionts in their intestines, two major groups of termites can be distinguished: higher termites and lower termites (Lange et al. 2005). Higher termites have a diverse diet and do not form a symbiosis with protozoa. The main orders of methanogenic archaea developing in their gut are *Methanobacteriales*, *Methanomassiliicoccales*, *Methanomicrobiales* and *Methanosarcinales* (Brune 2014). Non-methanogenic organisms affiliated with *Thaumarchaeota* and *Crenarchaeota* are also found more marginally. Lower termites feed on wood and maintain a symbiosis with ciliated protozoa (*Ciliophora*). These ciliates are themselves in mutualistic interactions with methanogenic archaea in the form of ecto- or endosymbiosis. During the cellulose degradation process, ciliates produce lactate, acetate, carbon dioxide and dihydrogen. H_2 and CO_2 are then used as a source of energy and carbon by hydrogenotrophic methanogenic archaea (Brune 2014). In lower termites, the main methanogenic *taxa* present are affiliated with the genus *Methanobrevibacter* (*Mbb.*). It is important to note that similar methane production mechanisms exist in other insects, such as cockroaches.

Ruminants are significant producers of methane, emitting 250–500 L of methane daily (Lange et al. 2005) (Figure 4.4(g)). Like lower termites, the rumen of ruminants is colonized by ciliated protozoa. The processes of cellulose degradation and methane production are similar. In the rumen of cows, *Methanobacteriaceae* can account for up to 92% of total archaea, and 20–50% belong to the species *Mbb. ruminantium* (Lange et al. 2005; Moissl-Eichinger et al. 2018). Other lineages of methanogenic archaea have also been identified in the bovine rumen: *Methanomicrobiales*, *Methanosarcinales* and *Methanomassiliicoccales* (Söllinger et al. 2016; Moissl-Eichinger et al. 2018). The representatives of these lineages are hydrogenotrophic or methylotrophic. Sequences of non-methanogenic archaea have also been found in the gastrointestinal tract of goats, affiliated with *Desulfurococcaceae* (*Crenarchaeota*) and the genus *Archaeoglobus* (*Archaeoglobi*) (Moissl-Eichinger et al. 2018).

4.3.4.3. *Human microbiomes*

The human body hosts a significant diversity of archaea. These microorganisms have specifically colonized various human microbiomes, including the digestive tract (mouth, gastrointestinal tract and colon), the respiratory system (nose and lungs), the reproductive system (vagina) and the skin surface (Figure 4.4(h)) (Dridi et al. 2011; Koskinen et al. 2017).

The primary archaeal clades in the human digestive tract belong to the *Euryarchaeota*. Six species or species-candidates have been isolated or cultured from human gastrointestinal samples: *Methanobrevibacter smithii* (*Methanobacteriales*), *Methanosphaera stadmanae* (*Methanobacteriales*), *Methanomassiliicoccus luminyensis* (*Methanomassiliicoccales*), *Haloferax massiliensis* (*Halobacteriales*), *Candidatus Methanomethylophylus alvus* and *Candidatus Methanomassiliicoccus intestinalis* (Koskinen et al. 2017). Additionally, sequences of *Crenarchaeota*

(*Desulfurococcales*, *Sulfolobales* and *Thermoproteales*), *Thaumarchaeota* (*Nitrososphaerales*) and *Woesearchaeota* have been detected in this complex environment (Koskinen et al. 2017). The functions of these *taxa* remain poorly characterized. Representing 70–90% of the archaea in the gastrointestinal tract, *Methanobrevibacter smithii* is the dominant archaeal species in the digestive tract, present in over 95% of the population. *Haloferax massiliensis* is the only known halophilic archaeal species associated with the human body (Khelaifia and Raoult 2016). Notably, the seventh order of methanogens, the *Methanomassiliicoccales*, was recently described from human feces (Dridi et al. 2012). This lineage is medically significant: its representatives are hydrogen-dependent methylotrophs that consume compounds like trimethylamine (TMA), produced by the microbiota in the intestine. TMA, resulting from the degradation of certain foods (eggs, soy, cabbage, red meat, seafood, etc.), is known to contribute to trimethylaminuria in individuals with enzymatic deficiencies or atherosclerosis. Utilizing *Methanomassiliicoccales* as probiotics could help prevent these conditions (Brugère et al. 2014). Conversely, other methanogenic archaea might be harmful to health. For instance, *Methanobrevibacter* spp. have been linked to weight gain, periodontitis and brain abscesses (Dridi et al. 2011; Koskinen et al. 2017; Moissl-Eichinger et al. 2018).

To date, archaeal species isolated from the oral cavity (mucous membranes and subgingival plaques) are limited to *Methanobacteriales Mbb. oralis* and *Mbb. smithii*. However, sequences affiliated with other *Euryarchaeota* (*Methanomassiliicoccales*, *Candidatus Mbb. massiliensis*) and *Thaumarchaeota* reveal a broader diversity of archaea in this environment (Moissl-Eichinger et al. 2018). The vaginal microbiota also includes archaea, with *Mbb. smithii* being the only known species in this environment.

In humans, archaeome represent approximately 4.2% of the prokaryotes on the skin (Probst et al. 2013). The two main archaeal *phyla* found on this organ are *Thaumarchaeota* and *Euryarchaeota*, with prevalences of 88% and 12% on the torso, respectively (Probst et al. 2013). Among the *Thaumarchaeota* detected, the genus *Nitrososphaera* may play a role in the oxidation of ammonia and urea in sweat. According to some authors (Moissl-Eichinger et al. 2018), this phenomenon could improve skin quality and reduce body odor. *Euryarchaeota* found on the skin are either halophilic archaea (*Halobacteriaceae*) or, intriguingly, methanogenic archaea (*Methanobrevibacter*, *Methanomethylovorans* and *Methanosarcina*).

Surprisingly, the archaeome in the nasal cavity differs significantly from those in the lungs. This body part hosts lineages found in both the digestive tract and on the skin: *Thaumarchaeota* and methanogenic *Euryarchaeota* (*Methanobrevibacter*). Currently, limited data are available on the archaeal diversity in the lungs.

According to some authors (Koskinen et al. 2017), *Candidatus Woesearchaeota* are the main (≈ 66% of the sequences) archaeal taxon in the lungs, followed by unknown lineages affiliated with DPANN (≈ 32%) and *Thaumarchaeota* (≈ 2%).

The roles and interactions of these lineages with their host are not yet elucidated.

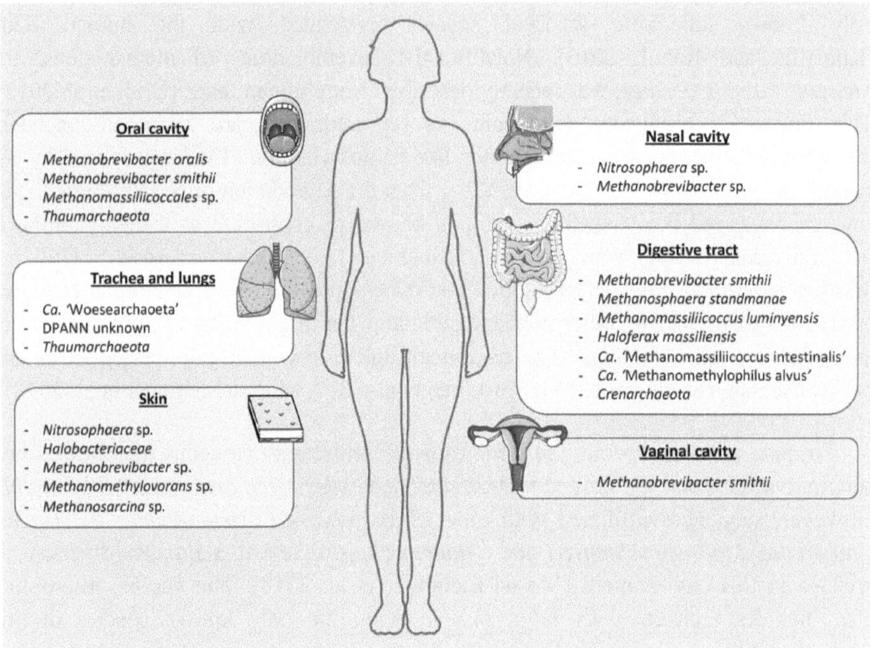

Figure 4.5. *Archaeal taxa of different microbiota of the human body (image modified from Sevier Medical Art)*

4.4. Archaea resistant to cultivation efforts

For decades, the description of new microbial *taxa* has relied on their isolation and physiological, metabolic and chemotaxonomic characterization. In the mid-1990s, the first genomes of cultivated archaeal strains were sequenced, starting with the methanogen *Methanocaldococcus jannaschii*, published in 1996 (Bult et al. 1996). Since then, genome sequencing has exponentially increased, with some genomes reconstructed from community genomes (=metagenomes) derived from environments with low diversity.

Phylum	MAGs/Bins	Environment	Hypothetical metabolism	References
Candidatus Methanofastidiosa sp. (WSA2)	ADurb1213_Bin02801 U1lsi0214_ Bin055 B15fssc0709_Meth_Bin003	Wastewater treatment bioreactors	Methanogenesis via the reduction of methyl-thiol compounds, using H_2 and CO as electron donors, and acetate, malonate, or propionate as carbon sources	(Nobu et al. 2016)
Candidatus Verstraetearchaeota	V1 V2 V3 V4 V5	Anaerobic digesters	Hydrogen-dependent methylotrophic methanogenesis; utilization of organic carbon	(Vanwonterghem et al. 2016)
Candidatus Syntrophoarchaeum sp.	GoM-Arch87-1 GoM-Arch87-2	Hydrothermal vent sediments	Butane oxidation	(Laso-Pérez et al. 2016)
Candidatus Altiarchaeum hamiconexum	CG Bin14	Subsurface carbon dioxide storage reservoir	Anaerobic glucose metabolism	(Emerson et al. 2016)
Candidatus Thorarchaeota sp.	WOR bin SMTZ-83 WOR bin SMTZ-45	Methane-sulfate transition zones in estuarine sediments	Acetogenesis Sulfide and thiosulfate reduction	(Seitz et al. 2016)

Phylum	MAGs/Bins	Environment	Hypothetical metabolism	References
Candidatus Hadesarchaea sp. (SAGMEG)	Yellowstone hot spring bin 45 WOR sediment bin DG-33 Yellowstone hot spring bin N21	Terrestrial hot springs or estuarine sediments	CO or H_2 oxidation Nitrite reduction to ammonia (DNRA)	(Baker et al. 2016)
Candidatus Theionarchaea sp.	WOR bin DG-70 WOR bin DG-70-1	Estuarine sediments	Degradation of proteins and amino acids to pyruvate and acetyl Co-A Acetate oxidation S° and $S_2O_3^{2-}$ reduction	(Lazar et al. 2017)
Candidatus Woesearchaeota sp.	AR20	Subterranean aquifer	Parasitic or symbiotic lifestyle	(Castelle et al. 2015)
Candidatus Diapherotrites *sp.*	AR10	Subterranean aquifer	Fermentation	(Castelle et al. 2015)
Candidatus Pacearchaeota sp.	AR13	Subterranean aquifer	Heterotrophy, fermentation and polysaccharide degradation	(Castelle et al. 2015)

Table 4.3. *Hypothetical metabolisms of archaeal lineages with no isolated representatives to date (Note: Chapter written in 2019)*

For a little over five years, the number of available archaeal genomes in databases has increased tenfold due to the tremendous rise of isolated individual cell genomics methods and metagenomics approaches. The reduction in sequencing costs, combined with advances in computational data processing methods, has enabled colossal progress in assembling genomes from metagenomes. These methods now allow the study of complex habitats characterized by very high specific richness, which were previously excluded from such analyses. Complete genomes have been reconstructed using these sophisticated approaches. However, most "metagenome-assembled genomes" (MAGs) are often incomplete or likely to comprise contigs derived from several phylogenetically close strains with sequences that are difficult to discriminate. Thus, the description of a MAG is generally accompanied by an evaluation of its quality, expressed by the completeness of the assembled genome and the degree of contamination in the assembly.

The acquisition of numerous MAGs from archaeal lineages without cultivated representatives has illuminated the phylogenetic position of these microbial groups and contributed to understanding the tree of life. It has also enabled the formulation of hypotheses about the metabolism of the *taxa* of these lineages currently without cultivated representatives. Table 4.3, though not exhaustive, lists some examples of archaeal groups for which metabolic and functional hypotheses have been proposed based on their genetic content. Metagenomics studies have notably revolutionized our knowledge of methanogenic archaeal lineages by demonstrating the significance of certain methanogenesis pathways (particularly the hydrogen-dependent methylotrophic pathway) in archaea, the existence of new pathways (methoxydotrophic methanogenesis) and establishing that methanogenic archaea are not exclusively affiliated with *Euryarchaeota* (Evans et al. 2019). It is highly likely that the metabolic hypotheses generated through MAGs will guide the cultivation and even the isolation of archaeal lineages known only through genomic sequences.

4.5. Challenges and success stories

While the diversity of archaeal *taxa* and their primary habitats are becoming fairly well known, many challenges remain in understanding the biology and ecology of archaea across different biomes. Future challenges include determining the abundance of archaea in their habitats and deciphering their physiology.

Indeed, data on the abundances of archaea in ecosystems remain very fragmented and insufficient to identify general trends. Similarly, the physiology of archaea known only from MAGs is purely hypothetical and needs to be validated through functional approaches. Today, although the scientific community agrees on the necessity of abandoning the axiom that only 1% of unicellular microorganisms without a nucleus can be cultured, the vast majority of bacteria and archaea remains uncultivated (Steen et al. 2019). The physiology of this multitude of uncultivated *taxa* is therefore unknown or merely hypothetical. However, implementing functional approaches, thoroughly analyzing the physiological potential of a strain and studying the involved metabolic pathways are more manageable when cultures (clonal or otherwise) of the *taxa* of interest are available. Cultivating uncultivated microorganisms remains a current and pressing issue.

The physiology of archaea available in culture is, by definition, partly known because the necessary nutrients and physico–chemical conditions for their growth are understood. However, it is important to note that new physiological properties are frequently discovered in strains which have been cultivated for a long time, including other physico–chemical combinations that allow their growth or new redox reactions for energy production, which have not been tested before. For example, *Euryarchaea* of the genus *Thermococcus* were first cultivated and described in 1983 (Zillig et al. 1983) and were long considered in the collective mind as sulfur-reducing chemo-organoheterotrophs. It was not until 2004 that some *taxa* of this genus were discovered to develop by carboxydotrophy (Sokolova et al. 2004).

Many similar examples exist. Over the years, and with the discovery of new microbial metabolic pathways, it has been found that *taxa* available in microorganism collections have a wider range of physiological properties than those described at the time of their isolation. The study of microbial physiology is far from revealing all its secrets and promises many exciting discoveries in the future.

Among successes in cultivating archaea which can serve as inspiration is the work of a Japanese team that, after 12 years of effort, cultivated the first representative of the Asgard *Lokiarchaea*, resolving a debate on the genetic content of these archaea (Imachi et al. 2020). The ancestors of these archaea could be linked to the emergence of eukaryotes. The strain cultivated in this study, *Candidatus Prometheoarchaeum syntrophicum* (strain MK-D1), is a small anaerobic mesophilic coccoid with long protrusions, degrading amino acids and peptides through syntrophic association with a microbial partner consuming the dihydrogen or formate produced. The success of this cultivation relied on the combination of advanced cultural approaches (development of a continuous culture reactor mimicking the in situ conditions of methane-rich sediments), a good understanding of microbial physiology (addition of cocktails of antibiotics and inhibitors of cocultured strains), molecular or analytical methods for monitoring *taxa* and substrates, and finally, a significant amount of patience and perseverance.

4.6. Conclusion

Archaea are omnipresent on earth and have colonized all terrestrial habitats where there is life. Different lineages of archaea inhabit a wide range of conditions, including temperature, salinity, pH, pressure, luminosity and niches characterized by variable water, dioxygen and nutrient contents.

Initially discovered in extreme environments, where they often dominate in terms of biomass, archaea also populate a wide range of more ordinary, non-extreme environments described in this chapter. In most biomes where they are present, archaea play crucial roles in ecosystem functioning, particularly in the flow of matter and energy, such as methanogenic archaea, which participate in the final stage of organic matter remineralization.

Archaea from extreme environments can be used as models for studying the origins of life on earth, with some of these environments corresponding to analogues of early Earth. The environmental conditions in which these archaea live can help define terrestrial habitability conditions and can be extrapolated to the study of extraterrestrial life.

Archaea also serve as relevant study models for understanding the diversity of microbial metabolic pathways and the diversity of oxidation-reduction reactions for energy production. Since many archaea hold records for living in extreme conditions, they are excellent models for studying the adaptation of organisms to environmental physico–chemical parameters.

NOTE.– This chapter was written in 2019. Since then, some previously uncultivated archaeal lineages have been cultivated, and archaeal taxonomy has undergone rearrangements and name changes.

4.7. References

Adam, P.S., Borrel, B., Brochier-Armanet, C., Gribaldo, S. (2017). The growing tree of Archaea: New perspectives on their diversity, evolution and ecology. *The ISME Journal*, 11(11), 2407–2425.

Akcil, A. and Koldas, S. (2006). Acid mine drainage (AMD): Causes, treatment and case studies. *Journal of Cleaner Production*, 14(12/13), 1139–114.

Alegria, T., Giles, R.C., Paterson, E., Robertson-Alertyn, S., Cesco, S., Mimmo, T., Pii, Y., Bulgarelle, D. (2016). Plant-microbiota interactions as a driver of the mineral turnover in the rhizosphere. *Advances in Applied Microbiology*, 95, 1–67.

Alpana, A., Vishwakarma, P., Adhya, T.K., Inubushi, K., Dubey, S.K. (2017). Molecular ecological perspective of methanogenic archaeal community in rice agroecosystem. *Science of the Total Environment*, 596–597, 136–146.

Andersen, R., Chapman, S.J., Artz, R.R.E. (2013). Microbial communities in natural and disturbed peatlands: A review. *Soil Biology and Biochemistry*, 57, 979–994.

Anthony, K.W., Daanen, R., Anthony, P., Schneider von Deimling, T., Ping, C.L., Chanton, J. P., Grosse, G. (2016). Methane emissions proportional to permafrost carbon thawed in Arctic lakes since the 1950s. *Nature Geoscience*, 9(9), 679.

Asakawa, S., Morii, H., Akagawa-Matsushita, M., Koga, Y., Hayano, K. (1993). Characterization of *Methanobrevibacter arboriphilicus* SA isolated from a paddy field soil and DNA-DNA hybridization among *M. arboriphilicus* strains. *International Journal of Systematic and Evolutionary Microbiology*, 43(4), 683–686.

Asakawa, S., Akagawa-Matsushita, M., Morii, H., Koga, Y., Hayano, K. (1995). Characterization of *Methanosarcina mazeii* TMA isolated from a paddy field soil. *Current Microbiology*, 31(1), 34–38.

Auguet, J.C., Barberan, A., Casamayor, E.O. (2010). Global ecological patterns in uncultured Archaea. *The ISME Journal*, 4(2), 182–190.

Bachran, M., Kluge, S., Lopez-Fernandez, M., Cherkouk, A. (2018). Microbial diversity in an arid, naturally saline environment. *Microbial Ecology*, 78(2), 494–505.

Baker, B.J. and Banfield, J.F. (2003). Microbial communities in acid mine drainage. *FEMS Microbiology Reviews*, 44(2), 139–152.

Baker, B.J., Tyson, G.W., Webb, R.I., Flanagan, J., Hugenholtz, P., Allen, E.E., Banfield, J.F. (2006). Lineages of acidophilic archaea revealed by community genomic analysis. *Science*, 314(5807), 1933–1935.

Baker, B.J., Saw, J.H., Lind, A.E., Lazar, C.S., Hinrichs, K.-U., Teske, A.P., Etterna, T.J.G. (2016). Genomic inference of the metabolism of cosmopolitan subsurface Archaea, *Hadesarchaea*. *Nature Microbiology*, 1, 16002.

Bakermans, C. (2012). Psychrophiles: Life in the cold. In *Extremophiles: Microbiology and Biotechnology*, Anitori, R.P. (ed.). Horizon Scientific Press, Hethersett.

Banerjee, R., Halder, A., Natta, A. (2016). Psychrophilic microorganisms: Habitats and exploitation potentials. *European Journal of Biotechnology and Bioscience*, 4(3), 16–24.

Bang, C. and Schmitz, R.A. (2015). Archaea associated with human surfaces: Not to be underestimated. *FEMS Microbiology Reviews*, 39, 631–648.

Beulig, F., Røy, H., McGlynn, S.E., Jørgensen, B.B. (2019). Cryptic CH4 cycling in the sulfate–methane transition of marine sediments apparently mediated by ANME-1 archaea. *The ISME Journal*, 13(2), 250–262.

Bhattacharyya, A., Majumder, N.S., Basak, P., Mukherji, S., Roy, D., Nag, S., Haldar, A., Chattopadhyay, D., Mitra, S., Bhattacharyya, M. et al. (2015). Diversity and distribution of archaea in the mangrove sediment of Sundarbans. *Archaea*, 2015(1), 968582. doi.org/10.1155/2015/968582.

Biddle, J.F., Lipp, J.S., Lever, M.A., Lloyd, K.G., Sorensen, K.B., Anderson, R., Fredricks, H.F., Elvert, M., Kelly, T.J., Schrag, D.P. et al. (2006). Heterotrophic Archaea dominate sedimentary subsurface ecosystems off Peru. *Proceedings of the National Academy of Sciences USA*, 103, 3846–3851.

Birrien, J.L., Zeng, X., Jebbar, M., Cambon-Bonavita, M.A., Querellou, J., Oger, P., Bienvenu, N., Xiao, X., Prieur, D. (2011). *Pyrococcus yayanosii* sp. nov., an obligate piezophilic hyperthermophilicarchaeon isolated from a deep-sea hydrothermal vent. *International Journal of Systematic and Evolutionary Microbiology*, 61, 2827–2881.

Blöchl, E., Rachel, R., Burggraf, S., Hafenbradl, D., Jannasch, H.W., Stetter, K.O. (1997). *Pyrolobus fumarii*, gen. and sp. nov., represents a novel group of archaea, extending the upper temperature limit for life to 113 degrees C. *Extremophiles*, 1(1), 14–21.

Bowers, K.J. and Wiegel, J. (2011). Temperature and pH optima of extremely halophilic archaea: A mini-review. *Extremophiles*, 15(2), 119–128.

Bowers, K.J., Mesbah, N.M., Wiegel, J. (2009). Biodiversity of poly-extremophilic bacteria: Does combining the extremes of high salt, alkaline pH and elevated temperature approach a physico–chemical boundary for life? *Saline Systems*, 5(9), 1–8.

Bowman, J.P. (2017). Genomics of psychrophilic bacteria and archaea. In *Psychrophiles: From Biodiversity to Biotechnology*, Margesin, R. (ed.). Springer, Cham.

Bräuer, S.L., Cadillo-Quiroz, H., Ward, R.J., Yavitt, J.B., Zinder, S.H. (2011). *Methanoregula boonei* gen. gov., sp. nov., an acidiphilic methanogen isolated from an acidic peat bog. *International Journal of Systematic and Evolutionary Microbiology*, 61(1), 45–52.

Brugère, J.F., Borrel, G., Gaci, N., Tottey, W., O'Toole, P.W., Malpuech-Brugère, C. (2014). Archaebiotics. *Gut Microbes*, 5(1), 5–10.

Brune, A. (2014). Symbiotic digestion of lignocellulose in termite guts. *Nature Reviews in Microbiology*, 12(3), 168–180.

Buée, M., De Boer, W., Martin, F., van Overbeek, L., Jurkevitch, E. (2009). The rhizosphere zoo: An overview of plant-associated communities of microorganisms, including phages, bacteria, archaea, and fungi, and of some of their structuring factors. *Plant and Soil*, 321(1), 189–212.

Bult, C.J., White, O., Olsen, G.J., Zhou, L., Fleischmann, R.D., Sutton, G.G., Blake, J.A., FitzGerald, L.M., Clayton, R.A., Gocayne, J.D. et al. (1996). Complete genome sequence of the methanogenic archaeon. *Methanococcus jannaschii*. *Science*, 273(5278), 1058–1073.

Cassarini, C., Zhang, Y., Lens, P.N. (2019). Pressure selects dominant anaerobic methanotrophic phylotype and sulfate reducing bacteria in coastal marine lake Greve-lingen sediment. *Frontiers in Environmental Sciences*, 6, 162.

Castelle, C.J., Wrighton, K.C., Thomas, B.C., Hug, L.A., Brown, C.T., Wilkins, M.J., Frischkorn, K.R., Tringe, S.G., Singh, A., Markillie, L.M. et al. (2015). Genomic expansion of domain archaea highlights roles for organisms from new phyla in anaerobic carbon cycling. *Current Biology*, 25(6), 690–701.

Cavicchioli, R. (2016). On the concept of a psychrophile. *The ISME Journal*, 10(4), 793.

Cayol, J.-L., Ollivier, B., Alazard, D., Amils, R., Godfroy, A., Piette, F., Prieur, D. (2015). The extreme conditions of life on the planet and exobiology. In *Environmental Microbiology: Fundamentals and Applications*, Bertrand, J.C., Caumette, P., Lebaron, P., Matheron, R., Normand, P., Sime-Ngando, T. (eds). Springer, Dordrecht.

Chaban, B., Ng, S.Y.M., Jarrel, K.F. (2006). Archaeal habitats – From the extreme to the ordinary. *Canadian Journal of Microbiology*, 52, 73–116.

Chen, S. and He, Q. (2016). Distinctive non-methanogen archaeal populations in anaerobic digestion. *Applied Microbiology and Biotechnology*, 100(1), 419–430.

Chen, L.X., Mendez-Garcia, C., Dombrowski, N., Servín-Garcidueñas, L.E., Eloe-Fadrosh, E.A., Fang, B.Z., Luo, Z.H. (2018). Metabolic versatility of small archaea *Micrarchaeota* and *Parvarchaeota*. *The ISME Journal*, 12(3), 756–775.

Chin, K.J., Lukow, T., Conrad, R. (1999). Effect of temperature on structure and function of the methanogenic archaeal community in an anoxic rice field soil. *Applied and Environmental Microbiology*, 65(6), 2341–2349.

Colman, D.R., Poudel, S., Hamilton, T.L., Havig, J.R., Selensky, M.J., Shock, E.L., Boyd, E.S. (2018). Geobiological feedbacks and the evolution of thermoacidophiles. *The ISME Journal*, 12(1), 225–236.

Cox, M.M. (2010). Rising from the ashes: DNA repair in *Deinococcus radiodurans*. *PLOS Genetics*, 6(1), e1000815.

Dalmasso, C., Oger, P., Selva, G., Courtine, D., L'Haridon, S., Garlaschelli, A., Roussel, E., Miyazaki, J., Reveillaud, J., Jebbar, M. et al. (2016). *Thermococcus piezophilus* sp. nov., a novel hyperthermophilic and piezophilic archaeon with a broad pressure range for growth, isolated from a deepest hydrothermal vent at the Mid-Cayman Rise. *Systematic and Applied Microbiology*, 39, 440–444.

Dargie, G.C., Lewis, S.L., Lawson, I.T., Mitchard, E.T.A., Page, S.E., Bocko, Y.E., Ifo, S.A. (2017). Age, extent and carbon storage of the central congo basin peatland complex. *Nature*, 542, 7639.

DeLong, E.F. and Pace, N. (2001). Environmental diversity of bacteria and archaea. *Systematic Biology*, 17, 219–232.

Dick, G.J. (2019). The microbiomes of deep-sea hydrothermal vents: Distributed globally, shaped locally. *Nature Reviews in Microbiology*, 50(4), 219–232.

Dridi, B., Raoult, D., Drancourt, M. (2011). Archaea as emerging organisms in complex human microbiomes. *Anaerobe*, 17(2), 56–63.

Dridi, B., Henry, M., Richet, H., Raoult, D., Drancourt, M. (2012). Age-related prevalence of *Methanomassiliicoccus luminyensis* in the human gut microbiome. *APMIS*, 120(10), 773–777.

Elshahed, M.S., Najar, F.Z., Roe, B.A., Oren, A., Dewers, T.A., Krumholz, L.R. (2004). Survey of archaeal diversity reveals an abundance of halophilic archaea in a low-salt, sulfide- and sulfur-rich spring. *Applied and Environmental Microbiology*, 70, 2230–2239.

Emerson, J.B., Thomas, B.C., Alvarez, W., Banfield, J.F. (2016). High CO_2 subsurface metagenomics. *Environmental Microbiology*, 18, 1686–1703.

Evans, P.N., Boyd, J.A., Leu, A.O., Woodcroft, B.J., Parks, D.H., Hugenholtz, P., Tyson, G.W. (2019). An evolving view of methane metabolism in the Archaea. *Nature Reviews in Microbiology*, 50(4), 219–232.

Feller, G. (2017). Cryosphere and psychrophiles: Insights into a cold origin of life? *Life*, 7(2), 25.

Flemming, H.C. and Wuertz, S. (2019). Bacteria and archaea on Earth and their abundance in biofilms. *Nature Reviews in Microbiology*, 17(4), 247–260.

Gopal, M. and Gupta, A. (2016). Microbiome selection could spur next-generation plant breeding strategies. *Frontiers in Microbiology*, 7, 1971.

Grant, W.D. (2004). Introductory chapter: Half a lifetime in soda lakes. In *Halophilic Microorganisms*, Ventosa, A. (ed.). Springer, Heidelberg.

Grant, W.D. and Jones, B.E. (2016). Bacteria, archaea and viruses of soda lake. In *Soda Lakes of East Africa*, Shagerl, M. (ed.). Springer, New York.

Grant, W.D., Mwatha, W.E., Jones, B.E. (1990). Alkaliphiles: Ecology, diversity and applications. *FEMS Microbiology Reviews*, 6(2/3), 255–269.

Großkopf, R., Stubner, S., Liesack, W. (1998). Novel euryarchaeotal lineages detected on rice roots and in the anoxic bulk soil of flooded rice microcosms. *Applied and Environmental Microbiology*, 64(12), 4983–4989.

Hallam, S.J., Konstantinidis, K.T., Putnam, N., Schleper, C., Watanabe, Y.I., Sugahara, J., Preston, C., de la Torre, J., Richardson, P.M., DeLong, E.F. (2006). Genomic analysis of the uncultivated marine crenarchaeote *Cenarchaeum Symbiosum*. *Proceedings of the National Academy of Sciences USA*, 103, 18296–18301.

Hallberg, K.B. and Johnson, D.B. (2001). Biodiversity of acidophilic prokaryotes. *Advances in Applied Microbiology*, 49, 37–84.

Hartmann, A., Schmid, M., van Tuinen, D., Berg, G. (2009). Plant-driven selection of microbes. *Plant and Soil*, 321(1/2), 235–257.

Hashimoto, T., Horikawa, D.D., Saito, Y., Kuwahara, H., Kozuka-Hata, H., Shin-I, T., Minakuchi, Y., Ohishi, K., Motoyama, A., Aizu, T. et al. (2016). Extremotolerant tardigrade genome and improved radiotolerance of human cultured cells by tardigrade-unique protein. *Nature Communications*, 20(7), 12808.

Hawkins, A.N., Johnson, K.W., Bräuer, S.L. (2014). Southern appalachian peatlands support high archaeal diversity. *Microbial Ecology*, 67(3), 587–602.

Hensel, R., Matussek, K., Michalke, K., Tacke, L., Tindall, B.J., Kohlhoff, M., Siebers, B., Dielenschneider, J. (1997). *Sulfophobococcus zilligii* gen. nov., spec. nov.: A novel hyperthermophilic archaeum isolated from hot alkaline springs of Iceland. *Systematic and Applied Microbiology*, 20(1), 102–110.

Hentschel, U., Piel, J., Degnan, S.M., Taylor, M.W. (2012). Genomic insights into the marine sponge microbiome. *Nature Reviews Microbiology*, 10(9), 641–654.

Hug, L.A., Baker, B.J., Anantharaman, K., Brown, C.T., Probst, A.J., Castell, C.J., Butterfield, C.N., Hernsdorf, A.W., Amano, Y., Ise, K. et al. (2016). A new view of the tree of life. *Nature Microbiology*, 1, 16048.

Imachi, H., Nobu, M.K., Nakahara, N., Morono, Y., Ogawara, M., Takaki, Y., Takano, Y., Uematsu, K., Ikuta, T., Ito, M. et al (2020). Isolation of an archaeon at the prokaryote–eukaryote interface. *Nature*, 577, 7791.

Itoh, H., Ishii, S., Shiratori, Y., Oshima, K., Otsuka, S., Hattori, M., Senoo, K. (2013). Seasonal transition of active bacterial and archaeal communities in relation to water management in paddy soils. *Microbes and Environments*, 28(3), 370–380.

Jannasch, H.W. and Taylor, C.D. (1984). Deep-sea microbiology. *Annual Review of Microbiology*, 38, 487–514.

Jarrell, K.F., Walters, A.D., Bochiwal, C., Borgia, J.M., Dickinson, T., Chong, J.P.J. (2011). Major players on the microbial stage: Why archaea are important. *Microbiology*, 157(4), 919–936.

Javor, B. (1984). Growth potential of halophilic bacteria isolated from solar salterns environments: Carbon sources and salt requirements. *Applied and Environmental Microbiology*, 48, 352–360.

Jebbar, M., Franzetti, B., Girard, E., Oger, P. (2015). Microbial diversity and adaptation to high hydrostatic pressure in deep-sea hydrothermal vents prokaryotes. *Extremophiles*, 19(4), 721–740.

Jiang, H., Dong, H., Yu, B., Ye, Q., Shen, J., Rowe, H., Zhang, C. (2008). Dominance of putative marine benthic Archaea in Qinghai Lake, north-western China. *Environmental Microbiology*, 10(9), 2355–2367.

Johnson, D.B. (1998). Biodiversity and ecology of acidophilic microorganisms. *FEMS Microbiology Ecology*, 27(4), 307–317.

Johnson, D.B. (2007). Physiology and ecology of acidophilic microorganisms. In *Physiology and Biochemistry of Extremophiles*, Gerday, C. and Glansdorff, N. (eds). ASM Press, Washington, DC.

Jolivet, E., L'Haridon, S., Corre, E., Forterre, P., Prieur, D. (2003). *Thermococcus gammatolerans* sp. nov., a hyperthermophilic archaeon from a deep-sea hydrothermal vent that resists ionizing radiation. *International Journal of Systematic and Evolutionary Microbiology*, 53(3), 847–851.

Jones, B.E., Grant, W.D., Duckworth, A.W., Owenson, G.G. (1998). Microbial diversity of soda lakes. *Extremophiles*, 2(3), 191–200.

Jung, K.W., Lim, S., Bahn, Y.S. (2017). Microbial radiation-resistance mechanisms. *Journal of Microbiology*, 55(7), 499–507.

Kambourova, M., Tomova, I., Boyadzhieva, I., Radchenkova, N., Vasileva-Tonkova, E. (2016). Unusually high archaeal diversity in a crystallizer pond, Pomorie Salterns, Bulgaria, revealed by phylogenetic analysis. *Archaea*, 7459679.

Ke, X., Lu, Y., Conrad, R. (2016). Different behaviour of methanogenic archaea and *Thaumarchaeota* in rice field microcosms. *FEMS Microbiology Ecology*, 87(1), 18–29.

Khelaifia, S. and Raoult, D. (2016). *Haloferax massiliensis* sp. nov., the first human-associated halophilic archaea. *New Microbes and New Infections*, 12, 96–98.

Kirschke, S., Bousquet, P., Ciais, P., Saunois, M., Canadell, J.G., Dlugokencky, E.J., Bergamaschi, P., Bergmann, D., Blake, D.R., Bruhwiler, L. et. al. (2013). Three decades of global methane sources and sinks. *Nature Geoscience*, 6(10), 813–823.

Koskinen, K., Pausan, M.R., Perras, A.K., Beck, M., Bang, C., Mora, M., Schilhabel, A., Schmitz, R., Moissl-Eichinger, C. (2017). First insights into the diverse human archaeome: Specific detection of Archaea in the gastrointestinal tract, lung, and nose and on skin. *mBio*, 8(6), e00824–17.

Kozubal, M.A., Romine, M., Jennings, R.D., Jay, Z.J., Tringe, S.G., Rusch, D.B., Beam, J.P., McCue, L.A., Inskeep, W.P. (2013). *Geoarchaeota*: A new candidate phylum in the Archaea from high-temperature acidic iron mats in Yellowstone National Park. *The ISME Journal*, 7(3), 622–634.

Kulkarni, S., Kusum, D., Amaraja, J. (2016). Alkaliphiles: Diversity and bioprospection. In *Microbial Diversity in the Genomic Era*, Surajit, D. and Hirak, R. (eds). Academic Press, Cambridge.

Kurr, M., Huber, R., König, H., Jannasch, H.W., Fricke, H., Trincone, A., Kristjansson, J.K., Stetter, K.O. (1991). *Methanopyrus kandleri*, gen. and sp. nov. represents a novel group of hyperthermophilic methanogens, growing at 110°C. *Archives of Microbiology*, 156(4), 239–247.

Lange, M., Westermann, P., Kiaer Ahring, B. (2005). Archaea in protozoa and metazoa. *Applied Microbiology and Biotechnology*, 66(5), 465–474.

Larsson, L., Olsson, G., Hoist, O., Karlsson, H.T. (1990). Pyrite oxidation by thermophilic archaebacteria. *Applied and Environmental Microbiology*, 56(3), 697–701.

Laskar, F., Das Purkayastha, S., Sen, A., Bhattacharya, M.K., Misra, B.B. (2018). Diversity of methanogenic archaea in freshwater sediments of lacustrine ecosystems. *Journal of Basic Microbiology*, 58, 101–119.

Laso-Pérez, R., Wegener, G., Knittel, K., Widdel, F., Harding, K.J., Krukenberg, V., Meier, D.V., Richter, M., Tegetmeyer, H.E., Riedel, D. et al. (2016). Thermophilic archaea activate butane via alkyl-coenzyme M formation. *Nature*, 539, 396–401.

Lazar, C.S., Baker, B.J., Seitz, K.W., Teske, A.P. (2017). Genomic reconstruction of multiple lineages of uncultured benthic archaea suggests distinct biogeochemical roles and ecological niches. *The ISME Journal*, 11, 1118–1129.

Lee, O.O., Wang, Y., Yang, J., Lafi, F., Al-Suwailem, A., Qian, P.-Y. (2011). Pyrosequencing reveals highly diverse and species-specific microbial communities in sponges from the Red Sea. *The ISME Journal*, 5, 650–664.

Liesack, W., Schnell, S., Revsbech, N.P. (2000). Microbiology of flooded rice paddies. *FEMS Microbiology Reviews*, 24(5), 625–645.

Liu, X., Pan, J., Liu, Y., Li, M., Gu, J.D. (2018). Diversity and distribution of Archaea in global estuarine ecosystems. *Science of the Total Environment*, 637/638, 349–358.

Lyu, Z., Shao, N., Akinyemi, T., Whitman, W.B. (2018). Methanogenesis. *Current Biology*, 28(13), R727–R732.

Ma, K., Qiu, Q., Lu, Y. (2010). Microbial mechanism for rice variety control on methane emission from rice field soil. *Global Change Biology*, 16(11), 3085–3095.

Madigan, M.T., Bender, K.S., Buckley, D.H., Sattley, W.M., Stahl, D.A. (eds) (2018). *Brock Biology of Microorganisms*, 15th edition. Pearson, London.

Margesin, R. and Miteva, V. (2011). Diversity and ecology of psychrophilic microorganisms. *Research in Microbiology*, 162(3), 346–361.

Marteinsson, V.T., Birrien, J.L., Reysenbach, A.L., Vernet, M., Marie, D., Gambacorta, A., Messner, P., Sleytr, U.B., Prieur, D. (1989). *Thermococcus barophilus* sp. nov., a new barophilic and hyperthermophilic archaeon isolated under high hydrostatic pressure from a deep-sea hydrothermal vent. *International Journal of Systematic Bacteriology*, 49(2), 351–359.

Mehta, D. and Satyanarayana, T. (2013). Diversity of hot environments and thermophilic microbes. In *Thermophilic Microbes in Environmental and Industrial Biotechnology*, Satyanarayana, T., Littlechild, J., Kawarabayasi, Y. (eds). Springer, New York.

Merino, N., Aronson, H.S., Bojanova, D.P., Feyhl-Buska, J., Wong, M.L., Zhang, S., Giovannelli, D. (2019a). Living at the extremes: Extremophiles and the limits of life in a planetary context. *Frontiers in Microbiology*, 10, 780.

Merino, N., Aronson, H.S., Bojanova, D.P, Feyhl-Buska, J., Wong, M.L., Zhang, S., Giovannelli, D. (2019b). Living at the extremes: Extremophiles and the limits of life in a planetary context. *Frontiers in Microbiology*, 10, 1785.

Miller, J.F., Shah, N.N., Nelson, C.M., Ludlow, J.M., Clark, D.S. (1988). Pressure and temperature effects on growth and methane production of the extreme thermophile *Methanococcus jannaschii*. *Applied and Environmental Microbiology*, 54, 3039–3042.

Moissl-Eichinger, C., Pausan, M., Taffner, J., Berg, G., Bang, C., Schmitz, R.A. (2018). Archaea are interactive components of complex microbiomes. *Trends in Microbiology*, 26(1), 70–85.

Müller, H., Berg, C., Landa, B.B., Auerbach, A., Moissl-Eichinger, C., Berg, G. (2015). Plant genotype-specific archaeal and bacterial endophytes but similar bacillus antagonists colonize Mediterranean olive tree. *Frontiers in Microbiology*, 6, 138.

Nachshon, S., Ben-Dov, E., Sivan, A., Kushmaro, A. (2008). Global distribution and diversity of coral-associated archaea and their possible role in the coral holobiont nitrogen cycle. *Environmental Microbiology*, 10(11), 2979–2990.

Narrowe, A.B., Angle, J.C., Daly, R.A., Stefanik, K.C., Wrighton, K.C., Miller, C.S. (2017). High-resolution sequencing reveals unexplored archaeal diversity in freshwater wetland soils. *Environmental Microbiology*, 19, 2192–2209.

Neori, A. and Agami, M. (2017). The functioning of rhizosphere biota in wetlands – A review. *Wetlands*, 37(4), 615–633.

Nguyen, L.N., Nguyen, A.Q., Nghiem, L.D. (2018). Microbial community in anaerobic digestion system: Progression in microbial ecology. In *Wastewater Treatment Technologies. Energy, Environment, and Sustainability*, Bui, X.T., Chiemchaisri, C., Fujioka, T., Varjani, S. (eds). Springer, Singapore.

Nobu, M.K., Narihiro, T., Kuroda, K., Mei, R., Liu, W.T. (2016). Chasing the elusive *Euryarchaeota* class WSA2: Genomes reveal a uniquely fastidious methyl-reducing methanogen. *The ISME Journal*, 10, 2478–2487.

Nordstrom, D.K., Alpers, C.N., Ptacek, C.J., Blowes, D.W. (2000). Negative pH and extremely acidic mine waters from Iron Mountain, California. *Environmental Science & Technology*, 34(2), 254–258.

Oger, P. (2018). Complete genome sequences of 11 type species from the *Thermococcus* genus of hyperthermophilic and piezophilic archaea. *Genome Announcements*, 6, e00037-18.

Oger, P. and Jebbar, M. (2010). The many ways of coping with pressure. *Research in Microbiology*, 161, 799–809.

Orcutt, B.N., Sylvan, J.B., Knab, N.J., Edwards, K.J. (2011). Microbial ecology of the dark ocean above, at, and below the seafloor. *Microbiology Molecular Biology Reviews*, 75(2), 361–422.

Oren, A. (2010). The dying Dead Sea: The microbiology of an increasingly extreme environment. *Lakes and Reservoirs*, 15(3), 215–222.

Oren, A. (2014). Halophilic archaea on Earth and in space: Growth and survival under extreme conditions. *Philosophical Transactions of the Royal Society A*, A372, 20140194.

Pannekens, M., Kroll, L., Müller, H., Mbow, F.T., Meckenstock, R.U. (2019). Oil reservoirs, an exceptional habitat for microorganisms. *New Biotechnology*, 49, 1–9.

Parkes, R.J., Cragg, B., Roussel, E., Webster, G., Weightman, A., Sass, H. (2014). A review of prokaryotic populations and processes in sub-seafloor sediments, including biosphere:geosphere interactions. *Marine Geology*, 352, 409–425.

Pikuta, E.V., Hoover, R.B., Tang, J. (2007). Microbial extremophiles at the limits of life. *Critical Reviews in Microbiology*, 33(3), 183–209.

Plumb, J.J., Haddad, C.M., Gibson, J.A.E., Franzmann, P.D. (2007). *Acidianus sulfidivorans* sp. nov., an extremely acidophilic, thermophilic archaeon isolated from a solfatara on Lihir Island, Papua New Guinea, and emendation of the genus description. *International Journal of Systematic and Evolutionary Microbiology*, 57(7), 1418–1423.

Pontefract, A., Zhu, T.F., Walker, V.K., Hepburn, H., Lui, C., Zuber, M.T., Ruvkun, G., Carr, C.E. (2017). Microbial diversity in a hypersaline sulfate lake: A terrestrial analog of ancient Mars. *Frontiers in Microbiology*, 8, 1819.

Postec, A., Quéméneur, M., Bes, M., Mei, N., Benaïssa, F., Payri, C., Pelletier, B., Monnin, C., Guentas-Dombrowsky, L., Ollivier, B. et al. (2015). Microbial diversity in a submarine carbonate edifice from the serpentinizing hydrothermal system of the Prony Bay (New Caledonia) over a 6-year period. *Frontiers in Microbiology*, 6, 857.

Probst, A.J., Auerbach, A.K., Moissl-Eichinger, C. (2013). Archaea on human skin. *PLoS ONE*, 8(6), e65388.

Purdy, K.J., Cresswell-Maynard, T.D., Nedwell, D.B., McGenity, T.J., Grant, W.D., Timmis, K.N., Embley, T.M. (2004). Isolation of haloarchaea that grow at low salinities. *Environmental Microbiology*, 6(6), 591–595.

Ranawat, P. and Rawat, S. (2017). Radiation resistance in thermophiles: Mechanisms and applications. *World Journal of Microbiology and Biotechnology*, 33(6), 112.

Raymond-Bouchard, I., Tremblay, J., Altshuler, I., Greer, C.W., Whyte, L.G. (2018). Comparative transcriptomics of cold growth and adaptive features of a eury- and steno-psychrophile. *Frontiers in Microbiology*, 31(9), 1565.

Reigstad, L.A., Jorgensen, S.L., Schleper, C. (2010). Diversity and abundance of *Korarchaeota* in terrestrial hot springs of Iceland and Kamchatka. *The ISME Journal*, 4(3), 346–356.

Reysenbach, A.L., Liu, Y., Banta, A.B., Beveridge, T.J., Kirshtein, J.D., Schouten, S., Tivey, M.K., Von Damm, K.L., Voytek, M.A. (2006). A ubiquitous thermoacido-philic archaeon from deep-sea hydrothermal vents. *Nature*, 442(7101), 444–447.

Roh, S.W., Kim, K.H., Nam, Y.D., Chang, H.W., Park, E.J., Bae, J.W. (2010). Investigation of archaeal and bacterial diversity in fermented seafood using barcoded pyrosequencing. *The ISME Journal*, 4, 1–16.

Santoro, A.E., Richter, R.A., Dupont, C.L. (2019). Planktonic marine archaea. Annual review of marine science. *Annual Review of Marine Science*, 711(1), 131–158.

Sato, T., Watanabe K., Yamamoto H., Yamamoto S., Kurosawa N. (2013). Archaeal community structures in the solfataric acidic hot springs with different temperatures and elemental compositions. *Archaea*, 723871, 1–10.

Schleper, S.C., Puehler, G., Holz, I., Gambacorta, A., Janekovic, D., Santarius, U., Klenk, H.P., Zillig, W. (1995). *Picrophilus* gen. nov., fam. nov.: A novel aerobic, heterotrophic, thermoacidophilic genus and family comprising archaea capable of growth around pH 0. *Journal of Bacteriology*, 177(24), 7050–7059.

Seckbach, C. and Rampelotto, P.H. (2015). Polyextremophiles. In *Microbial Evolution Under Extreme Conditions*, Bakermans, C. (ed.). De Gruyters, Berlin.

Seitz, K.W., Lazar, C.S., Hinrichs, K.U., Teske, A.P., Baker, B.J. (2016). Genomic reconstruction of a novel, deeply branched sediment archaeal phylum with pathways for acetogenesis and sulfur reduction. *The ISME Journal*, 10(7), 1696–705.

Siddiqui, K.S., Williams, T.J., Wilkins, D., Yau, S., Allen, M.A., Brown, M.V., Lauro, F.M., Cavicchioli, R. (2013). Psychrophiles. *Annual Review of Earth and Planetary Sciences*, 41, 87–115.

Sievert, S. and Vetriani, C. (2012). Chemoautotrophy at deep-sea vents: Past, present, and future. *Oceanography*, 25, 218–233.

Skousen, J.G., Ziemkiewicz, P.F., McDonald, L.M. (2019). Acid mine drainage formation, control and treatment: Approaches and strategies. *The Extractive Industries and Society*, 6(1), 241–249.

Sokolova, T.G., Jeanthon, C., Kostrikina, N.A., Chernyh, N.A., Lebedinsky, A.V., Stacke-brandt, E., Bonch-Osmolovskaya, E.A. (2004). The first evidence of anaerobic CO oxidation coupled with H_2 production by a hyperthermophilic archaeon isolated from a deep-sea hydrothermal vent. *Extremophiles*, 8(4), 317–323.

Söllinger, A., Schwab, C., Weinmaier, T., Loy, A., Tveit, A.T., Schleper, C., Urich, T. (2016). Phylogenetic and genomic analysis of *Methanomassiliicoccales* in wetlands and animal intestinal tracts reveals clade-specific habitat preferences. *FEMS Microbiology Ecology*, 92(1), fiv149.

Sorokin, D.Y., Messina, E., Smedile, F., Roman, P., Sinninghe Damsté, J.P., Kublanov, I.V., Samarov, N.I., Toschavok, S.V., La Cono, V., Yakimov, M.M. (2017). Discovery of anaerobic lithoheterotrophic haloarchaea, ubiquitous in hypersaline habitats. *The ISME Journal*, 11, 1245–1260.

Steen, A.D., Crits-Christoph, A., Carini, P., DeAngelis, K.M., Fierer, N., Lloyd, K.G., Thrash, J.C. (2019). High proportions of bacteria and archaea across most biomes remain uncultured. *The ISME Journal*, n10.1038.

Takai, K., Nakamura, K., Toki, T., Tsunogai, U., Miyazaki, M., Miyazaki, J., Hirayama, H., Nakagawa, S., Nunoura, T., Horikoshi, K. (2008). Cell proliferation at 122°C and isotopically heavy CH_4 production by a hyperthermophilic methanogen under high-pressure cultivation. *Proceedings of the National Academy of Sciences USA*, 105(31), 10949–10954.

Taylor, M.W., Radax, R., Steger, D., Wagner, M. (2007). Sponge-associated microorganisms: Evolution, ecology, and biotechnological potential. *Microbiology and Molecular Biology Reviews*, 71(2), 295–347.

Tindall, B.J., Ross, H.N.M., Grant, W.D. (1984). *Natronobacterium* gen. nov. and *Natronococcus* gen. nov., two new genera of haloalkaliphilic archaebacteria. *Systematic and Applied Microbiology*, 5(1), 41–57.

Tully, B.J. (2019). Metabolic diversity within the globally abundant Marine Group II *Euryarchaea* offers insight into ecological patterns. *Nature Communications*, 10(1), 271.

Vanwonterghem, I., Evans, P.E., Parks, D.H., Jensen, P.D., Woodcroft, B.J., Hugenholtz, P., Tyson, G.W. (2016). Methylotrophic methanogenesis discovered in the archaeal phylum *Verstraetearchaeota*. *Nature Microbiology*, 16170, 1–9.

Vorholt, J.A. (2012). Microbial life in the phyllosphere. *Nature Reviews Microbiology*, 10, 12.

Woese, C.R. and Fox, G.E. (1977). Phylogenetic structure of the prokaryotic domain: The primary kingdoms. *Proceedings of the National Academy of Sciences USA*, 74(11), 5088–5090.

Wrede, C., Dreier, A., Kokoschka, S., Hoppert, M. (2012). Archaea in symbioses. *Archaea*, 596846.

Xu, Y., Zhou, P., Tian, X. (1999). Characterization of two novel haloalkaliphilic archaea, *Natronorubrum* bangense gen. nov. sp. nov. and *Natronorubrum tibetense* gen. nov., sp. nov. *International Journal of Systematic Bacteriology*, 49, 261–266.

Yang, Y., Dai, Y., Wu, Z., Xie, S., Liu, Y. (2016). Temporal and spatial dynamics of archaeal communities in two freshwater lakes at different trophic status. *Frontiers in Microbiology*, 7, 451.

Yavitt, J.B., Yashiro, E., Cadillo-Quiroz, H., Zinder, S.H. (2012). Methanogen diversity and community composition in peatlands of the central to northern Appalachian Mountain region, North America. *Biogeochemistry*, 109(1), 117–131.

Yim, K. J., Kwon, J., Cha, I.-T., Oh, K.-S., Lee, H.-W., Rhee, J.-K., Song, E.-J., Rho, J.R., Seo, M.L., Choi, J.-S. et al. (2015). Occurrence of viable, red-pigmented haloarchaea in the plumage of captive flamingoes. *Scientific Reports*, 5, e16425.

Zeng, X., Birrien, J.L., Fouquet, Y., Cherkashov, G., Jebbar, M., Querellou, J., Oger, P., Cambon-Bonavita, M.A., Xiao, X., Prieur, D. (2009). *Pyrococcus* CH1, an obligate piezophilic hyperthermophile: Extending the upper pressure-temperature limits for life. *The ISME Journal*, 3, 873–876.

Zhang, T., Fang, H.H.P., Ko, B.C.B. (2003). Methanogen population in a marine biofilm corrosive to mild steel. *Applied Microbiology and Biotechnology*, 63, 101–106.

Zhang, R., Hedrich, S., Ostertag-Henning, C., Schippers, A. (2018). Effect of elevated pressure on ferric iron reduction coupled to sulfur oxidation by biomining microorganisms. *Hydrometallurgy*, 178, 215–223.

Zillig, W., Holz, I., Janekovic, D., Schäfer, W., Reiter, W.D. (1983). The archaebacterium *Thermococcus celer* represents, a novel genus within the thermophilic branch of the archaebacteria. *Systematic and Applied Microbiology*, 4(1), 88–94.

5

Methanogenic Archaea

Tristan WAGNER[1], Laurent TOFFIN[2] and Guillaume BORREL[3]

[1] *Max Planck Institute for Marine Microbiology, Bremen, Germany*

[2] *Univ Brest, Ifremer, CNRS, Unité Biologie et Ecologie des Ecosystèmes Marins Profonds, F-29280, Plouzané, France*

[3] *Institut Pasteur, Université Paris Cité, UMR CNRS 6047, Evolutionary Biology of the Microbial Cell, Paris, France*

5.1. Diversity of methanogens and their environments

Since Alessandro Volta's discovery in November 1776 that "the air of the marshes is flammable and burns very slowly with a beautiful blue flame" (Moletta et al. 2008), the process of methanogenesis involving so-called "methanogenic" microorganisms has garnered significant interest in both the scientific and public communities. The methane generated by these microorganisms is not only a powerful greenhouse gas but also a fuel that meets the energy needs of our modern society.

5.1.1. *Methane sources and sinks*

Methane (CH_4) is the simplest organic molecule and the most reduced form of carbon. Despite its molecular simplicity, the physical-chemical properties of methane represent one of the major challenges of our modern society. This fuel, capable of meeting our electrical and heating needs, is also the second-largest contributor to global warming due to human activities. For several decades, atmospheric methane concentrations have been increasing at a rate of about 1–2% per year (Reeburgh 2007).

For a color version of all figures in this chapter, see www.iste.co.uk/clouet/archaea1.zip.

Biology of Archaea 1,
coordinated by Béatrice CLOUET-D'ORVAL, Bruno FRANZETTI and Philippe OGER.

The main natural sources of atmospheric methane include wetlands (e.g. peatlands and marshes), lake sediments and enteric digestion of animals such as termites (Table 5.1). It is estimated that these natural sources are responsible for one-third of methane emissions. Anthropogenic contributions include enteric fermentation of farmed ruminants, rice paddies and landfilling of waste. These contributions account for about three-quarters of total emissions (Reeburgh 2007; Conrad 2009).

Sources	Production ($Tg \cdot an^{-1}$)	Net emission ($Tg \cdot an^{-1}$)
Natural sources		
Wetlands	142	115
Termites	44	20
Oceans and lakes	85.3	10
Hydrates	10	5
Anthropogenic sources		
Rice fields	577	100
Ruminants	80	80
Waste landfilling	62	40
Biomass fermentation	55	55
Coal mines	35	35
Gas production	58	40
Others	54	25
Total	1,202.3	525

Table 5.1. *Estimated methane production and net emissions on land (Reeburgh 2007). Tg: one teragram equals 10^9 kilograms*

Globally, abiogenic methane represents a minor share of emissions since it is formed either at high temperatures by magmatic processes in volcanic and geothermal zones or at low temperatures (< 100°C) through reactions between rock, water and gases in continental regimes (Etiope and Lollar 2013). Of the estimated 500–600 teragrams of methane entering the atmosphere annually (Ehhalt et al. 2001), more than 70% (350–400 Tg) is derived from microbial activity. On average, about 1–2% of the carbon fixed by photosynthesis each year is processed through methanogenesis (Hedderich and Whitman 2006). Many researchers are focused on understanding the processes of degradation of sedimentary organic matter leading up to the final stage of methane production.

Microorganisms involved in the methane cycle (production or consumption) and the alteration of their activity in response to anthropogenic disturbances are particularly significant because they have a global and lasting impact on our environment.

Atmospheric methane can be naturally trapped and stored in ocean sediments as methane hydrate or clathrate (one volume of hydrate contains and can release about 160 volumes of methane gas). Since the stability zone of gas hydrates depends on ocean warming, the additional methane flux from hydrates could contribute to ocean acidification, oxygen consumption in the water column, instability of the upper part of continental slopes and a potential increase in methane emissions to the atmosphere (Weinstein et al. 2016).

5.1.2. *Taxonomic and metabolic diversity*

5.1.2.1. *Expanding taxonomic diversity*

To date, microorganisms dependent on methane production to support their energy needs belong exclusively to the domain Archaea. The first methanogens were isolated in the mid-20th century, and today, there are about 200 characterized species. These species have long been divided into five major orders: *Methanobacteriales*, *Methanococcales*, *Methanopyrales*, *Methanomicrobiales* and *Methanosarcinales*. These orders are often grouped into two categories based on their evolutionary relationships: class I methanogens, which include the first three orders mentioned above, and class II methanogens, which include the other two orders (Figure 5.1).

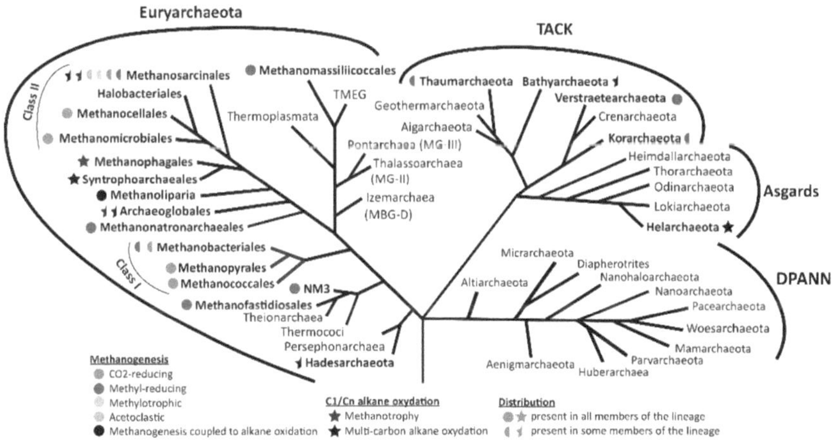

Figure 5.1. *Phylogenetic tree of the domain Archaea*

COMMENT ON FIGURE 5.1.– *The lineages in bold contain one or more members with enzymes linked to methanogenesis. Circles indicate lineages that contain methanogens, while stars indicate lineages that contain alkanotrophs. The colors indicate the type of metabolism. Complete symbols mean that all members of the lineage have the metabolism, whereas half-symbols mean that only some members have the metabolism. "*C_1*" refers to methane, the simplest alkane. "*$C_{n>1}$*" refers to multi-carbon alkanes (e.g. butane, which is a four-carbon chain (*C_4*)).*

A new order has recently been added to class II methanogens, the *Methanocellales* (Sakai et al. 2008). In the early 2010s, after more than half a century of studies on methanogen diversity in a wide variety of environments, a new order of methanogens not belonging to classes I and II was discovered within the intestines of humans and several other animals. This order was named *Methanomassiliicoccales* (Mihajlovski et al. 2008; Borrel et al. 2012; Dridi et al. 2012; Paul et al. 2012; Poulsen et al. 2013). More recently, the use of metagenomic methods has made it possible to overcome biases related to culture and the use of PCR (polymerase chain reaction) primers to explore the diversity of life. Within 3 years, these approaches have led to the discovery of eight additional orders of methanogens. Beyond expanding the diversity of methanogens, these discoveries have also changed our understanding of methane metabolism (see Table 5.2 and section 5.3.3).

5.1.2.2. *Limited diversity of energetic substrates*

Unlike other archaea or *bacteria* that generally possess varied energy metabolisms based on a wide range of substrates, methanogens are limited to a single energy metabolism: methanogenesis. Additionally, many methanogens are highly specialized and are able to use only one type of substrate. Hydrogenotrophs use hydrogen (H_2) to reduce CO_2 or methyl compounds to methane; acetoclastic methanogens use acetate, and methylotrophs use methyl compounds (such as methanol, methylamines, or methylthiol) in the absence of hydrogen (Table 5.2).

Most described methanogenic species reduce CO_2 to methane, contributing to one-third of biological methane emissions. These species belong to *Methanobacteriales*, *Methanococcales*, *Methanopyrales*, *Methanomicrobiales*, *Methanocellales* and to a lesser extent, *Methanosarcinales* (Figure 5.1). *Methanosarcinales* contain methanogens with the greatest metabolic flexibility, largely due to the presence of electron transport via a membrane cofactor (see section 5.3.2). This order is the only one to contain acetoclastic species, which are responsible for the majority of methane emissions into the atmosphere. These species dominate habitats where hydrogenotrophic methanogens lower the partial pressure of hydrogen sufficiently to create the conditions necessary for acetate formation. Acetoclastic methanogens are particularly dominant in anaerobic digesters, rice paddies, peat bogs and lake sediments, contributing up to two-thirds of biological methane emissions.

The *Methanosarcinales* order is also the only one containing methylotrophic species. These methanogens are common in marine sediments (King 1984) and sulfate-rich hypersaline environments (McGenity 2010). Unlike hydrogen and acetate utilization, methanogens are more competitive than sulfate-reducing bacteria in utilizing methylated compounds. Some of these methylated substrates (e.g. trimethylamine (TMA)) originate from the degradation of osmoprotectants produced in large quantities by bacteria in hypersaline environments. Although methylotrophic methanogenesis can be a significant source of methane in certain marine environments, its overall contribution to the global methane budget remains minor (Liu and Whitman 2008). More recently, a team of researchers characterized a strain, *Methermicoccus shengliensis*, from coal deposits (Mayumi et al. 2016). This species is capable of producing methane from methoxylated aromatic compounds found in coal via the methylotrophic methanogenesis pathway. This metabolism could play an important role in the production of natural gas and provides new insights into potential microbial mechanisms in the deep biosphere. These microbial representatives are suspected to be responsible for the formation of explosive methane pockets in coal mines.

Metabolism	Reaction	$\Delta G^{0'}$ (kJ·mol^{-1} of product formed)
Glucose oxidation	$C_6H_{12}O_6 + 6O_2 \rightarrow 6CO_2 + 6H_2O$	– 2,870 kJ·mol^{-1} of glucose consumed
Hydrogenotrophic methanogenesis	$CO_2 + 4H_2 \rightarrow CH_4 + 2H_2O$	– 131 kJ·mol^{-1} of methane formed
Acetoclastic methanogenesis	$CH_3COOH \rightarrow CH_4 + CO_2$	– 36 kJ·mol^{-1} of methane formed
Methylotrophic methanogenesis	$4CH_3OH \rightarrow 3CH_4 + CO_2 + 2H_2O$	– 106.5 kJ·mol^{-1} of methane formed
Methyl-reducing hydrogenotrophic	$CH_3OH + H_2 \rightarrow CH_4 + H_2O$	– 112.5 kJ·mol^{-1} of methane formed

Table 5.2. *Methanogenesis pathways and Gibbs free energies under standard biological conditions (constant pressure of 1 atm, constant temperature of 298 K (25°C), concentrations of dissolved reactants [i] = 1 mol·L^{-1}, and pH 7) for methanogenesis reactions. For comparison, the Gibbs free energy associated with aerobic glucose oxidation is given*

Methanogenesis that uses hydrogen to reduce methyl compounds to methane has long been considered restricted to a few species present in digestive environments (Miller and Wolin 1985; Sprenger et al. 2000). However, eight of the nine orders of methanogens discovered since 2010 consist of microorganisms exclusively capable of this metabolism. Some of these new orders have been identified in the digestive systems of humans and other animals (Borrel et al. 2020), sediments, hypersaline environments (Sorokin et al. 2017), digesters (Nobu et al. 2016; Vanwonterghem et al. 2016) and hot springs (Borrel et al. 2019). Due to their recent discovery, the environmental distribution and global significance of these methanogens remain largely unknown.

This metabolic diversity plays a central role in microbial processes, suggesting that methanogenic communities must interact with other partners to stimulate consortia or competitors and ensure their development.

5.1.3. *Ecological diversity of methanogens*

Despite their sensitivity to oxygen (strict anaerobes), methanogens occupy a wide range of ecological niches (Borrel et al. 2011; Serrano-Silva et al. 2014; Vigneron et al. 2015; Borrel et al. 2020), including some of the most extreme environments such as deep marine and terrestrial sediments, coal deposits, hot springs, petroleum hydrocarbon reservoirs, granitic aquifers, mafic and ultramafic hydrothermal systems, hypersaline sediments, permafrost soils, glacial lakes and pack ice (Liu and Whitman 2008) (Table 5.3).

As a result, methanogens tolerate incredibly varied physicochemical environmental conditions and hold numerous records among known microorganisms.

For instance, *Methanopyrus kandleri* holds the record for the highest temperature tolerated by a living organism (122°C), and the *Methanococcales* species *Methanocaldococcus villosus* holds the speed record, moving at 500 body lengths per second, compared to *Escherichia coli*, which moves at 20 body lengths per second, or the cheetah at 14 body lengths per second (Herzog and Wirth 2012).

Methanogenic strains isolated in the laboratory demonstrate growth at temperatures ranging from 0 to 122°C, pH levels from 3 to 10, salinities from 0 up to halite saturation (e.g. salt marshes with > 5 M NaCl, such as marine halophilic *Methanohalophilus*) and hydrostatic pressures up to 75 MPa, equivalent to a depth of about 7,500 m of water (Table 5.3).

Phenotype	Taxonomy	Growth range (°C)	Salinity (M)	Optimal pressure (MPa)	pH	Environment	References
Hyperthermophilic	*Methanothermus fervidus*	61–(83)–97	ND	ND	6.5	Volcanic source and Icelandic mud	(Stetter et al. 1981)
	Methanothermus sociabilis	55–(88)–97	ND	ND	5.5–(6.5)–7.5	Solfataras in Iceland	(Lauerer et al. 1986)
	Methanotorris igneus	45–(88)–91	0.15–(0.30)–0.92	ND	5–(5.7)–7.5	Submarine vent at the Kolbeinsey Ridge, Iceland, at 106 m	(Burggraf et al. 1990)
	Methanocaldococcus infernus	55–(85)–91	0.21–(0.43)–0.85	ND	5.25–(6.5)–7.0	Logatchev hydrothermal vent, Mid-Atlantic Ridge, 3,000 m	(Jeanthon et al. 1998)
	Methanocaldococcus vulcanius	49–(80)–89	0.1–(0.43)–0.96	ND	5.25–(6.5)–7.0	Logatchev hydrothermal vent, East Pacific Rise, 2,600 m	(Jeanthon et al. 1999)
	Methanocaldococcus fervens	48–(85)–92	0.51	ND	5.5–(6.5)–7.6	Warm hydrothermal marine sediments (Gulf of California) at 2,300 m	(Jeanthon et al. 1999)

Phenotype	Taxonomy	Growth range (°C)	Salinity (M)	Optimal pressure (MPa)	pH	Environment	References
Hyperthermophilic and piezophilic	*Methanocaldococcus jannaschii*	50–(85)–86	0.17–(0.50)–0.85	75	5.2–(6.0)–7.0	Hydrothermal fluid, East Pacific Rise, 2,600 m	(Jones et al. 1983; Miller et al. 1988)
	Methanopyrus kandleri	84–(98)–110 optimal at 40 MPa, 105°C, survival up to 122°C	0.034–(0.34)–0.68	40	5.5–(6.5)–7.0	Hydrothermal marine sediments, Gulf of California and Kolbeinsey Ridge, Iceland	(Kurr et al. 1991; Takai et al. 2008)
Psychrophilic and psychrotolerant	*Methanogenium frigidum*	0–(15)–17	0.35–(ND)–0.6	ND	6.5–(7.5–7.8)–7.9	Sediment of Ace Lake, Vestfold Hills, Antarctica	(Franzmann et al. 1997)
	Methanococcoides burtonii	1.7–(23)–30	(0.2)–0.5	ND	6.8–(7.7)–8.2	Anoxic hypolimnion of Ace Lake, Antarctica	(Franzmann et al. 1992)
	Methanosarcina lacustris	1–(25)–35	ND	ND	4.5–(7.0)–8.5	Anoxic lake sediment (Switzerland)	(Simankova et al. 2001)
	Methanosarcina baltica	4–(21)–27	0.17–(0.34)–1.2	ND	4–(6.5)–8.5	Anoxic surface sediment at 241 m Gotland Deep, Baltic Sea	(von Klein et al. 2002)

Phenotype	Taxonomy	Growth range (°C)	Salinity (M)	Optimal pressure (MPa)	pH	Environment	References
Psychrophilic and psychrotolerant	*Methanosarcina acetivorans*	10–(40)–50	(0.1–0.4)	ND	6.5	Marine sediments of Scripps Canyon, La Jolla, California	(von Klein et al. 2002)
	Methanocorpusculum aggregans	10–(35)–40	(0.034)	ND	6.5–7.0	Sewage sludge digester, Bourg-en-Bresse, France	(Ollivier et al. 1985)
	Methanomethylovorans hollandica	12–(34–37)–40	(0–0.04)	ND	6.5–7.0	Eutrophic pond sediment (Dekkerswald Institute campus), Nijmegen, Netherlands	(Lomans et al. 1999)
	Methanogenium marinum	5–(25)	0.25–(1.1)–1.25	ND	5.5–(6.0–6.6)–7.5	Skan Bay sediment, Alaska	(Chong et al. 2002)
Acidotolerant	*Candidatus* Methanoregula booneii	10–(37)–40	0	ND	(4.7)	Acidic (pH 4.0–4.5) and ombrotrophic site near Ithaca, NY, USA	(Bräuer et al. 2006)
	Methanosphaerula palustris	14–(30)–37	0.035	ND	4.8–(5.5)–6.5	Minerotrophic peatlands, Michigan Hollow, NY	(Cadillo-Quiroz et al. 2008)
	Methanobacterium espanolae	15–(35)–50	ND	ND	4.7–(5.6–6.2)–7.7	Anoxic zone in mud from the Eddy Forest Products Ltd. bleaching plant (Espanola, ON, Canada)	(Patel et al. 1990)

Phenotype	Taxonomy	Growth range (°C)	Salinity (M)	Optimal pressure (MPa)	pH	Environment	References
Alkalotolerant	*Methanolobus oregonensis*	10–(35)–45	0.1–(0.48)–1.6	ND	8.1–(8.6)–9.1	Subterranean saline and alkaline aquifer mud (West Alkali Lake, south-central Oregon)	(Liu et al. 1990)
	Methanonatronarchaeum thermophilum	30–(50)–60	3.0–(4.0)–4.8	ND	8.2–(9.6)–10.2	Sediments from hypersaline lakes in the Kulunda Steppe (Altai, Russia)	(Sorokin et al. 2018)
	Methanolobus taylorii	5–(37)–45	0.2–(0.5)–1.2	ND	5.5–(8.0)–9.2	Estuarine sediment from San Francisco Bay	(Oremland Boone 1994)
Halophilic and halotolerant	*Methanohalobium evestigatum*	ND	4.3	ND	7–7.5	Sediments of saline lagoons in Sivash	(Zhilina Zavarzin 1987)
	Methanohalophilus mahii	35–(37)–45	0.5–(1.2)–3.5	0.1	7.0–(7.5)–7.8	Sediment from the Great Salt Lake, UT (thermal lake)	(Paterek Smith 1988)

Phenotype	Taxonomy	Growth range (°C)	Salinity (M)	Optimal pressure (MPa)	pH	Environment	References
Halophilic and halotolerant	*Methanohalophilus halophilus*	18–(26–36)–42	0.26–(1.20–1.54)–2.57	0.1	6.5–(7.4)	Marine cyanobacterial mat	(Zhilina 1983; Guan et al. 2019)
	Methanohalophilus portucalensis	>25–(42)–48	0.9–(2)–3.5	0.1	6.6–8	Sediments of a saltern in Figueira da Foz, Portugal	(Boone et al. 1993)
Piezophilic	*Methanococcus thermolithotrophicus*	17–(65)–70	0.22–(0.68)–1.42	50	6.2	Marine sediments heated by hydrothermal fluids, Italy	(Bernhardt et al. 1988)
Piezophilic and halophilic	*Methanohalophilus vulcani*	12–(30)–37	0.77–(2.3)–4.1	35 MPa at 14°C	6.5–(7.5)–8.2	Deep hypersaline basin in the Mediterranean Sea	(L'Haridon et al. 2020)

Table 5.3. *Methanogens in cultures from international collections and metabolic characteristics (temperature, pH, salinity and pressure). The numbers in parentheses indicate the optimal growth condition. MPa: Mega Pascal = 10^6 Pascal; ND: Not Determined*

The distribution of methanogenic populations is influenced by ecological interactions and combined physicochemical factors that regulate biochemical and bioenergetic activities.

In addition to extreme environmental conditions (temperature, salinity and pH), methanogenic populations are distributed based on the availability of energy sources. They are primarily constrained by oxygenation conditions (less than 10 ppm, parts per million, 10^{-6} of oxygen) and the absence of sulfate or other inorganic electron acceptors (e.g. nitrate and ferric iron) in their environments due to competition with bacteria, such as sulfate reducers.

5.2. Interactions of methanogens with their environment

5.2.1. *Competition for substrates*

Acetate, hydrogen and formate serve as substrates for a variety of microbial metabolisms that compete with methanogens, thereby limiting their development. In anoxic systems, primary competitors for acetate and/or hydrogen are microbes that oxidize these substrates using a range of electron acceptors. Predominant terminal electron acceptors include sulfate (SO_4^{2-}), nitrate (NO_3^-), manganese (Mn^{4+}) and ferric iron (Fe^{3+}). Most methanogenic species isolated from marine sediments utilize H_2/CO_2, formate and methyl compounds exclusively. Hydrogenotrophic methanogens, which use H_2 to reduce CO_2 to methane, face disadvantageous competition from SO_4^{2-}-reducing bacteria, which utilize hydrogen faster and more effectively for sulfate reduction. This thermodynamic and kinetic competition favors sulfate-reducing bacteria over methanogens in hydrogen and acetate utilization (Lovley and Klug 1983; Robinson and Tiedje 1984). This competition is a significant factor controlling methanogenesis in marine and intertidal sediments where sulfate concentrations are high (Mountfort and Asher 1981; Oremland and Polcin 1982). Sulfate present in seawater is transported by diffusion into the surface sediments, where it is reduced by sulfate-reducing bacteria. In the deeper layers, methanogenesis is more significant than sulfate reduction.

Conversely, acetogenic bacteria using H_2 to reduce CO_2 to acetate do not benefit from favorable thermodynamics, thus allowing hydrogenotrophic methanogens to dominate in many habitats.

5.2.2. *Ecological and syntrophic interactions*

Several methanogens are capable of autotrophic growth, utilizing simple compounds such as H_2 and CO_2 as their sole energy source (see Table 5.2). This capability provides them direct access to unique ecological niches, such as submarine hydrothermal vents (Kelley et al. 2005), which are fueled by abiotic hydrogen generation through the reaction of water with mantle rocks (serpentinization). Globally, however, these abiotic sources of substrates are a minority, with methanogens primarily relying on H_2/CO_2, acetate and methylated compounds resulting from organic matter degradation. In anaerobic environments, the degradation of organic matter involves the collaborative action of diverse microbial communities, with methanogens playing a pivotal role in the final stages (Reeburgh 2007).

This widely accepted model involves four metabolic stages:

– *Hydrolytic and fermentative bacteria* play crucial roles in the initial stages of organic matter degradation. Hydrolysis breaks down complex organic polymers into soluble monomers. This step involves various strict and facultative anaerobic bacteria whose composition depends on the qualitative and quantitative characteristics of the available organic matter.

– *Acidogenesis* carried out by acidogenic bacteria further degrades simple monomers into volatile fatty acids (acids with a carbon chain composed of 2–10 carbon atoms, e.g. propionic or butyric acid) and alcohols (e.g. ethanol).

– During acetogenesis, acetogenic bacteria convert volatile fatty acids and alcohols into hydrogen, CO_2 and acetate. However, when hydrogen partial pressure increases, this conversion becomes thermodynamically unfavorable. Thus, the growth of acetogenic microorganisms and complete organic matter degradation rely on other organisms utilizing hydrogen.

– In environments with low sulfate concentrations, hydrogen is consumed by methanogenic archaea (which reduce CO_2 or methylated compounds to methane) and homoacetogenic bacteria (which reduce CO_2 to acetate). These organisms are referred to as hydrogenotrophs.

These four microbial groups must form a balanced ecosystem to ensure that most of the generated reducing equivalents end up as methane. Thus, the methanogenic process is dependent on syntrophic microbial partnerships. Syntrophy is a symbiotic process where different organisms cooperate to degrade compounds (such as polysaccharides, proteins and fats) that they cannot degrade individually. The symbiotic association between acidogens and hydrogenotrophs benefits both parties:

hydrogen produced by acetogens serves as an energy substrate for hydrogenotrophs, making acetogenesis reactions thermodynamically favorable (exergonic) (Schink 1997; Sa 2005). An increase in hydrogen partial pressure in wastewater treatment plants inhibits these reactions, leading to malfunction. Hydrogen-dependent methanogens quickly deplete H_2 stocks produced by fermentative bacteria, maintaining low concentrations that favor organic matter degradation.

Inter-species hydrogen transfer also occurs in symbiotic relationships between methanogens and eukaryotic cells. For instance, methanogens act as endosymbionts within anaerobic eukaryotes (e.g. protists), where hydrogenosomes (specific organelles of anaerobic eukaryotes) oxidize pyruvate into H_2, CO_2 and acetate to produce ATP. This close interaction between hydrogenosomes and endosymbiotic methanogens (such as *Methanobrevibacter* and *Methanocorpusculum* genera) results in methane generation from hydrogenosome-derived products (Lind et al. 2018).

5.2.3. *Human–methanogen association*

In humans, the predominant methanogenic species is *Methanobrevibacter smithii*. This species is sometimes transmitted from mother to child at birth, potentially being the first methanogenic archaea to colonize our intestines. This hydrogenotroph utilizes H_2/CO_2 and formate generated during bacterial fermentation reactions. Other methanogenic species in the human gut reduce methylated compounds with hydrogen. Some of these species, belonging to the order *Methanomassiliicoccales*, use TMA (Brugère et al. 2014; Borrel et al. 2017), a molecule produced by fermentative bacteria from choline, betaine and carnitine (found in eggs and red meat). The depletion of TMA by methanogens could have a fundamental role on health, as oxidized TMA (known as TMAO) is involved in the development of cardiovascular diseases (atherosclerosis) (Zhu et al. 2016) and liver and kidney diseases (Tang et al. 2015). TMA, meanwhile, is implicated in social disability related to a metabolic disorder (trimethylaminuria) causing a strong odor (Mackay et al. 2011). The term "Archaebiotic" has even been used to describe the use of these methanogens to correct trimethylaminuria or prevent cardiovascular diseases (Brugère et al. 2014). To date, there are no known pathogenic archaea, and methanogens are no exception. Nevertheless, methanogenic archaea have been identified in an increasing number of infections involving multiple microorganisms. These observations suggest that methanogens, through syntrophic relationships (described in section 5.2.2), could promote the development of certain anaerobic pathogenic bacteria (Borrel et al. 2020).

5.3. Bioenergetics and biochemistry of methanogenesis

5.3.1. *Energy extremophiles*

Generating methane from molecules as simple as CO_2 or methanol poses a thermodynamic challenge. Animals "burn" sugars (see Table 5.2, aerobic respiration) through enzymatic processes involving glycolysis, the Krebs cycle and the respiratory chain. The energy transferred during these reactions leads to the formation of ATP, the cellular energy currency. The complete oxidation of one glucose molecule can yield approximately 32–36 moles of ATP. In comparison, methanogenesis using hydrogen and CO_2 can theoretically generate about 2 ATP per molecule of methane produced.

However, hydrogenotrophic methanogens typically live in environments with 0.01 bar of H_2, far from the theoretical 1 bar of H_2 shown in Table 5.2. Under such conditions, the energy transfer is so minimal that two molecules of methane are required to form one molecule of ATP. Therefore, methanogens are described as "energy extremophiles", capable of living at the thermodynamic edge of life.

With such low ATP production, the rapid growth of these microorganisms seems unlikely. Yet, in laboratory conditions, where methanogens can be grown at very high H_2 partial pressures, the growth of some species surpasses that of *Escherichia coli* growing on glucose in the presence of oxygen (the most common model of bacteria using aerobic respiration). This division time (25 min) in a minimal mineral medium is only possible due to adapted metabolic strategies. The enzymes catalyzing the reactions of methanogenesis have had billions of years to evolve and perfect themselves to meet the physiological needs of the microorganism. One notable feature of such a system is the use of cofactors specialized in single-carbon unit chemistry (C1 = CHO- or CH- or CH_2- or CH_3-).

5.3.2. *Cofactors used in methanogenesis*

Some scientists suggest that methanogenesis is restricted to archaea because the transmission of all enzymes involved in the biosynthesis of cofactors is too extensive for all of the necessary genes to be transferred to bacteria. More than 200 genes would be needed for methanogenesis that reduces CO_2 to methane (Kaster et al. 2011a). However, some of these cofactors (e.g. F_{420} and H_4MPT) are found in other species of archaea or even bacteria. The coenzymes and metallocofactor necessary for methanogenesis can be divided into two classes based on their functions: C1 unit carriers and electron carriers necessary for redox reactions. Most of these coenzymes and cofactors are also used in reverse methanogenesis reactions (or anaerobic methanotrophy, see section 5.4).

C1 unit carriers:

– *Methanofuran* (see Figure 5.2), capable of binding a formyl group (CHO-) on its furan ring. It also exists in other archaea, such as Archaeoglobales, where it is involved in the degradation of lactate to CO_2 (Gorris et al. 1991). Methylotrophic bacteria also use a methanofuran derivative to generate formate during the formaldehyde oxidation process (Hemmann et al. 2019).

– *Tetrahydromethanopterin* (H_4MPT), which carries the formyl group, is involved in carbon reduction (Figure 5.2). This cofactor is similar to tetrahydrofolate, which is used by other organisms as a C1 carrier (Chistoserdova et al. 1998). H_4MPT is also present in methylotrophic bacteria where it oxidizes the C1 unit (these bacteria use the reverse pathway of methanogenesis; their H_4MPT does not contain phosphate and hydroxyglutarate groups) as well as in sulfate-reducing archaea (Gorris et al. 1991). Methanogens of the orders *Methanosarcinales* and *Methanococcales* use an analog of H_4MPT for their central metabolism called tetrahydrosarcinapterin (H_4SPT), corresponding to H_4MPT extended by an additional glutamate (Buchenau and Thauer 2004). Other derivatives, such as tatiopterin, found in *Methanogenium tationis*, also exist (Raemakers-Franken et al. 1989).

– *5-hydroxybenzimidazolyl cobamide*, a close analogue of vitamin B_{12} (cobalamin). Its cobalt atom binds a methyl group.

– *2-mercaptoethanesulfonate*, also called *coenzyme M*, carries C1 on its thiol in the form of methyl. This methyl group is eventually released by cofactor F_{430}, the most reduced tetrapyrrole known to date, which contains nickel at its center (Thauer 2019).

Electron carriers:

– *Cofactor F_{420}*, with a structure related to flavins (such as vitamin B_2, also called riboflavin) (Figure 5.2). It plays a role similar to NADH/NADPH in carrying electrons from central metabolism. Due to its more negative redox potential compared to NADH/NADPH, methanogenesis reactions are facilitated and accelerated. This cofactor is also responsible for the fluorescence of cells exposed to a wavelength of 420 nm (Figure 5.3).

– *Ferredoxin*, a protein containing iron-sulfur-based metal centers capable of transferring low-potential electrons.

– *Coenzyme B* (Figure 5.2), which uses its thiol function for the methane generation reaction (see below).

– *Methanophenazine*, a quinone analogue with a membrane anchor capable of transporting protons and electrons (Figure 5.2).

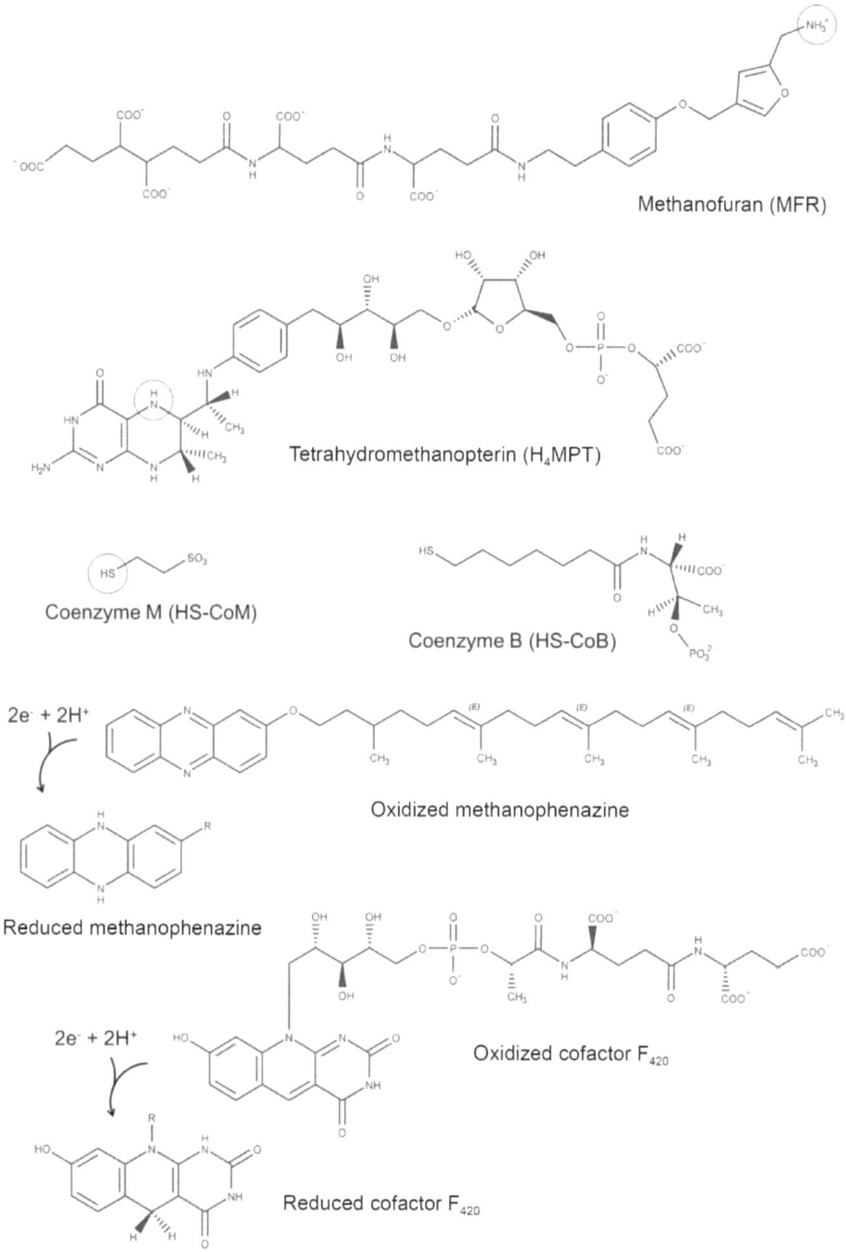

Figure 5.2. *Chemical formulas of cofactors used in methanogenesis. A gray circle indicates the branching of the C1 unit to the carrier*

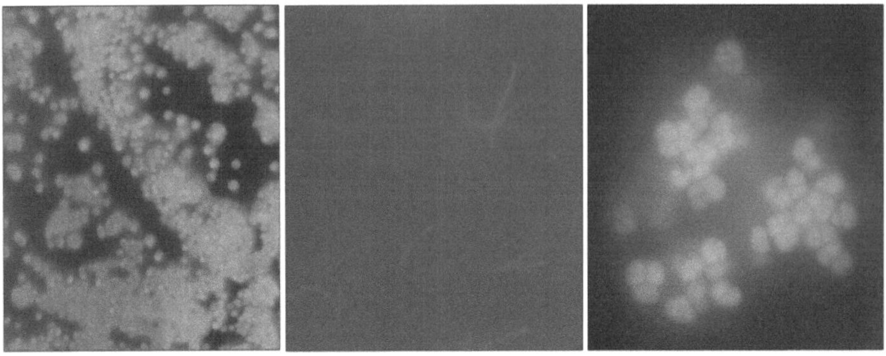

a) b) c)

Figure 5.3. *Example of the intrinsic fluorescence phenomenon in methanogens. This fluorescence is due to the cofactor F_{420}*

COMMENT ON FIGURE 5.3.– *(a) Methanothermococcus thermolithotrophicus is a hydrogenotrophic marine thermophilic methanogen. (b) Methanothermobacter marburgensis is a hydrogenotrophic terrestrial thermophile. (c) Methanosarcina barkeri is a hydrogenotrophic, acetotrophic and methylotrophic terrestrial mesophile.*

5.3.3. *Different types of methanogenesis*

5.3.3.1. *Hydrogenotrophic methanogenesis reducing carbon dioxide (or CO_2-reducing methanogenesis)*

Figure 5.4 illustrates the various reactions involved in hydrogenotrophic methanogenesis (Shima et al. 2020), potentially one of the most ancient metabolisms, capable of reducing CO_2 using H_2. The two gases diffuse easily through the membrane.

– *Capture of CO_2 as a formyl-methanofuran (reaction 1)*: formyl-methanofuran dehydrogenase (Fwd or Fmd, reaction 1) is named after the reverse reaction: the oxidation of formyl-methanofuran. This enzymatic complex accumulates electrons on its metal center containing tungsten or molybdenum (this metallocofactor is a tungstopterin or molybdopterin). Only a very low redox potential electron carried by ferredoxins can effectively force the reduction of CO_2 to formate. The formate is transferred inside the enzymatic complex through a tunnel. It then reacts with the second metal center, which catalyzes the condensation of formate on methanofuran, creating formyl-methanofuran (Wagner et al. 2016) and a water molecule.

– *Transfer of formyl to H_4MPT (reaction 2)*: this step is catalyzed by the enzyme formyl-transferase (Ftr), releasing methanofuran.

– *Condensation of formyl-H_4MPT (reaction 3)*: this is carried out by the cyclohydrolase (Mch), where formyl-H_4MPT is dehydrated to methenyl-H_4MPT.

– *Reduction of methenyl to methyl (reactions 4, 5, 5' and 6)*: both successive reduction reactions require an electron donor. When cells have sufficient nickel (Huang et al. 2020), a nickel-iron hydrogenase (F_{420}-reducing hydrogenase, Frh) reduces F_{420} with H_2 (reaction 4). The reduced F_{420} ($F_{420}H_2$) is then used by F_{420}-dependent methylene-H_4MPT dehydrogenase (Mtd, reaction 5) and methylene-H_4MPT reductase (Mer, reaction 6) to form methyl-H_4MPT. This methyl-H_4MPT can be used for cellular carbon assimilation, a reaction that requires the CO-dehydrogenase/Acetyl-CoA synthase complex (Thauer 2012). $F_{420}H_2$ is also used in other anabolic reactions.

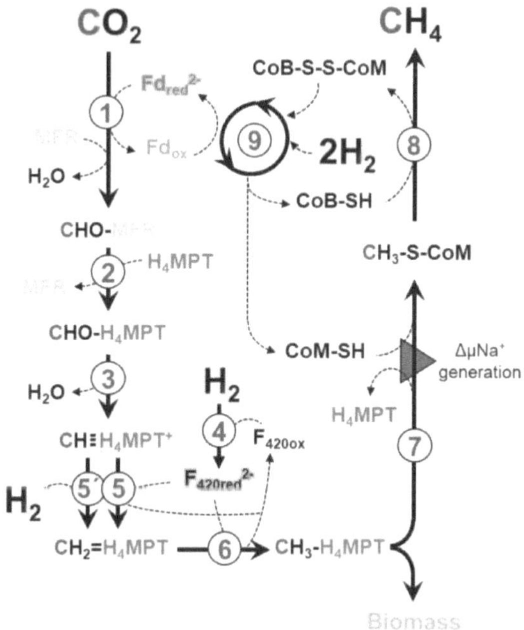

Figure 5.4. *Metabolism of hydrogenotrophic methanogens. MFR, methanofuran; H_4MPT, tetrahydromethanopterin; F_{420}, cofactor F_{420}; Fd, ferredoxin; CoB-SH, coenzyme B; CoM-SH, coenzyme M. The pink triangle represents the membrane-bound methyl transferase that is responsible for establishing the chemiosmotic gradient by the translocation of sodium ions (Na^+) outside the cell*

In the absence of nickel, methanogens utilize an iron-only hydrogenase (Hmd) to reduce methenyl-H_4MPT (reaction 5'). The F_{420} is then reduced by Mtd through its reverse reaction.

– *Translocation of sodium ions across the membrane via the formation of methyl-S-coenzyme M (reaction 7)*: this reaction is responsible for energy conservation. The transfer of the methyl group from H_4MPT to coenzyme M is highly favorable. Instead of dissipating this energy, methanogenesis couples the methyl transfer reaction to a mechanical motion that enables the translocation of sodium ions across the membrane (Gottschalk and Thauer 2001), leading to the accumulation of sodium ions on the cell surface. The membrane enzyme complex orchestrating this reaction contains 5-hydroxybenzimidazolyl cobamide. The methyl group is transferred to the cobalt of the cofactor and then further transferred to coenzyme M. The second transfer reaction coordinates the translocation of two sodium ions. ATP synthase forms ATP from ADP and inorganic phosphate with the mechanical force provided by the passage of sodium ions into the cell.

– *Generation of methane (reaction 8)*: methyl-coenzyme M reductase (MCR) contains the cofactor F_{430}. This cofactor catalyzes the formation of methane and the oxidation of coenzyme M and coenzyme B to heterodisulfide (CH_3-S-CoM+ HS-CoB ➔ CoM-S-S-CoB+CH_4).

– *Regeneration of cofactors by electron bifurcation (reaction 9)*: the coupling of methane generation and CO_2 fixation (called the R.P.G. effect after its discoverer Robert P. Gunsalus) was observed as early as 1977 (Gunsalus and Wolfe 1977), but it was not experimentally proven until 2012 (Kaster et al. 2011b). This step involves a recent biochemical concept called electron bifurcation (Buckel and Thauer 2018). Heterodisulfide reductase captures electrons from hydrogen and transfers them to a flavin carried by the enzyme (Wagner et al. 2017a, 2018). Flavins can receive two electrons of the same redox potential and then separate them into two different potentials: one high and one low (Figure 5.5). The high-potential electron is used to reduce the heterodisulfide, and the low-potential one is transferred to ferredoxin, which is delivered directly or indirectly to the formyl-methanofuran dehydrogenase (reaction 1). Thus, methanogenesis is not linear but cyclical, with the last step coupled to the first.

Variants of CO_2-reducing hydrogenotrophic methanogenesis exist in which CO_2 and H_2 are replaced by another substrate, such as formate. Under these conditions, formate dehydrogenases replace hydrogenases. Formate is oxidized to CO_2, and the electrons are transferred to F_{420} (reaction 4), heterodisulfide and ferredoxin (Costa et al. 2010). Other electron donors can also be used to reduce CO_2 to methane, such as ethanol, isopropanol, 2-butanol and even ferrous iron (Kurth et al. 2020).

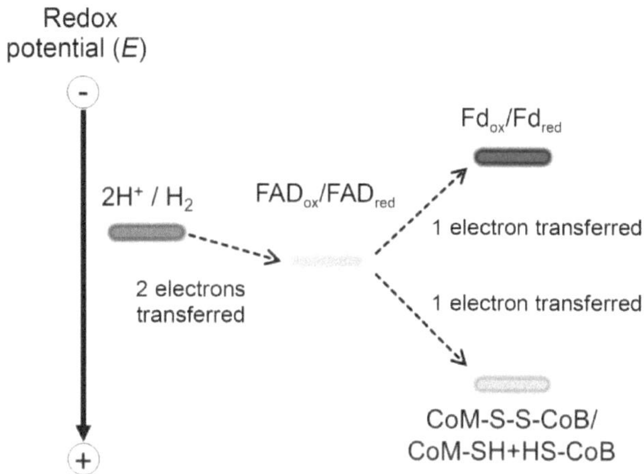

Figure 5.5. *Concept of electron bifurcation illustrated by the reaction of heterodisulfide reductase*

COMMENT ON FIGURE 5.5.– *The reaction uses hydrogen as the electron donor. FAD stands for flavin adenine dinucleotide, Fd for ferredoxin and CoM-S-S-CoB for heterodisulfide. The bifurcation of electrons functions like an energy wheel (noria), coupling an endergonic reaction with an exergonic reaction. This process is also used in many systems, such as the respiratory chain of aerobic organisms.*

5.3.3.2. *Acetoclastic methanogenesis*

The conversion of acetate (one of the products of fermentation, see section 5.2.2) into methane and CO_2 is illustrated in Figure 5.6.

– *Acetate transport and activation (reaction 1)*: acetate is internalized and then captured as acetyl-coenzyme A (acetyl-CoA) through phosphorylation (reaction 1) (Welte and Deppenmeier 2014). Acetyl-CoA is central to carbon metabolism as most of cellular metabolites can be synthesized from this precursor. Acetyl-CoA is also used in catabolism for these methanogens.

– *Conversion of acetyl-CoA to CO_2 and methyl-coenzyme M (Reaction 2)*: acetyl-CoA is metabolized by the CO-dehydrogenase/acetyl-CoA decarbonylase complex (reaction 2), a "primordial" enzyme considered part of the LUCA proteome (Adam et al. 2019), the ancestor of all living cells. Acetyl-CoA decarbonylase converts acetyl-CoA into CoA, methyl-H_4MPT and carbon monoxide (CO). CO, which is toxic to the cell, does not diffuse out of the enzyme. It is sequestered in an internal tunnel up to the CO-dehydrogenase subunit to be oxidized to CO_2. The very low redox potential electrons extracted during CO oxidation are transferred to ferredoxin and will be used in reaction 5, involving the regeneration of the heterodisulfide. A carbonic anhydrase, strongly expressed under acetoclastic conditions, may drive the reaction by favorably evacuating the CO_2 (Karrasch et al. 1989; Smith and Ferry 2000).

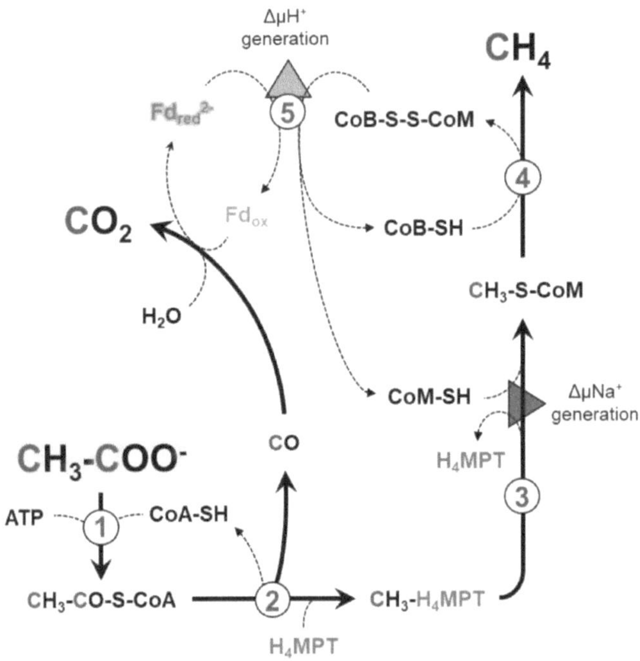

Figure 5.6. *Metabolism of acetoclastic methanogens*

COMMENT ON FIGURE 5.6.– *H_4MPT, tetrahydromethanopterin; Fd, Ferredoxin; CoM, coenzyme M; CoB, coenzyme B. The pink triangle represents the membrane-bound (Na^+) methyl transferase pump. The green triangle represents two enzymes, membrane methanophenazine reductase and membrane heterodisulfide reductase, involved in proton (H^+) translocation.*

– *Transfer of the methyl group to coenzyme M and generation of methane (reactions 3 and 4)*: as previously described, the membrane methyl transferase transfers the methyl group from methyl-H_4MPT to coenzyme M, forming the sodium gradient (reaction 3). Then, MCR generates methane and heterodisulfide (reaction 4).

– *Generation of a membrane proton gradient by reduction of the heterodisulfide (reaction 5)*: the heterodisulfide is not reduced by electron bifurcation, as in hydrogenotrophs, but by a membrane-coupled heterodisulfide reductase using methanophenazine as an electron donor. Depending on the methanogen, methanophenazine is reduced either directly (Suharti et al. 2014) or indirectly (Kulkarni et al. 2018) by ferredoxin from CO oxidation. This electron transfer is coupled to the translocation of ions (H^+ or Na^+, depending on the methanogen) via a membrane pump. The translocation of ions across the membrane enables ATP synthase to generate ATP.

This methanogenesis pathway generates the least cellular energy (see Table 5.2) due to the ATP investment required during acetate capture.

5.3.3.3. *Methylotrophic methanogenesis*

Methylated substrates used by methylotrophic methanogens can be very diverse (section 5.1.2.2). In this example, we will consider the case of methanol (Figure 5.7).

– *Activation of methanol to methyl-coenzyme M and formation of methane (reactions 1–3)*: the first enzyme transfers the methyl group from methanol to a protein carrying 5-hydroxybenzimidazolyl cobamide. The methyl group bound to cobalt is then recognized by a second methyltransferase (Hagemeier et al. 2006; Sollinger and Urich 2019; Kurth et al. 2020), which branches it on coenzyme M. From methyl-coenzyme M and coenzyme B, MCR forms methane and heterodisulfide.

– *The oxidative methyl branch (reactions 4–12)*: in the absence of an electron donor, such as hydrogen or CO, the heterodisulfide accumulates without being recycled. Methylotrophs oxidize the methyl group bound to coenzyme M to CO_2 and use the electrons obtained from this oxidation to reduce the heterodisulfide, thereby regenerating coenzyme M and coenzyme B. This oxidation is possible because the reactions are near equilibrium. The oxidation of one methyl molecule to CO_2 generates three pairs of electrons, which are used to reduce three heterodisulfides, which explains why three methane and one CO_2 are generated from four methanol.

The first step consumes the sodium ion gradient to transfer the methyl group bound on the coenzyme M to H_4MPT (reaction 4). Methyl-H_4MPT can be used to form acetyl-CoA and enter central carbon metabolism. The complete oxidation of the methyl group relies on the same enzymes and reactions as hydrogenotrophic metabolism (reactions 5 to 9). Oxidation of reactions 5 and 6 are coupled to F_{420} reduction and reaction 9 to ferredoxin reduction. Reduced ferredoxin will enable the translocation of protons or sodium ions (varies depending on the methanogen species, reaction 10), and the transfer of electrons from reduced F_{420} to methanophenazine will allow the proton translocation (reaction 11). This reduced methanophenazine will reduce back the heterodisulfide produced during methanogenesis thanks to a membrane heterodisulfide reductase (Sollinger and Urich 2019; Kurth et al. 2020) (reaction 12).

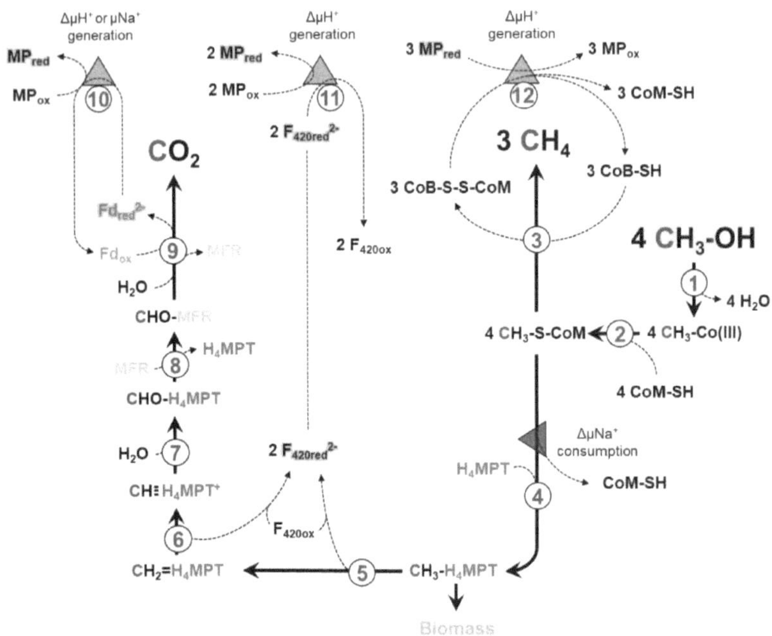

Figure 5.7. *Metabolism of methylotrophic methanogens*

COMMENT ON FIGURE 5.7.– *H_4MPT, tetrahydromethanopterin; Fd, ferredoxin; MP, methanophenazine; CH_3-Co(III), methyl group bound to vitamin B_{12} analogue; CoB, coenzyme B; CoM, coenzyme M. The pink triangle represents membrane-bound methyl transferase. The green triangle represents two enzymes, a membrane methanophenazine reductase (reactions 10 and 11, being selective of ferredoxin or F_{420}, respectively) and a membrane heterodisulfide reductase (reaction 12) involved in proton (H^+) translocation.*

This metabolism produces more energy than hydrogenotrophic and acetoclastic methanogenesis (Table 5.2) and offers high metabolic plasticity through the use of different methyl donors. A battery of specific methyl transferases (reactions 1 and 2) allows the specific use of methanol, TMAs, methylthiols and many others.

5.3.3.4. *Hydrogenotrophic methanogenesis reducing methylated compounds (or reductive methyl methanogenesis)*

Some methanogens can also utilize methylated compounds with external electron donors, such as hydrogen. *Methanosphaera stadtmanae* (van de Wijngaard et al. 1991; Fricke et al. 2006; Thauer et al. 2008) employs this specific strategy to reduce the heterodisulfide and acquire its cellular energy (Figure 5.8).

The methyl group from methanol is converted to methyl-coenzyme M (reactions 1 and 2), which is subsequently metabolized to methane along with the production of the heterodisulfide (reaction 3). The soluble bifurcating heterodisulfide reductase/hydrogenase facilitates the recycling of heterodisulfide using H_2 as a reducing agent. Due to electron bifurcation, this reaction is coupled with the reduction of ferredoxin (reaction 4). The reduced ferredoxin would be reoxidized by a membrane pump that translocates sodium ions while generating H_2 (Ehb complex, reaction 5).

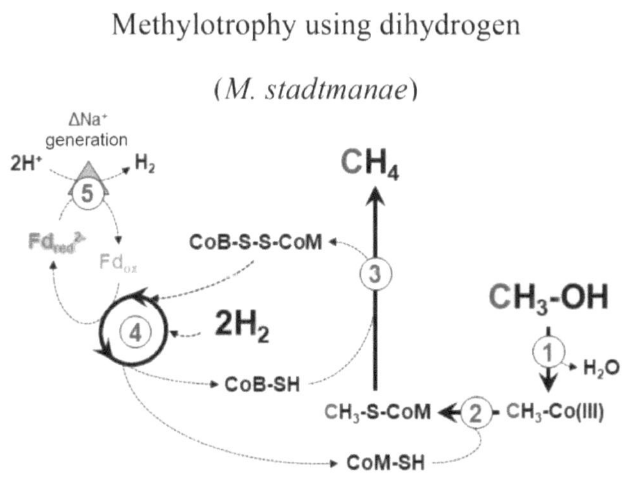

Figure 5.8. *Metabolism of M. stadtmanae. Fd: ferredoxin; CH_3-Co(III): methyl group bound to a vitamin B_{12} analogue; CoM: coenzyme M; CoB: coenzyme B. The green arrow represents a membrane-bound hydrogenase pumping sodium ions outside the cell*

This type of methanogenesis has long been considered anecdotal, limited to a few species of the genera *Methanosphaera* and *Methanimicrococcus*. However, recent studies have shown that there are as many taxonomic orders of methanogens capable of this metabolism as there are orders of methanogens capable of reducing CO_2 to methane (Borrel et al. 2019). These new lineages utilize a wide variety of enzymes, based on either a soluble heterodisulfide reductase/hydrogenase, membrane-embedded soluble cytochromes, or other strategies (Borrel et al. 2019). Due to the recent discovery of these methanogens and the lack of cultures for most of these new lineages, many of the involved enzymes currently have only predicted functions. Finally, as with the reduction of CO_2 to methane, other electron donors besides hydrogen can be used to reduce methylated compounds to methane, such as ethanol (Hoedt et al. 2016) or formate (Sorokin et al. 2017).

5.3.4. *MCR, the unique enzyme capable of generating biogenic methane*

The formation of catabolic methane depends on the MCR and its associated heterodisulfide reductase (either soluble or membrane-bound). MCR has been the focus of extensive research due to its remarkable chemistry and significant role in the carbon cycle. In 1997, the first structure of the enzyme was elucidated, providing an atomic-scale view and detailed coordination of its cofactor F_{430} (Ermler et al. 1997). MCR consists of three peptide chains (α, β and γ) that oligomerize into a dimer of heterotrimer. The active site, located at the dimeric interface, is accessible via a hydrophilic tunnel that guides the substrates. The tunnel and the active site contain post-translational modifications on the α chain, which vary among different methanogens. For example, *Methanothermobacter marburgensis*, a terrestrial species found in wastewater treatment plants, contains N^1-methylhistidine, S-methylcysteine, 5-(*S*)-methylarginine, 2-(*S*)-methylglutamine, thioglycine and didehydroaspartate (Figure 5.9).

Methanotorris formicicus, a marine hyperthermophile, lacks the S-methylcysteine and didehydroaspartate but contains a 6-hydroxytryptophan (Wagner et al. 2017b). Although some of these modifications are common across all studied MCRs, their functions remain unclear. Recent investigations have demonstrated that thioglycine, methylarginine and methylcysteine are not essential but may play roles in certain physiological processes and temperature tolerance (Lyu et al. 2020; Nayak et al. 2020). One current hypothesis suggests that, as the most abundant cytoplasmic protein in methanogenic archaea (constituting about 10% of the protein content), MCR represents the rate-limiting step of methanogenesis and, consequently, cellular energy supply. Over billions of years of evolution, the catalytic speed has been enhanced through mutations, yet the traditional set of 20 amino acids was proven to be insufficient. New modifications, though not critical for activity, may have improved the enzyme's properties, providing ecological advantages. A structure of the enzyme in its active form could shed light on the role of these modifications, particularly their impact during catalysis. F_{430} and its Ni are the key elements of catalysis, as well as the Achilles' heel of methanogens. Their extreme sensitivity to oxidants inactivates the cofactor, as evidenced by spectroscopic studies (Thauer 2019). An ATP-dependent system reactivates the cofactor in vivo. The reaction mechanism for methane formation has been debated for several years, requiring the development of various techniques to trap and observe intermediates (Thauer 2019). Spectroscopy studies (Wongnate et al. 2016) have proposed the following reaction mechanism. (1) Methyl-coenzyme M binds to Ni(I) of the cofactor F_{430}. Ni(I) attacks the sulfur of coenzyme M, transitioning to Ni(II) and activating the methyl group into a radical form. (2) Coenzyme B is attacked by the methyl radical, donating its proton. (3) Methane is formed, and the radical sulfur of coenzyme B binds to the sulfur of coenzyme M. (4) The resulting heterodisulfide transfers the electron back to Ni(II), regenerating Ni(I).

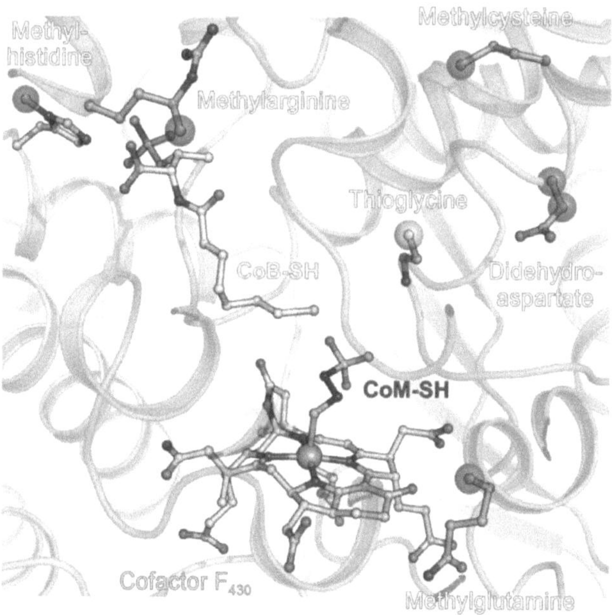

Figure 5.9. *Catalytic core of MCR from Methanothermobacter marburgensis (PDB code 5A0Y). The α, α', β and γ chains are depicted as ribbons and loops, colored green, chartreuse, cyan and orange, respectively. The modified residues, coenzymes and cofactor F_{430} are shown in a ball-and-stick model*

5.4. Anaerobic methanotrophs and anaerobic oxidation of multi-carbon alkanes

Natural gas sources releasing methane are home to a fascinating consortium: an association between methanotrophic archaea (capable of oxidizing methane into CO_2) and sulfate-reducing bacteria. Numerous studies worldwide have finally uncovered the intimate relationship of this anaerobic prokaryotic couple: the archaea directly transmit the electrons produced during methane oxidation to the bacteria. The electrons recovered by the bacterium enable it to reduce sulfate (SO_4^{2-}) to sulfide (S^{2-}), thereby allowing respiration.

The prevailing hypothesis is that electron transfer is mediated by biological electrical wiring (McGlynn et al. 2015, 2017; Wegener et al. 2015; Bhattarai et al. 2019). It was a significant surprise when researchers discovered that genomes of methanotrophic archaea contain genes coding for MCR and other enzymes from the methanogenesis pathway. The current hypothesis is that these enzymes work in reverse to convert methane into methyl-coenzyme M and oxidize it into CO_2. The electrons released during these reactions are transferred to the membrane and then to the bacterium via a chain of cytochromes, which drives membrane pumps, ensuring ion translocation and establishing a proton/sodium gradient to obtain cellular energy. Subsequently, other methanotrophic archaea belonging to different lineages were discovered, some capable of using nitrate, iron or manganese as final electron acceptors in the absence of a bacterial partner.

Recently, consortia have been enriched using various multi-carbon alkanes (e.g. ethane, propane and butane) instead of methane. Some of these archaea are specific to the oxidation of ethane (Chen et al. 2019; Hahn et al. 2020), while others specialize in propane and butane (Laso-Pérez et al. 2016). These multi-carbon alkanes may be similarly activated by an alkyl-coenzyme M reductase (ACR), evolutionarily linked to the MCR enzyme of methanogens but with different characteristics (Hahn et al. 2021).

Genome reconstruction from petroleum samples and hydrocarbon-rich marine sediments has highlighted a new class of archaea called *Candidatus Methanoliparia*. Members of this class possess both the MCR enzyme of methanogens and the ACR enzyme involved in the utilization of long-chain alkanes.

The metabolic prediction of *Ca. Methanoliparia* suggests they are capable of degrading long-chain alkanes or even fatty acids anaerobically to form methane without a syntrophic partner (Borrel et al. 2019; Laso-Pérez et al. 2019; Zhou et al. 2022).

This metabolism of *Ca. Methanoliparia* represents a completely new type of methanogenesis (Borrel et al. 2019; Laso-Pérez et al. 2019; Zhou et al. 2022). Such a new *terra incognita* excites the scientific community, revealing not only new metabolic pathways and unexplored enzymes but also a gateway for biotechnology: converting methanogens into hydrocarbon producers.

5.5. Evolution of methanogenesis

5.5.1. *An ancestral metabolism*

A body of evidence suggests that methanogenesis is likely one of the oldest metabolic processes. Traces of methane with isotopic signatures indicating biological origins have been found in 3.46-billion-year-old quartz crystals (Ueno et al. 2006). Additionally, phylogenomic studies have shown that several key enzymes of methanogenesis were probably present in the last common ancestor of Archaea (Borrel et al. 2016; Adam et al. 2019; Borrel et al. 2019), and for some, even in LUCA (Adam et al. 2018), the last universal common ancestor of all life forms. This ancestral nature is also supported by the fact that many enzymes from the methanogenesis pathway serve for cellular energy production and the transformation of inorganic carbon into organic carbon. These early methanogens could have emerged in hot environments (Hua et al. 2019), such as hydrothermal vents, which could have provided all of the simple elements needed for producing cellular energy and biomass (Chapelle et al. 2002). The ancestral nature of methanogenesis likely had significant implications for the evolution of the domain Archaea. Since not all modern archaea are methanogenic, it suggests that their current metabolisms could have evolved from methanogenesis (Borrel et al. 2016). In some lineages, like the *Halobacteria*, the transition to an aerobic lifestyle based on the utilization of complex substrates was enabled by the transfer of a substantial number of genes from bacteria (Nelson-Sathi et al. 2012; Becker et al. 2014).

Due to their ancestral nature and the potent greenhouse gas effect of methane, methanogens influenced climate evolution long before the current global warming. Several hypotheses based on geochemical models and field observations suggest that through their activity, methanogens prevented the Earth from becoming a "snowball" during a period when solar activity was 30% lower than today. Notably, glaciation periods 2.2–2.5 billion years ago (Evans et al. 1997) have been associated with declines in methanogen metabolic activity.

5.5.2. *Metabolic adaptations*

Methanogens have endured several major geochemical upheavals on Earth. The first significant event was a decrease in biologically available nickel at the end of the Archean period (2.7 billion years ago) (Konhauser et al. 2009). As discussed in section 5.3.3, several key reactions in methanogenesis, including methane generation by MCR, as well as nickel-iron hydrogenases and the CO-dehydrogenase/acetyl-CoA synthase complex, depend on nickel. This reduction in bioavailable nickel was likely a critical limiting factor for the development of methanogens at that time (Konhauser et al. 2009). Modern methanogens have evolved strategies to cope with this deficiency, such as utilizing iron-only hydrogenase (Huang et al. 2020). The second crisis faced by methanogens was the "Great Oxygenation Event" (2.4 billion years ago), associated with the development of cyanobacteria performing oxygenic photosynthesis. During this period, the previously reducing environment became oxidatively toxic for all anaerobic organisms, including methanogens. Oxygen irreversibly inhibits several key enzymes of the methanogens, confining them to anaerobic ecological niches protected from oxygen. Today, only certain aquatic environments (e.g. sediments, wet soils), deep habitats and animal digestive systems are devoid of oxygen. The penetration of oxygen into sediments or wet soils depends on the availability of easily degradable organic matter (labile). In environments rich in labile organic matter, oxygen penetrates only a few millimeters, being consumed by aerobic organisms. However, in some sediments that are poor in organic matter, such as those in the South Pacific, oxygen can penetrate several kilometers into the sediments. Adaptation to oxidative stress became essential for species occasionally exposed to oxygen. Consequently, most methanogen genomes contain a suite of enzymes, such as superoxide dismutase, catalase and F_{420} oxidase, capable of effectively eliminating oxygen and preventing the formation of free radicals (Lyu and Lu 2018; Engilberge et al. 2020). This tolerance enables them to persist in habitats with fluctuating environmental conditions (e.g. periodically aerated sediments or soils).

5.6. The impact of methanogens in our modern society

Due to their ability to grow on simple inorganic compounds under extreme temperature conditions and in the absence of oxygen and light, methanogens are used to explore the possibilities of extraterrestrial life. Methane concentrations in the Martian atmosphere are not fully explained by known physical–chemical processes. Although the probability is low, it is not entirely ruled out that this methane could be the product of past or still active life forms. Moreover, hydrogen and methane have been detected in the hydrothermal plumes of Enceladus, an icy moon of Saturn

(Waite et al. 2017). Several research teams have shown that methanogenic species can develop under conditions mimicking those on Mars (Morozova et al. 2007; Kral et al. 2016) or Enceladus (Taubner et al. 2018). These significant findings suggest that methanogens could potentially develop on planets other than Earth, but it is crucial to note that they do not confirm that methane observed on other planets originates from extraterrestrial life forms. Additionally, for Mars, these experiments also test the possibility that methanogens could be the first Earth organisms to colonize the Red Planet in the future.

The impact of methanogenesis on our society is pervasive. Methane is viewed both as a concerning greenhouse gas and as the "green" biofuel of the future. Most of us maintain a symbiotic relationship with these organisms *sensu stricto*, as they are an integral part of our microbiota (see section 5.2.3) and through their active role in recycling our biological waste in wastewater treatment plants (see section 5.2.2). Without them, fermentation products (methylamines, acetate, hydrogen, etc.) would accumulate, slowing or blocking the initial stages of degradation. This bioconversion is currently harnessed in biogas plants, where methane is recovered to heat our homes or generate electricity. Industries are capitalizing on these properties by cultivating hydrogenotrophic methanogens in large fermenters to convert CO_2 into methane.

The enzymes involved in methanogenesis are a source of inspiration for chemists and are valuable resources for biotechnology. For example, methanogen hydrogenases can act as catalysts for producing or capturing H_2 (Huang et al. 2020). Understanding how these enzymes function could lead to their domestication for use as biochemical tools or to inhibit them. For instance, the three-dimensional structure of MCR has enabled the modeling of highly specific inhibitory molecules that can block methanogens without affecting the surrounding microbiota. One such man-made molecule, called 3-nitro-oxypropanol, has already demonstrated remarkable effectiveness against various rumen methanogens. When administered to cattle, the amount of methane emitted by the animals drops drastically and their weight slightly increases (Duin et al. 2016).

Methanogens occupy most of the anaerobic niches on our planet and play an indisputable key role in the global carbon cycle. Given that methanogenesis is a very ancient metabolism, its cofactors and fundamental reactions have been optimized over billions of years of evolution, continually adapting enzymatic machinery. Every day, scientists discover new species of methanogens through enrichment and isolation or by analyzing metagenomic sequences from various environments. Recently, bioinformatics analyses have revealed that the methanogens known and isolated to date represent only a minority of all methanogens, hinting at the potential discovery of new enzymes, new metabolisms and functional roles within ecosystems.

5.7. References

Adam, P.S., Borrel, G., Gribaldo, S. (2018). Evolutionary history of carbon monoxide dehydrogenase/acetyl-CoA synthase, one of the oldest enzymatic complexes. *Proc. Natl. Acad. Sci. USA*, 115(6), E1166–E1173.

Adam, P.S., Borrel, G., Gribaldo, S. (2019). An archaeal origin of the Wood-Ljungdahl H_4MPT branch and the emergence of bacterial methylotrophy. *Nat. Microbiol.*, 4(12), 2155–2163.

Becker, E.A., Seitzer, P.M., Tritt, A., Larsen, D., Krusor, M., Yao, A.I., Wu, D.Y., Madern, D., Eisen, J.A., Darling, A.E., Facciotti, M.T. (2014). Phylogenetically driven sequencing of extremely halophilic archaea reveals strategies for static and dynamic osmo-response. *PloS Genetics*, 10(11), 1–23. doi: 10.1371/journal.pgen.1004784.

Bernhardt, G., Jaenicke, R., Lüdemann, H.D., König, H., Stetter, K.O. (1988). High pressure enhances the growth rate of the thermophilic archaebacterium *Methanococcus Thermolithotrophicus* without extending its temperature range. *Applied and Environmental Microbiology*, 54(5), 1258–1261.

Bhattarai, S., Cassarini, C., Lens, P.N.L. (2019). Physiology and distribution of archaeal methanotrophs that couple anaerobic oxidation of methane with sulfate reduction. *Microbiology and Molecular Biology Reviews*, 83(3), 1–31. doi: 10.1128/mmbr.00074-18.

Boone, D.R., Mathrani, I.M., Liu, Y., Menaia, J.A.G.F., Mah, R.A., Boone, J.E. (1993). Isolation and characterization of *Methanohalophilus portucalensis* sp. nov. and DNA reassociation study of the genus *Methanohalophilus*. *International Journal of Systematic Bacteriology*, 43(3), 430–437.

Borrel, G., Jezequel, D., Biderre-Petit, C., Morel-Desrosiers, N., Morel, J.P., Peyret, P., Fonty, G., Lehours, A.C. (2011). Production and consumption of methane in freshwater lake ecosystems. *Res. Microbiol.*, 162(9), 832–847.

Borrel, G., Harris, H.M.B., Tottey, W., Mihajlovski, A., Parisot, N., Peyretaillade, E., Peyret, P., Gribaldo, S., O'Toole, P.W., Brugère, J.F. (2012). Genome sequence of "*Candidatus* Methanomethylophilus alvus" Mx1201, a methanogenic archaeon from the human gut belonging to a seventh order of methanogens. *Journal of Bacteriology*, 194(24), 6944–6945.

Borrel, G., Adam, P.S., Gribaldo, S. (2016). Methanogenesis and the Wood-Ljungdahl pathway: An ancient, versatile, and fragile association. *Genome Biology and Evolution*, 8(6), 1706–1711.

Borrel, G., Mccann, A., Deane, J., Neto, M.C., Lynch, D.B., Brugère, J.F., O'Toole, P.W. (2017). Genomics and metagenomics of trimethylamine-utilizing Archaea in the human gut microbiome. *ISME J.*, 11(9), 2059–2074.

Borrel, G., Adam, P.S., Mckay, L.J., Chen, L.-X., Sierra-Garcia, I.N., Sieber, C.M.K., Letourneur, Q., Ghozlane, A., Andersen, G.L., Li, W.-J. et al. (2019). Wide diversity of methane and short-chain alkane metabolisms in uncultured archaea. *Nat. Microbiol.*, 4(4), 603–613.

Borrel, G., Brugère, J.F., Gribaldo, S., Schmitz, R.A., Moissl-Eichinger, C. (2020). The host-associated archaeome. *Nat. Rev. Microbiol.*, 18(11), 622–636.

Bräuer, S.L., Cadillo-Quiroz, H., Yashiro, E., Yavitt, J.B., Zinder, S.H. (2006). Isolation of a novel acidiphilic methanogen from an acidic peat bog. *Nature*, 442(7099), 192–194.

Brugère, J.F., Borrel, G., Gaci, N., Tottey, W., O'Toole, P.W., Malpuech-Brugère, C. (2014). Archaebiotics: Proposed therapeutic use of archaea to prevent trimethylaminuria and cardiovascular disease. *Gut Microbes*, 5(1), 5–10.

Buchenau, B. and Thauer, R.K. (2004). Tetrahydrofolate-specific enzymes in *Methanosarcina barkeri* and growth dependence of this methanogenic archaeon on folic acid or *p*-aminobenzoic acid. *Arch. Microbiol.*, 182(4), 313–325.

Buckel, W. and Thauer, R.K. (2018). Flavin-based electron bifurcation, a new mechanism of biological energy coupling. *Chemical Reviews*, 118(7), 3862–3886.

Burggraf, S., Fricke, H., Neuner, A., Kristjansson, J., Rouvier, P., Mandelco, L., Woese, C.R., Stetter, K.O. (1990). *Methanococcus igneus* sp. nov., a novel hyperthermophilic methanogen from a shallow submarine hydrothermal system. *Systematic and Applied Microbiology*, 13(3), 263–269.

Cadillo-Quiroz, H., Yashiro, E., Yavitt, J.B., Zinder, S.H. (2008). Characterization of the archaeal community in a minerotrophic fen and terminal restriction fragment length polymorphism-directed isolation of a novel hydrogenotrophic methanogen. *Applied and Environmental Microbiology*, 74(7), 2059–2068.

Chapelle, F.H., O'Neill, K., Bradley, P.M., Methé, B.A., Ciufo, S.A., Knobel, L.L., Lovley, D.R. (2002). A hydrogen based subsurface microbial community dominated by methanogens. *Nature*, 415(6869), 312–315.

Chen, S.C., Musat, N., Lechtenfeld, O.J., Paschke, H., Schmidt, M., Said, N., Popp, D., Calabrese, F., Stryhanyuk, H., Jaekel, U. et al. (2019). Anaerobic oxidation of ethane by archaea from a marine hydrocarbon seep. *Nature*, 568(7750), 108–111.

Chistoserdova, L., Vorholt, J.A., Thauer, R.K., Lidstrom, M.E. (1998). C1 transfer enzymes and coenzymes linking methylotrophic bacteria and methanogenic Archaea. *Science*, 281(5373), 99–102.

Chong, S.C., Liu, Y.T., Cummins, M., Valentine, D.L., Boone, D.R. (2002). Methanogenium marinum sp nov., a H_2-using methanogen from Skan Bay, Alaska, and kinetics of H_2 utilization. *Antonie Van Leeuwenhoek International Journal of General and Molecular Microbiology*, 81(1–4), 263–270.

Conrad, R. (2009). The global methane cycle: Recent advances in understanding the microbial processes involved. *Environ Microbiol. Rep.*, 1(5), 285–292.

Costa, K.C., Wong, P.M., Wang, T., Lie, T.J., Dodsworth, J.A., Swanson, I., Burn, J.A., Hackett, M., Leigh, J.A. (2010). Protein complexing in a methanogen suggests electron bifurcation and electron delivery from formate to heterodisulfide reductase. *Proc. Natl. Acad. Sci. USA*, 107(24), 11050–11055.

Dridi, B., Fardeau, M.L., Ollivier, B., Raoult, D., Drancourt, M. (2012). *Methanomassiliicoccus luminyensis* gen. nov., sp nov., a methanogenic archaeon isolated from human faeces. *International Journal of Systematic and Evolutionary Microbiology*, 62, 1902–1907.

Duin, E.C., Wagner, T., Shima, S., Prakash, D., Cronin, B., Yanez-Ruiz, D.R., Duval, S., Rumbeli, R., Stemmler, R.T., Thauer, R.K., Kindermann, M. (2016). Mode of action uncovered for the specific reduction of methane emissions from ruminants by the small molecule 3-nitrooxypropanol. *Proc. Natl. Acad. Sci. USA*, 113(22), 6172–6177.

Ehhalt, D., Prather, M., Dentener, F., Derwent, R., Dlugokencky, E., Holland, E., Isaksen, I., Katima, J., Kirchhoff, V., Matson, P., Midgley, P., Wang, M. (2001). Atmospheric chemistry and greenhouse gases. In *Climate Change 2001: The Scientific Basis*, Houghton, J.T., Ding, Y., Griggs, D.J., Noguer, M., van der Linden, P.J., Dai, X., Maskell, K. and Johnson, C.A. (eds). Cambridge University Press, Cambridge.

Engilberge, S., Wagner, T., Carpentier, P., Girard, E., Shima, S. (2020). Krypton-derivatization highlights O_2-channeling in a four-electron reducing oxidase. *Chem. Commun. (Camb)*, 56(74), 10863–10866.

Ermler, U., Grabarse, W., Shima, S., Goubeaud, M., Thauer, R.K. (1997). Crystal structure of methyl-coenzyme M reductase: The key enzyme of biological methane formation. *Science*, 278(5342), 1457–1462.

Etiope, G. and Lollar, B.S. (2013). Abiotic methane on earth. *Reviews of Geophysics*, 51(2), 276–299.

Evans, D.A., Beukes, N.J., Kirschvink, J.L. (1997). Low-latitude glaciation in the Palaeoproterozoic era. *Nature*, 386(6622), 262–266.

Franzmann, P.D., Springer, N., Ludwig, W., De Macario, E.C., Rohde, M. (1992). A methanogenic archaeon from Ace Lake, Antarctica, *Methanococcoides burtonii* sp. nov. *Systematic and Applied Microbiology*, 15(4), 573–581.

Franzmann, P.D., Liu, Y.T., Balkwill, D.L., Aldrich, H.C., De Macario, E.C., Boone, D.R. (1997). *Methanogenium frigidum* sp. nov., a psychrophilic, H_2-using methanogen from Ace Lake, Antarctica. *International Journal of Systematic Bacteriology*, 47(4), 1068–1072.

Fricke, W.F., Seedorf, H., Henne, A., Krüer, M., Liesegang, H., Hedderich, R., Gottschalk, G., Thauer, R.K. (2006). The genome sequence of *Methanosphaera stadtmanae* reveals why this human intestinal archaeon is restricted to methanol and H_2 for methane formation and ATP synthesis. *J. Bacteriol.*, 188(2), 642–658.

Gorris, L.G., Voet, A.C., Van Der Drift, C. (1991). Structural characteristics of methanogenic cofactors in the non-methanogenic archaebacterium *Archaeoglobus fulgidus*. *Biofactors*, 3(1), 29–35.

Gottschalk, G. and Thauer, R.K. (2001). The Na^+-translocating methyltransferase complex from methanogenic archaea. *Biochim. Biophys. Acta.*, 1505(1), 28–36.

Guan, Y., Ngugi, D.K., Vinu, M., Blom, J., Alam, I., Guillot, S., Ferry, J.G., Stingl, U. (2019). Comparative genomics of the genus *Methanohalophilus*, including a newly isolated strain from Kebrit Deep in the Red Sea. *Frontiers in Microbiology*, 10(839), 1–11.

Gunsalus, R.P. and Wolfe, R.S. (1977). Stimulation of CO_2 reduction to methane by methyl-coenzyme M in extracts of *Methanobacterium*. *Biochem. Biophys. Res. Commun.*, 76(3), 790–795.

Hagemeier, C.H., Kruer, M., Thauer, R.K., Warkentin, E., Ermler, U. (2006). Insight into the mechanism of biological methanol activation based on the crystal structure of the methanol-cobalamin methyltransferase complex. *Proceedings of the National Academy of Sciences of the United States of America*, 103(50), 18917–18922.

Hahn, C.J., Laso-Pérez, R., Vulcano, F., Vaziourakis, K.M., Stokke, R., Steen, I.H., Teske, A., Boetius, A., Liebeke, M., Amann, R. et al. (2020). "*Candidatus* Ethanoperedens," a thermophilic genus of archaea mediating the anaerobic oxidation of ethane. *MBio*, 11(2), 1–18. doi: 10.1128/mbio.00600-20.

Hahn, C.J., Lemaire, O.N., Kahnt, J., Engilberge, S., Wegener, G., Wagner, T. (2021). Crystal structure of a key enzyme for anaerobic ethane activation. *Science*, 373(6550), 118–121.

Hedderich, R. and Whitman, W.B. (2006). Physiology and biochemistry of the methane-producing archaea. In *Prokaryotes: A Handbook on the Biology of Bacteria*. Springer, Berlin, Heidelberg.

Hemmann, J.L., Wagner, T., Shima, S., Vorholt, J.A. (2019). Methylofuran is a prosthetic group of the formyltransferase/hydrolase complex and shuttles one-carbon units between two active sites. *Proceedings of the National Academy of Sciences of the United States of America*, 116(51), 25583–25590.

Herzog, B. and Wirth, R. (2012). Swimming behavior of selected species of archaea. *Applied and Environmental Microbiology*, 78(6), 1670–1674.

Hoedt, E.C., Cuív, P.O., Evans, P.N., Smith, W.J., Mcsweeney, C.S., Denman, S.E., Morrison, M. (2016). Differences down-under: Alcohol-fueled methanogenesis by archaea present in Australian macropodids. *ISME J.*, 10(10), 2376–2388.

Hua, Z.S., Wang, Y.L., Evans, P.N. Qu, Y.-N., Goh, K.M., Rao, Y.-Z., Qi, Y.-L., Li, Y.-X., Huang, M.-J., Jiao, J.-Y. et al. (2019). Insights into the ecological roles and evolution of methyl-coenzyme M reductase-containing hot spring Archaea. *Nature Communications*, 10, 1–11.

Huang, G.F., Wagner, T., Ermler, U., Shima, S. (2020). Methanogenesis involves direct hydride transfer from H_2 to an organic substrate. *Nature Reviews Chemistry*, 4(4), 213–221.

Jeanthon, C., L'Haridon, S., Reysenbach, A.L., Vernet, M., Messner, P., Sleytr, U.B., Prieur, D. (1998). *Methanococcus infernus* sp. nov., a novel hyperthermophilic lithotrophic methanogen isolated from a deep-sea hydrothermal vent. *International Journal of Systematic Bacteriology*, 48, 913–919.

Jeanthon, C., L'Haridon, S., Reysenbach, A.L., Corre, E., Vernet, M., Messner, P., Sleytr, U.B., Prieur, D. (1999). *Methanococcus vulcanius* sp. nov., a novel hyperthermophilic methanogen isolated from East Pacific Rise, and identification of *Methanococcus* sp. DSM 4213(T) as *Methanococcus fervens* sp. nov. *International Journal of Systematic Bacteriology*, 49, 583–589.

Jones, W.J., Leigh, J.A., Mayer, F., Woese, C.R., Wolfe, R.S. (1983). *Methanococcus jannaschii sp. nov.*, an extremely thermophilic methanogen from a submarine hydrothermal vent. *Archives of Microbiology*, 136(4), 254–261.

Karrasch, M., Bott, M., Thauer, R.K. (1989). Carbonic anhydrase activity in acetate grown *Methanosarcina barkeri*. *Archives of Microbiology*, 151(2), 137–142.

Kaster, A.K., Goenrich, M., Seedorf, H., Liesegang, H., Wollherr, A., Gottschalk, G., Thauer, R.K. (2011a). More than 200 genes required for methane formation from H_2 and CO_2 and energy conservation are present in *Methanothermobacter marburgensis* and *Methanothermobacter thermautotrophicus*. *Archaea*, 973848.

Kaster, A.K., Moll, J., Parey, K., Thauer, R.K. (2011b). Coupling of ferredoxin and heterodisulfide reduction via electron bifurcation in hydrogenotrophic methanogenic archaea. *Proc. Natl. Acad. Sci. USA*, 108(7), 2981–2986.

Kelley, D.S., Karson, J.A., Fruh-Green, G.L., Yoerger, D.R., Shank, T.M., Butterfield, D.A., Hayes, J.M., Schrenk, M.O., Olson, E.J., Proskurowski, G. et al. (2005). A serpentinite-hosted ecosystem: The lost city hydrothermal field. *Science*, 307(5714), 1428–1434.

King, G.M. (1984). Utilization of hydrogen, acetate, and noncompetitive substrates by methanogenic bacteria in marine-sediments. *Geomicrobiology Journal*, 3(4), 275–306.

Konhauser, K.O., Pecoits, E., Lalonde, S.V., Papineau, D., Nisbet, E.G., Barley, M.E., Arndt, N.T., Zahnle, K., Kamber, B.S. (2009). Oceanic nickel depletion and a methanogen famine before the great oxidation event. *Geochimica et Cosmochimica Acta*, 73(13), A678–A678.

Kral, T.A., Goodhart, T.H., Harpool, J.D., Hearnsberger, C.E., Mccracken, G.L., Mcspadden, S.W. (2016). Sensitivity and adaptability of methanogens to perchlorates: Implications for life on Mars. *Planetary and Space Science*, 120, 87–95.

Kulkarni, G., Mand, T.D., Metcalf, W.W. (2018). Energy conservation via hydrogen cycling in the methanogenic archaeon *Methanosarcina barkeri*. *Mbio*, 9(4), 1–10.

Kurr, M., Huber, R., König, H., Jannasch, H.W., Fricke, H., Trincone, A., Kristjansson, J.K., Stetter, K.O. (1991). *Methanopyrus kandleri*, gen. and sp. nov. represents a novel group of hyperthermophilic methanogens, growing at 110°C. *Archives of Microbiology*, 156(4), 239–247.

Kurth, J.M., Op Den Camp, H.J.M., Welte, C.U. (2020). Several ways one goal-methanogenesis from unconventional substrates. *Appl. Microbiol. Biotechnol.*, 104(16), 6839–6854.

L'Haridon, S., Haroun, H., Corre, E., Roussel, E., Chalopin, M., Pignet, P., Balière, C., La Cono, V., Jebbar, M., Yakimov, M. et al. (2020). *Methanohalophilus profundi* sp. nov., a methylotrophic halophilic piezophilic methanogen isolated from a deep hypersaline anoxic basin. *Syst. Appl. Microbiol.*, 43(5), 126107.

Laso-Pérez, R., Wegener, G., Knittel, K., Widdel, F., Harding, K.J., Krukenberg, V., Meier, D.V., Richter, M., Tegetmeyer, H.E., Riedel, D. et al. (2016). Thermophilic archaea activate butane via alkyl-coenzyme M formation. *Nature*, 539(7629), 396–401.

Laso-Pérez, R., Hahn, C., Van Vliet, D.M., Tegetmeyer, H.E., Schubotz, F., Smit, N.T., Pape, T., Sahling, H., Bohrmann, G., Boetius, A. et al. (2019). Anaerobic degradation of non-methane alkanes by "*Candidatus methanoliparia*" in hydrocarbon seeps of the Gulf of Mexico. *Mbio*, 10(4), 1–20.

Lauerer, G., Kristjansson, J.K., Langworthy, T.A., König, H., Stetter, K.O. (1986). *Methanothermus sociabilis* sp. nov., a second species within the *Methanothermaceae* growing at 97°C. *Systematic and Applied Microbiology*, 8(1/2), 100–105.

Lind, A.E., Lewis, W.H., Spang, A., Guy, L., Embley, T.M., Ettema, T.J.G. (2018). Genomes of two archaeal endosymbionts show convergent adaptations to an intracellular lifestyle. *ISME Journal*, 12(11), 2655–2667.

Liu, Y. and Whitman, W.B. (2008). Metabolic, phylogenetic, and ecological diversity of the methanogenic archaea. *Incredible Anaerobes: From Physiology to Genomics to Fuels*, 1125, 171–189.

Liu, Y., Boone, D.R., Choy, C. (1990). *Methanohalophilus oregonense* sp. nov., a methylotrophic methanogen from an alkaline, saline aquifer. *International Journal of Systematic Bacteriology*, 40(2), 111–116.

Lomans, B.P., Maas, R., Luderer, R., Den Camp, H.J.M.O., Pol, A., Van Der Drift, C., Vogels, G.D. (1999). Isolation and characterization of *Methanomethylovorans hollandica* gen. nov., sp. nov., isolated from freshwater sediment, a methylotrophic methanogen able to grow on dimethyl sulfide and methanethiol. *Applied and Environmental Microbiology*, 65(8), 3641–3650.

Lovley, D.R. and Klug, M.J. (1983). Sulfate reducers can out-compete methanogens at fresh-water sulfate concentrations. *Applied and Environmental Microbiology*, 45(1), 187–192.

Lyu, Z. and Lu, Y.H. (2018). Metabolic shift at the class level sheds light on adaptation of methanogens to oxidative environments. *ISME Journal*, 12(2), 411–423.

Lyu, Z., Shao, N.N., Chou, C.W., Shi, H., Patel, R., Duin, E.C., Whitman, W.B. (2020). Posttranslational methylation of arginine in methyl coenzyme M reductase has a profound impact on both methanogenesis and growth of *Methanococcus maripaludis*. *Journal of Bacteriology*, 202(3), 1–18.

Mackay, R.J., Mcentyre, C.J., Henderson, C., Lever, M., George, P.M. (2011). Trimethylaminuria: Causes and diagnosis of a socially distressing condition. *Clin. Biochem. Rev.*, 32(1), 33–43.

Mayumi, D., Mochimaru, H., Tamaki, H., Yamamoto, K., Yoshioka, H., Suzuki, Y., Kamagata, Y., Sakata, S. (2016). Methane production from coal by a single methanogen. *Science*, 354(6309), 222–225.

McGenity, T.J. (2010). Methanogens and methanogenesis in hypersaline environments. In *Handbook of Hydrocarbon and Lipid Microbiology*, Timmis, K.N. (eds). Springer, Berlin, Heidelberg.

McGlynn, S.E. (2017). Energy metabolism during anaerobic methane oxidation in ANME archaea. *Microbes Environ.*, 32(1), 5–13.

McGlynn, S.E., Chadwick, G.L., Kempes, C.P., Orphan, V.J. (2015). Single cell activity reveals direct electron transfer in methanotrophic consortia. *Nature*, 526(7574), 531–U146.

Mihajlovski, A., Alric, M., Brugère, J.F. (2008). A putative new order of methanogenic Archaea inhabiting the human gut, as revealed by molecular analyses of the mcrA gene. *Research in Microbiology*, 159(7/8), 516–521.

Miller, T.L. and Wolin, M.J. (1985). *Methanosphaera stadtmaniae* gen. nov.: A species that forms methane by reducing methanol with hydrogen. *Archives of Microbiology*, 141(2), 116–122.

Miller, J.F., Shah, N.N., Nelson, C.M., Ludlow, J.M., Clark, D.S. (1988). Pressure and temperature effects on growth and methane production of the extreme thermophile *Methanococcus jannaschii*. *Appl Environ Microbiol*, 54(12), 3039–3042.

Moletta, M., Wery, N., Delgenes, J.P., Godon, J.J. (2008). Microbial characteristics of biogas. *Water Science and Technology*, 57(4), 595–599.

Morozova, D., Möhlmann, D., Wagner, D. (2007). Survival of methanogenic archaea from Siberian permafrost under simulated Martian thermal conditions. *Orig. Life Evol. Biosph.*, 37(2), 189–200.

Mountfort, D.O. and Asher, R.A. (1981). Role of sulfate reduction versus methanogenesis in terminal carbon flow in polluted intertidal sediment of Waimea Inlet, Nelson, New Zealand. *Applied and Environmental Microbiology*, 42(2), 252–258.

Nayak, D.D., Liu, A.D., Agrawal, N., Rodriguez-Carerro, R., Dong, S.H., Mitchell, D.A., Nair, S.K., Metcalf, W.W. (2020). Functional interactions between posttranslationally modified amino acids of methyl-coenzyme M reductase in *Methanosarcina acetivorans*. *PLos Biology*, 18(2), 1–23.

Nelson-Sathi, S., Dagan, T., Landan, G., Janssen, A., Steel, M., Mcinerney, J.O., Deppenmeier, U., Martin, W.F. (2012). Acquisition of 1,000 eubacterial genes physiologically transformed a methanogen at the origin of Haloarchaea. *Proceedings of the National Academy of Sciences of the United States of America*, 109(50), 20537–20542.

Nobu, M.K., Narihiro, T., Kuroda, K., Mei, R., Liu, W.T. (2016). Chasing the elusive *Euryarchaeota* class WSA2: Genomes reveal a uniquely fastidious methyl-reducing methanogen. *ISME Journal*, 10(10), 2478–2487.

Ollivier, B.M., Mah, R.A., Garcia, J.L., Robinson, R. (1985). Isolation and characterization of *Methanogenium aggregans* sp. nov. *International Journal of Systematic Bacteriology*, 35(2), 127–130.

Oremland, R.S. and Boone, D.R. (1994). *Methanolobus taylorii* sp. nov., a new methylotrophic, estuarine methanogen. *International Journal of Systematic Bacteriology*, 44(3), 573–575.

Oremland, R.S. and Polcin, S. (1982). Methanogenesis and sulfate reduction – Competitive and noncompetitive substrates in estuarine sediments. *Applied and Environmental Microbiology*, 44(6), 1270–1276.

Patel, G.B., Sprott, G.D., Fein, J.E. (1990). Isolation and characterization of *Methanobacterium espanolae* sp. nov., a mesophilic, moderately acidiphilic methanogen. *International Journal of Systematic Bacteriology*, 40(1), 12–18.

Paterek, J.R. and Smith, P.H. (1988). *Methanohalophilus mahii* gen. nov., sp. nov., a methylotrophic halophilic methanogen. *International Journal of Systematic Bacteriology*, 38(1), 122–123.

Paul, K., Nonoh, J.O., Mikulski, L., Brune, A. (2012). "*Methanoplasmatales*," thermoplasmatales-related archaea in termite guts and other environments, are the seventh order of methanogens. *Applied and Environmental Microbiology*, 78(23), 8245–8253.

Poulsen, M., Schwab, C., Jensen, B.B., Engberg, R.M., Spang, A., Canibe, N., Højberg, O., Milinovich, G., Fragner, L., Schleper, C. et al. (2013). Methylotrophic methanogenic Thermoplasmata implicated in reduced methane emissions from bovine rumen. *Nature Communications*, 4, 1428.

Raemakers-Franken, P.C., Voncken, F.G., Korteland, J., Keltjens, J.T., Van Der Drift, C., Vogels, G.D. (1989). Structural characterization of tatiopterin, a novel pterin isolated from *Methanogenium tationis*. *Biofactors*, 2(2), 117–122.

Reeburgh, W.S. (2007). Oceanic methane biogeochemistry. *Chemical Reviews*, 107(2), 486–513.

Robinson, J.A. and Tiedje, J.M. (1984). Competition between sulfate-reducing and methanogenic bacteria for H_2 under resting and growing conditions. *Archives of Microbiology*, 137(1), 26–32.

Sa, D. (2005). *Mémento technique de l'eau*, volume 1 and 2. Édition Degrémont SUEZ, Rueil-Malmaison.

Sakai, S., Imachi, H., Hanada, S., Ohashi, A., Harada, H., Kamagata, Y. (2008). *Methanocella paludicola* gen. nov., sp. nov., a methane-producing archaeon, the first isolate of the lineage 'Rice Cluster I', and proposal of the new archaeal order *Methanocellales* ord. nov. *International Journal of Systematic and Evolutionary Microbiology*, 58, 929–936.

Schink, B. (1997). Energetics of syntrophic cooperation in methanogenic degradation. *Microbiology and Molecular Biology Reviews*, 61(2), 262.

Serrano-Silva, N., Sarria-Guzmán, Y., Dendooven, L., Luna-Guido, M. (2014). Methanogenesis and methanotrophy in soil: A review. *Pedosphere*, 24(3), 291–307.

Shima, S., Huang, G., Wagner, T., Ermler, U. (2020). Structural basis of hydrogenotrophic methanogenesis. *Annu. Rev. Microbiol.*, 74, 713–733.

Simankova, M.V., Parshina, S.N., Tourova, T.P., Kolganova, T.V., Zehnder, A.J.B., Nozhevnikova, A.N. (2001). *Methanosarcina lacustris* sp. nov., a new psychrotolerant methanogenic archaeon from anoxic lake sediments. *Systematic and Applied Microbiology*, 24(3), 362–367.

Smith, K.S. and Ferry, J.G. (2000). Prokaryotic carbonic anhydrases. *FEMS Microbiol. Rev.*, 24(4), 335–366.

Sollinger, A. and Urich, T. (2019). Methylotrophic methanogens everywhere – Physiology and ecology of novel players in global methane cycling. *Biochem. Soc. Trans.*, 47(6), 1895–1907.

Sorokin, D.Y., Makarova, K.S., Abbas, B., Ferrer, M., Golyshin, P.N., Galinski, E.A., Ciordia, S., Mena, M.C., Merkel, A.Y., Wolf, Y.I. et al. (2017). Discovery of extremely halophilic, methyl-reducing euryarchaea provides insights into the evolutionary origin of methanogenesis. *Nature Microbiology*, 2(8), 1–11.

Sorokin, D.Y., Merkel, A.Y., Abbas, B., Makarova, K.S., Rijpstra, W.I.C., Koenen, M., Sinninghe Damsté, J.S., Galinski, E.A., Koonin, E.V., Van Loosdrecht, M.C.M. (2018). *Methanonatronarchaeum thermophilum* gen. nov., sp. nov. and '*Candidatus* Methanohalarchaeum thermophilum', extremely halo(natrono)philic methyl-reducing methanogens from hypersaline lakes comprising a new euryarchaeal class *Methanonatronarchaeia* classis nov. *Int. J. Syst. Evol. Microbiol.*, 68(7), 2199–2208.

Sprenger, W.W., Van Belzen, M.C., Rosenberg, J., Hackstein, J.H., Keltjens, J.T. (2000). *Methanomicrococcus blatticola* gen. nov., sp. nov., a methanol- and methylamine-reducing methanogen from the hindgut of the cockroach *Periplaneta americana*. *Int. J. Syst. Evol. Microbiol.*, 50(6), 1989–1999.

Stetter, K.O., Thomm, M., Winter, J., Wildgruber, G., Huber, H., Zillig, W., Jané-Covic, D., König, H., Palm, P., Wunderl, S. (1981). *Methanothermus fervidus*, sp. nov., a novel extremely thermophilic methanogen isolated from an Icelandic hot spring. *Zentralblatt für Bakteriologie Mikrobiologie und Hygiene I Abteilung Originale C-Allgemeine Angewandte Und Okologische Mikrobiologie*, 2(2), 166–178.

Suharti, S., Wang, M., De Vries, S., Ferry, J.G. (2014). Characterization of the RnfB and RnfG subunits of the Rnf complex from the archaeon *Methanosarcina acetivorans*. *PLoS One*, 9(5), e97966.

Takai, K., Nakamura, K., Toki, T., Tsunogai, U., Miyazaki, M., Miyazaki, J., Hirayama, H., Nakagawa, S., Nunoura, T., Horikoshi, K. (2008). Cell proliferation at 122°C and isotopically heavy CH_4 production by a hyperthermophilic methanogen under high-pressure cultivation. *Proceedings of the National Academy of Sciences of the United States of America*, 105(31), 10949–10954.

Tang, W.H.W., Wang, Z.N., Kennedy, D.J., Wu, Y.P., Buffa, J.A., Agatisa-Boyle, B., Li, X.M.S., Levison, B.S., Hazen, S.L. (2015). Gut microbiota-dependent trimethylamine N-oxide (TMAO) pathway contributes to both development of renal insufficiency and mortality risk in chronic kidney disease. *Circulation Research*, 116(3), 448–455.

Taubner, R.S., Pappenreiter, P., Zwicker, J., Smrzka, D., Pruckner, C., Kolar, P., Bernacchi, S., Seifert, A.H., Krajete, A., Bach, W. et al. (2018). Biological methane production under putative Enceladus-like conditions. *Nat. Commun.*, 9(1), 748.

Thauer, R.K. (2012). The Wolfe cycle comes full circle. *Proc. Natl. Acad. Sci. USA*, 109(38), 15084–15085.

Thauer, R.K. (2019). Methyl (alkyl)-coenzyme M reductases: Nickel F-430-containing enzymes involved in anaerobic methane formation and in anaerobic oxidation of methane or of short chain alkanes. *Biochemistry*, 58(52), 5198–5220.

Thauer, R.K., Kaster, A.K., Seedorf, H., Buckel, W., Hedderich, R. (2008). Methanogenic archaea: Ecologically relevant differences in energy conservation. *Nat. Rev. Microbiol.*, 6(8), 579–591.

Ueno, Y., Yamada, K., Yoshida, N., Maruyama, S., Isozaki, Y. (2006). Evidence from fluid inclusions for microbial methanogenesis in the early Archaean era. *Nature*, 440(7083), 516–519.

Van De Wijngaard, W.M., Creemers, J., Vogels, G.D., Van Der Drift, C. (1991). Methanogenic pathways in *Methanosphaera stadtmanae*. *FEMS Microbiol. Lett.*, 64(2/3), 207–211.

Vanwonterghem, I., Evans, P.N., Parks, D.H., Jensen, P.D., Woodcroft, B.J., Hugenholtz, P., Tyson, G.W. (2016). Methylotrophic methanogenesis discovered in the archaeal phylum *Verstraetearchaeota*. *Nat. Microbiol.*, 1, 16170.

Vigneron, A., L'Haridon, S., Godfroy, A., Roussel, E.G., Cragg, B.A., Parkes, R.J., Toffin, L. (2015). Evidence of active methanogen communities in shallow sediments of the Sonora Margin cold seeps. *Applied and Environmental Microbiology*, 81(10), 3451–3459.

Von Klein, D., Arab, H., Völker, H., Thomm, M. (2002). *Methanosarcina baltica*, sp. nov., a novel methanogen isolated from the Gotland Deep of the Baltic Sea. *Extremophiles*, 6(2), 103–110.

Wagner, T., Ermler, U., Shima, S. (2016). The methanogenic CO_2 reducing-and-fixing enzyme is bifunctional and contains 46 [4Fe-4S] clusters. *Science*, 354(6308), 114–117.

Wagner, T., Koch, J., Ermler, U., Shima, S. (2017a). Methanogenic heterodisulfide reductase (HdrABC-MvhAGD) uses two noncubane [4Fe-4S] clusters for reduction. *Science*, 357(6352), 699–702.

Wagner, T., Wegner, C.E., Kahnt, J., Ermler, U., Shima, S. (2017b). Phylogenetic and structural comparisons of the three types of methyl coenzyme M reductase from *Methanococcales* and *Methanobacteriales*. *J. Bacteriol.*, 199(16), 1–15.

Wagner, T., Watanabe, T., Shima, S. (2018). Hydrogenotrophic methanogenesis. In *Biogenesis of Hydrocarbons*, Stams, A.J.M. and Sousa, D.Z. (eds). Springer, Berlin.

Waite, J.H., Glein, C.R., Perryman, R.S., Teolis, B.D., Magee, B.A., Miller, G., Grimes, J., Perry, M.E., Miller, K.E., Bouquet, A. et al. (2017). Cassini finds molecular hydrogen in the Enceladus plume: Evidence for hydrothermal processes. *Science*, 356(6334), 155–159.

Wegener, G., Krukenberg, V., Riedel, D., Tegetmeyer, H.E., Boetius, A. (2015). Intercellular wiring enables electron transfer between methanotrophic archaea and bacteria. *Nature*, 526(7574), 587–U315.

Weinstein, A., Navarrete, L., Ruppel, C., Weber, T.C., Leonte, M., Kellermann, M.Y., Arrington, E.C., Valentine, D.L., Scranton, M.I., Kessler, J.D. (2016). Determining the flux of methane into Hudson Canyon at the edge of methane clathrate hydrate stability. *Geochemistry Geophysics Geosystems*, 17(10), 3882–3892.

Welte, C. and Deppenmeier, U. (2014). Bioenergetics and anaerobic respiratory chains of aceticlastic methanogens. *Biochim. Biophys. Acta*, 1837(7), 1130–1147.

Wongnate, T., Sliwa, D., Ginovska, B., Smith, D., Wolf, M.W., Lehnert, N., Raugei, S., Ragsdale, S.W. (2016). The radical mechanism of biological methane synthesis by methyl-coenzyme M reductase. *Science*, 352(6288), 953–958.

Zhilina, T.N. (1983). New obligate halophilic methane-producing bacterium. *Microbiology*, 52(3), 290–297.

Zhilina, T.N. and Zavarzin, G.A. (1987). *Methanohalobium evestigatus* gen-nov, sp-nov, extremely halophilic methane forming archebacterium. *Doklady Akademii Nauk SSSR*, 293(2), 464–468.

Zhou, Z., Zhang, C.J., Liu, P.F., Fu, L., Laso-Pérez, R., Yang, L., Bai, L.P., Li, J., Yang, M., Lin, J.Z. et al. (2022). Non-syntrophic methanogenic hydrocarbon degradation by an archaeal species. *Nature*, 601(7892), 257–262.

Zhu, W.F., Gregory, J.C., Org, E., Buffa, J.A., Gupta, N., Wang, Z., Li, L., Fu, X., Wu, Y., Mehrabian, M. et al. (2016). Gut microbial metabolite TMAO enhances platelet hyperreactivity and thrombosis risk. *Cell*, 165(1), 111–124.

6

Hyperthermophilic Archaeal Viruses

Diana BAQUERO[1], Mart KRUPOVIC[1], Claire GESLIN[2] and David PRANGISHVILI[1,3]

[1] *Institut Pasteur, Université Paris Cité, France*

[2] *Univ Brest, Ifremer, BEEP, F-29280, Plouzané, France*

[3] *Ivane Javakhishvili TSU, Tbilisi, Georgia*

6.1. Introduction

Extreme thermal environments represent a habitat for viruses with remarkable morphotypical diversity and unique genome content, many of which have not been described in moderate temperature environments. The viruses which have been isolated from terrestrial and marine hydrothermal environments with temperatures above 75°C infect hyperthemophilic archaea and can be considered as hyperthermophiles themselves. The hyperthermophilic hosts are members of the phylum *Euryarchaeota*, in particular the genera *Thermococcus* and *Pyrococcus*, and members of the phylum *Crenarchaeota*, in particular the genera *Acidianus*, *Aeropyrum*, *Metallosphaera*, *Pyrobaculum*, *Stygiolobus*, *Sulfolobus* and *Thermoproteus*. The hyperthermophilic viruses represent an integral but unique part of the virosphere, with many members displaying exotic morphologies resembling bottles, droplets, lemons, etc. Along with the extraordinary diversity of morphotypes, another outstanding feature of archaeal viruses is the uniqueness of their genome content, with ~75% of the genes lacking detectable homologues in sequence databases (Krupovic et al. 2018). Based on their virion morphologies and genomic properties, the characterized hyperthermophilic archaeal viruses are

Biology of Archaea 1,
coordinated by Béatrice CLOUET-D'ORVAL, Bruno FRANZETTI and Philippe OGER.

classified currently into 13 families, but several virus species, which likely represent new virus families, remain unclassified (see Figure 6.1).

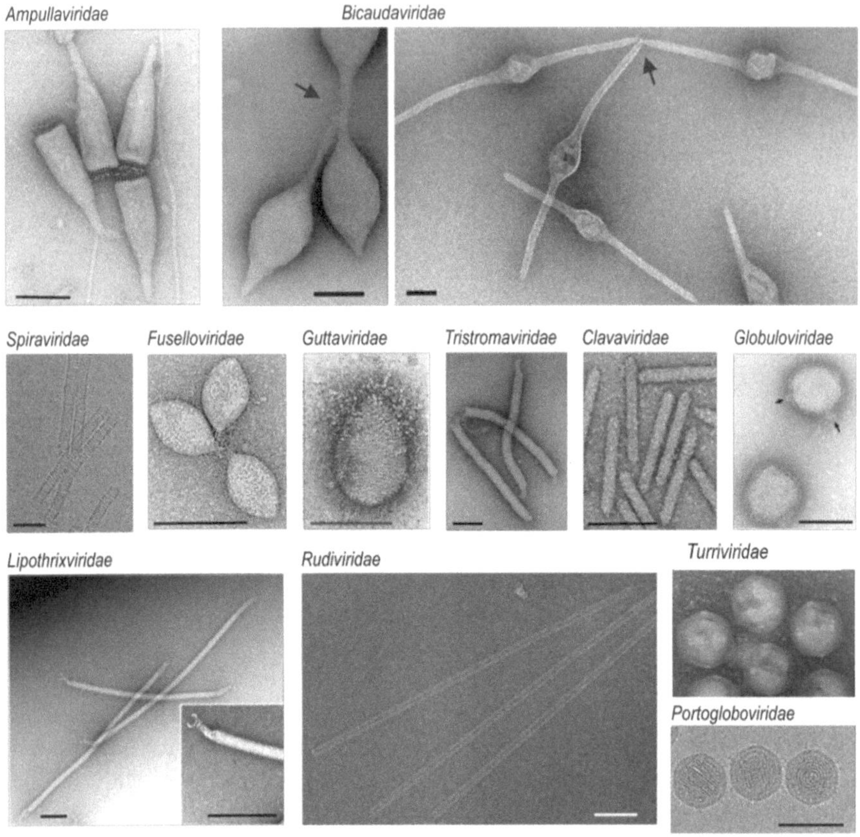

Figure 6.1. *Electron micrographs of hyperthermophilic archaeal viruses. Image modified from Prangishvili et al. (2017)*

6.2. Morphological and structural diversity

6.2.1. *Viruses with unique morphologies: families Ampullaviridae, Spiraviridae and Guttaviridae*

Certain families of hyperthermophilic viruses comprise members with structural features which are completely atypical among viruses infecting bacteria or eukaryotes. Virions of Acidianus bottled-shaped virus (ABV), the sole isolated member of the

Ampullaviridae family, exhibit a distinctive and exceptional bottle-shaped morphology (Figure 6.1) (Häring et al. 2005a; Prangishvili and Krupovic 2018). The virions display a complex architecture in which the linear dsDNA genome is condensed into a cone-shaped nucleocapsid – due to binding of multiple copies of the major capsid protein (MCP) – that is further encased in a lipid-containing envelope. The bottle-shaped virions possess a broad end with short, thick filaments inserted into a structural disc and a second pointed end that is involved in attachment to host cells. The virus ABV replicates in members of the genus *Acidianus* order *Sulfolobales*, phylum *Crenarchaeaota*). Notably, metagenomic analysis has revealed the presence of members of the *Ampullaviridae* family in several distantly located geothermal environments (Gudbergsdóttir et al. 2016).

The Aeropyrum coil-shaped virus (ACV), the sole member of the family *Spiraviridae,* has non-enveloped, cylindrical virions and infects hyperthermophilic archaea from the genus *Aeropyrum* (order *Desulfurococcales*, phylum *Crenarchaeaota*) (Mochizuki et al. 2012; Prangishvili et al. 2020). The virions are formed by the condensation of the circular ssDNA genome and capsid proteins into a rope-like structure that is further condensed into a spring-like coil. Virions of ACV carry single appendages protruding from both termini of the helical structure.

The *Guttaviridae* family includes *Sulfolobus neozealandicus* droplet-shaped virus (SNDV) and *Aeropyrum pernix* ovoid virus 1 (APOV1), infecting hyperthermophilic archaea from the crenarchaeal orders *Sulfolobales* and *Desulfurococcales*, respectively (Arnold et al. 2000a; Mochizuki et al. 2011; Prangishvili et al. 2018). Guttaviruses have slightly pleomorphic enveloped virions with circular dsDNA genomes. The droplet-shaped virions of SNDV carry a bunch of thin fibers attached to the pointed end of the particles (Figure 6.1). Conversely, the ovoid virions of APOV1 do not exhibit attached fibers.

6.2.2. *Filamentous viruses: families Rudiviridae, Lipothrixviridae, Tristromaviridae and Clavaviridae*

Unlike bacterial and eukaryotic filamentous viruses that have ssDNA and ssRNA genomes, filamentous archaeal viruses possess dsDNA genomes. They have been grouped into the families *Rudiviridae*, *Lipothrixviridae*, *Tristromaviridae* and *Clavaviridae*, and infect hyperthermophilic hosts from the orders *Sulfolobales*, *Desulfurococcales* and *Thermoproteales*, of the phylum *Crenarchaeota*. Members of the *Lipothrixviridae and Tristromaviridae* families are enveloped, whereas members of the other two families do not contain a lipid envelope.

The stiff, rod-shaped virions of the *Rudiviridae* are formed by multiple copies of a homodimer of the MCP and linear dsDNA (Prangishvili et al. 1999). Multiple copies of the MCP dimer wrap around DNA, condense it into a hollow helical tube and maintain it in an A-form, as revealed by cryo-electron microscopy reconstruction of the virion of *Sulfolobus* rod-shaped virus 2 (SIRV2) (DiMaio et al. 2015). At each end, the virions are decorated with three terminal fibers that mediate the interaction of the viral particles with the host cell (Quemin et al. 2013).

Viruses from the *Lipothrixviridae* family have flexible virions decorated at each terminus with thin filaments, varying in number and arrangement for different species; exceptionally, the termini of *Acidianus* filamentous virus 1 (AFV1) are decorated with claw-like structures (Figure 6.1), which fold upon interaction with host cell appendages (Arnold et al. 2000b; Bettstetter et al. 2003; Häring et al. 2005b; Vestergaard et al. 2008). The virion nucleoprotein core is built from multiple copies of the two MCP, which are similar in structure to each other as well as to the single MCP of the *Rudiviridae*. Same as in the latter case, the structural unit is a dimer of the MCPs, multiple copies of which bind linear dsDNA, condense it into a helical tube and maintain it in an A-form. However, in this case, the helical nucleocapsid is covered by a membrane containing lipids selectively acquired from the pool of host lipids. Specifically, those glycerol dibiphytanyl glycerol tetraether (GDGT) lipids are selected, which do not have cyclopentane rings and thus are rather flexible and can assume an extraordinary U-shaped, horseshoe conformation (Kasson et al. 2017). Due to the pronounced similarities in their structural and genomic properties, which suggest relatively close evolutionary relationships, the families *Rudiviridae* and *Lipothrixiridae* have been unified in the order *Ligamenvirales* (Prangishvili and Krupovic 2012).

Members of the *Tristromaviridae* have flexible filamentous virions, which, as in the case of the virions of *Liporthixviridae*, consist of an envelope and an inner helical core built from linear dsDNA and two MCPs. However, here, the contact between these two structural units is mediated by a protein matrix, consisting of the third type of the MCP (Rensen et al. 2016). Cryo-EM reconstruction of the virion of Pyrobaculum filamentous virus 2 (PFV2) has revealed that the structure of the helical nucleoprotein is similar in all three families of the filamentous dsDNA viruses, *Rudiviridae, Lipothrixviridae* and *Tristromaviridae*: the hollow, helical nucleoprotein is formed by the MCP dimers, which wrap around the linear dsDNA genome, condensing it and maintaining in an A-form. Moreover, MCPs with four helix bundle folds are highly similar in structure in these three families, despite lack of any significant sequence similarity, and apparently share a common origin (Wang et al. 2020). Due to the similar DNA topology and the homology of the MCPs, it was proposed to unify the family *Tristromaviridae* together with the order

Ligamenvirales into the first class-rank taxon for archaeal viruses, "*Tokiviricetes*" (Wang et al. 2020).

Aeropyrum pernix bacilliform virus 1 (APBV1), the single known member of the family *Clavaviridae,* has non-enveloped bacilliform virions. The two asymmetric ends, one pointed and the other rounded, most likely play a role in DNA packaging and delivery. The virions are built from circular dsDNA and multiple copies of a single MCP (Mochizuki et al. 2010). Cryo-EM tomography reconstruction has revealed that the DNA is tightly packed in the capsid as a left-handed superhelix and held in place by the interactions with positively charged residues of the MCP. The MCP carries two α-helices connected by a β-hairpin and is heavily glycosylated. Tight packaging of the MCP molecules is ensured by extended hydrophobic interfaces, and likely contributes to the extreme thermostability of the capsid. The double-stranded DNA is tightly packed in the capsid as a left-handed superhelix and held in place by the interactions with positively charged residues of the MCP (Ptchelkine et al. 2017).

6.2.3. *Spherical and icosahedral viruses: families Globuloviridae, Ovaliviridae, Portogloboviridae and Turriviridae*

Members of the three families *Turriviridae, Portogloboviridae* and *Ovaliviridae* infect hyperthermophilic acidophilic hosts from the order *Sulfolobales*, whereas members of the family *Globuloviridae* infect neutrophilic hyperthermophiles of the order *Thermoproteales*, both in the phylum *Crenarchaeota*. Although portogloboviruses, ovaliviruses and globuloviruses carry their dsDNA genomes in a form of helical nucleocapsid, members of *Turriviridae* package their dsDNA in a naked form.

The spherical virions of the *Globuloviridae* are covered with a lipid-containing envelope, which encases a superhelical core consisting of linear dsDNA and multiple copies of the MCPs. Multiple protrusions, spherical in their appearance, are found on the virion surface; however, their role remains unknown (Ahn et al. 2006; Häring et al. 2004; Baquero et al. 2020). Observation of partially disrupted virions of Pyrobaculum spherical virus (PSV) suggested that there is no regular arrangement of the superhelical nucleoprotein filament inside the virion. Exceptionally, the newly characterized *Thermoproteus* spherical piliferous virus 1 (TSPV1) has numerous unusual filaments decorating its surface, which can extend tens of micrometres from the virion and are presumed to be involved in host interactions (Hartman et al. 2020).

In contrast to the globuloviruses, the virions of the sole member of the family *Ovaliviridae*, *Sulfolobus* ellipsoid virus 1 (SEV1), are formed by nucleoprotein filaments that are regularly arranged. The circular dsDNA is covered by multiple copies of the MCPs and condensed into a unique spherical nucleoprotein coil, which is surrounded by a lipid membrane decorated with protruding spikes. Notably, the viral DNA wraps around the longitudinal axis of the virion to form a multilayered disk-like structure with a central hole; sixteen of these structures stack together to form an unusual spool-like capsid (Wang et al. 2018).

Even more complex is the structure of virions of *Sulfolobus* polyhedral virus 1 (SPV1), the sole member of the *Portogloboviridae* family (Liu et al. 2017). The arrangement of the core, similar to that of ovaliviruses, represents a spherical nucleoprotein coil surrounded by a lipid membrane. However, in this case, the virion is further encased by an outer icosahedral protein shell. Cryo-EM reconstruction at near-atomic resolution revealed that the shell is built from two capsid proteins with distinct single jelly-roll fold. The minor capsid protein forms pentamers and the MCP forms hexamers that assemble into an icosahedral shell. The virion reconstruction revealed that also in the portogloboviruses, the dsDNA is in an A-form, as is the case in virions from the families *Rudiviridae, Lipothrixviridae* and *Tristromaviridae.* The observation of A-form DNA in filamentous and spherical virions from different families strongly suggests that A-DNA may be the prevalent storage form of DNA in extreme geothermal environments. The detailed analysis of a SPV1 virion structure suggested a most unusual model of morphogenesis: the circular dsDNA is folded as a long linear double filament with loops at each end, which is eventually spooled into concentric shells, giving rise to the globular structure of the core (Wang et al. 2019). The fact that the circular dsDNA in SPV1 is fully covered with a viral protein facilitates such self-assembly process, due to attractive interactions between layers of the concentric shells of the globular core.

Virions of the members of the *Turriviridae* are similar in their overall design to the virions of the *Portogloboviridae*, that is, these are icosahedral particles with spikes protruding from the vertices of the polyhedron and an internal lipid membrane layer (Rice et al. 2004; Happonen et al. 2010; Veesler et al. 2013). However, there is a major difference between these two types of virions in the ways of genome packaging. By contrast to the portogloboviruses, and in fact to all other hyperthemophilic viruses, the DNA is present in a naked state as in icosahedral bacterial viruses with an inner lipid layer from the families *Tectiviridae* and *Corticovirida*e. The structural similarity between these bacterial and archaeal viruses is extended to the structures of the MCPs, which carry the double-jelly roll fold.

Moreover, due to the repulsion between adjacent layers of DNA, the naked DNA is packed under high pressure and requires a special energy-dependent molecular machinery, the genome packaging ATPase, which is homologous in the turriviruses and bacterial tecti- and corticoviruses.

6.2.4. *Fusiform viruses: families Fuselloviridae and Bicaudoviridae*

Spindle-shaped viruses are widespread in archaea-dominated terrestrial and marine environments. The isolated and characterized hyperthermophilic archaea viruses with spindle-shaped morphology have been assigned to two families *Fuselloviridae* and *Bicaudaviridae.* While fuselloviruses have tailless lemon-shaped virions decorated with terminal fibers at one of the two pointed ends, bicaudaviruses develop tails of different lengths at one or both ends of the spindle-shaped particles. Despite the shared morphology, the two families are evolutionarily unrelated, and their virions are constructed from unrelated MCPs, suggesting convergent evolution (Krupovic et al. 2014).

Known members of the *Fuselloviridae* isolated from terrestrial hot springs, for example, the most extensively studied representative of the family, *Sulfolobus* spindle-shaped virus 1 (SSV1), infect members of the order *Sulfolobales*, phylum *Crenarchaeota* (Palm et al. 1991; Schleper et al. 1992; Stedman et al. 2003; Redder et al. 2009; Zhang et al. 2020). Two unclassified viruses, *Pyrococcus abyssi* virus 1 (PAV1) and *Thermococcus prieurii* virus 1 (TPV1), morphologically resembling fuselloviruses, have been isolated from *deep-sea hydrothermal vents* (Geslin et al. 2003, 2007; Gorlas et al. 2012). These viruses infect members of the order *Thermococcales*, phylum *Euryarchaeota.* Notably, despite morphological resemblance, the genomes of PAV1 and TPV1 are not recognizably similar to those of fuselloviruses. Accordingly, the two viruses are likely to be classified into a separate family in the future.

Members of the family *Bicaudaviridae* also infect hosts from the order *Sulfolobales*. Notably, the bicaudavirus *Acidianus* two-tailed virus (ATV) was shown to develop tails extracellularly, after extrusion from the host cell, in conditions close to those in natural habitat, that is, at temperatures exceeding 80°C (Häring et al. 2005a; Prangishvili et al. 2006). Putative relatives, which share a considerable fraction of genes with the bicaudavirus ATV, have been described, but remain unclassified. Some of them also develop one or two tails extracellularly (Uldahl et al. 2016).

6.3. Genomic features of hyperthermophilic archaeal viruses

6.3.1. *Genome content*

All characterized hyperthermophilic archaeal viruses possess DNA genomes, which can be either ssDNA or dsDNA, linear or circular. The average genome sizes are rather small (20–30 kb) with the clavavirus APBV1 having one of the smallest known dsDNA genomes (5.5 kb). Notably, however, the sole representative of the *Spiraviridae* family, ACV, possesses one of the largest genomes (24.9 kb) among currently described ssDNA genomes. Viruses with linear genomes use different strategies to protect and replicate the genome termini, such as covalently closed hairpins, covalently attached terminal proteins and terminal inverted repeats (TIRs) (Prangishvili et al. 2017; Krupovic et al. 2018). Although no RNA viruses infecting archaea have been isolated, the existence of such viruses has been postulated based on metagenomic sequencing of samples from archaea-dominated acidic hot springs in the Yellowstone National Park, USA (Bolduc et al. 2012, 2015).

Along with the extraordinary diversity of morphotypes, another outstanding feature of hyperthermophilic archaeal viruses is the uniqueness of their genome content, with a very low proportion of genes with recognizable homologues in public databases. Indeed, family-specific comparison of viral proteomes against the available sequences revealed that a fraction of ~85% of the crenarchaeal virus proteins do not have identifiable homologs when an E-value threshold of <1e-5 is used (Krupovic et al. 2018). In line with this observation, the analysis of the evolutionary relationships between all dsDNA viruses using a bipartite network approach, which traces connections between viral genomes through shared gene families, revealed that hyperthermophilic archaeal viruses are largely disconnected from the global dsDNA virosphere (see Figure 6.2). Moreover, the families themselves are largely disconnected from each other and share just a small number of common genes, suggesting that they have evolved independently of one another (Iranzo et al. 2016a, 2016b). Most of the common genes shared between the families are involved in transcriptional regulation and glycosylation, and could be independently acquired by the viruses from their respective hosts (Krupovic et al. 2018). None of the viruses encodes an RNA polymerase and only members of two families, *Ampullaviridae* and *Ovaliviridae*, encode recognizable DNA polymerases (Peng et al. 2007; Wang et al. 2018).

Reconstruction of the evolutionary history of the individual virus families is hindered by scarcity of known members in the families. Only one or two members have been described in the families *Ampullaviridae, Bicaudaviridae, Clavaviridae, Spiraviridae, Guttaviridae, Tristomaviridae, Portogoboviridae* and *Turriviridae.*

Even in the most populated families, *Fuselloviridae, Rudiviridae* and *Lipothrixviridae*, the number of known members is rather limited, precluding tracing any general evolutionary trends.

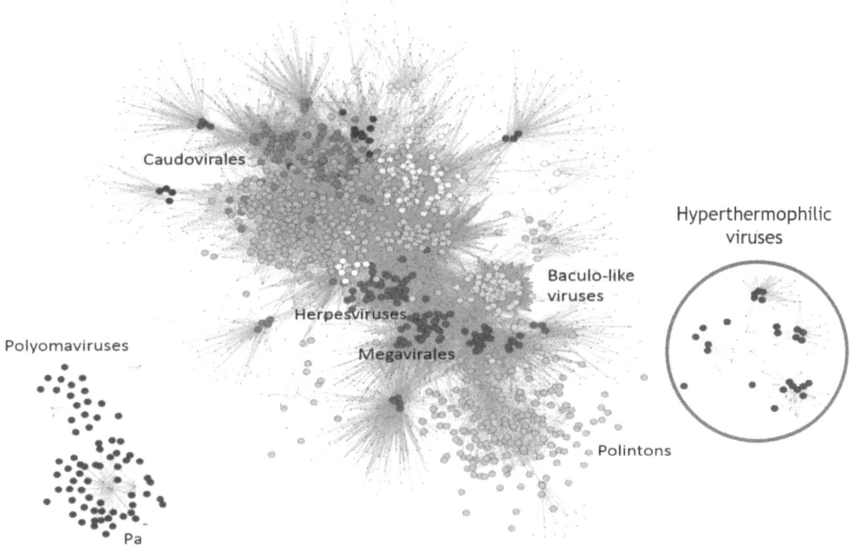

Figure 6.2. *Bipartite network approach of all dsDNA viruses. Image reproduced from Iranzo et al. (2016a). For a color version of this figure, see www.iste.co.uk/clouet/archaea1.zip*

6.3.1.1. *Genomes of fuselloviruses*

Comparative genomics show that members of the *Fuselloviridae* family share a set of 12 not contiguous "core" genes, suggesting a common evolutionary history, despite differences in the geographical context. Among the "core" genes, structural proteins, predicted DnaA-like ATPase and transcriptional regulators are found (Goodman and Stedman 2018). Intriguingly, nearly half of the predicted ORFs of SSV1, the type of virus of the *Fuselloviridae*, were shown to tolerate insertions or deletions without affecting the viral infectivity (Iverson et al. 2017). A high frequency of homologous recombination has been reported between fuselloviruses, which most likely occur in the protein-coding genes rather than intergenic regions (Redder et al. 2009). Furthermore, it has been suggested that these recombination events possibly occur between pairs of fuselloviruses infecting the same host (Zhang et al. 2020).

6.3.1.2. *Genomes of rudiviruses*

Rudiviruses represent one of the most extensively studied families of archaeal viruses and many of their proteins have been structurally and functionally characterized (Prangishvili et al. 2013). The linear genome of rudiviruses terminate in covalently closed hairpin structures. In addition, rudiviral genomes encompass TIRs that differ in size and sequence (Prangishvili and Krupovic 2012). Such structural design of the genome implies a unique mechanism of genome replication that involves strand displacement along with multiple reinitiation events on the parental template (Martínez-Alvarez et al. 2016). Similar to the fuselloviruses, members of the *Rudiviridae* family share a set of core proteins with critical roles in the virus cycle such as structural proteins, transcription factors with ribbon–helix–helix motifs, glycosyltransferases, acetyltransferases, Holliday junction resolvase, methyltransfrases, dUTPase, ssDNA-binding proteins, Rep protein, ssDNA annealing ATPase and Cas4-like nuclease (Krupovic et al. 2018). Interestingly, rudiviruses accumulate new genes with no similarity to any sequence in public databases at the termini of the linear genomes, providing a distribution of conserved and non-conserved genes across the genome (Krupovic et al. 2011).

6.3.1.3. *Genomes of lipothrixviruses*

Comparative genomic analysis has revealed that some members of the *Lipothrixviridae* possess TIRs with regularly spaced direct repeats and multiple overlapping ORFs of up to 23 bp. Conserved genes among members of the family include those encoding the two MCPs, transcriptional regulators, glycosyltransferases, helicases, a nuclease, a methyltransferase and a protein phosphatase. Interestingly, it has been shown that linear genomes of different species of lipothrixviruses can undergo intergenomic recombination. For example, a 13-kb segment of the AFV7 genome appears to have originated from a recombination event with a genomic region of the lipothrixvirus AFV2, resulting in possible phenotypic changes (Vestergaard et al. 2008).

The genomes of rudiviruses and lipothrixviruses display a considerable variation in gene order and content, in contrast to the conservation of the gene synteny of tailed dsDNA viruses infecting bacteria and archaea. The significant number of homologous genes shared by members of the *Rudiviridae* and *Lipothrixviridae* families is consistent with the common ancestry of the two families, unified in the order *Ligamenvirales*, which has also been suggested by the structural similarity of their MCPs (Goulet et al. 2009a; Prangishvili and Krupovic 2012; DiMaio et al. 2015; Kasson et al. 2017).

6.3.2. *Structural genomics*

As most of the proteins of hyperthermophilic archaeal viruses are refractory to bioinformatic analysis, structural genomics have been performed to promote the functional characterization of *Fuselloviridae*, *Bicaudoviridae*, *Rudiviridae*, *Lipothixviridae*, *Globuloviridae* and *Turriviridae* proteins (Krupovic et al. 2018). The structures of these proteins were determined by X-ray crystallography, nuclear magnetic resonance (NMR) spectroscopy and cryo-electron microscopy. Some structures of viral proteins display unique folds with no homologs in databases, hindering the assignment of putative functions. This is the case of the thermo- and acid-stable protein AFV1-99, for which a 3D crystal structure revealed a new α/β fold, which is conserved in linear genomes of lipothrixviruses and rudiviruses (Goulet et al. 2009b). Subsequently, it has been demonstrated that a homolog of AFV1-99 encoded by SIRV2 acts as an anti-CRISPR protein, which blocks type I-D CRISPR-Cas system of the host (He et al. 2018). Another unique fold is exemplified by ORF114 of SIRV1, which is a member of a protein family conserved in viruses of the families *Rudiviridae*, *Lipothrixviridae*, *Bicaudaviridae* and *Turriviridae*. The structure of SIRV1 ORF114 revealed a unique fold comprising five stranded beta-sheets flanked with loops and three helices on both sides. Although initial in vitro assays suggested that this protein might be involved in DNA binding (Larson et al. 2007), it has been recently demonstrated that the protein is a viral anti-CRISPR (Acr) enzyme that rapidly degrades cyclic tetra-adenylate (cA4) messenger, thereby blocking the type III CRISPR-Cas system (Athukoralage et al. 2020).

In other cases, the structural information was crucial to characterize functionally viral proteins. For example, the X-ray structure of P157, encoded by AFV1, revealed a novel fold that remotely resembles nucleotidyltransferases; in vitro experiments confirmed a nuclease activity of the protein on linear double-stranded DNA (Goulet et al. 2010a). The resolution of the crystal structure of the protein encoded by ORF119 from the rudivirus SIRV1 facilitated its assignment as a member of the replication initiator (Rep) superfamily of proteins, which are involved in the initiation of the rolling-circle replication in viruses with ssDNA genome (Oke et al. 2011). Similarly, the high-resolution structure of A197, encoded by turrivirus STIV, revealed the presence of a canonical DXD motif and a putative catalytic base, both hallmarks of the glycosyltransferase superfamily (Larson et al. 2006).

Many archaeal virus proteins have recognizable structural elements that are characteristics of DNA-binding domains of transcription factors, such as helix–turn–helix (HTH), zinc-finger motifs and ribbon–helix–helix (RHH) (Sheppard and Werner 2017). Crystallography and NMR studies have favored the functional characterization of these putative transcription factors, supporting the idea that they resemble those in bacteria (Sheppard and Werner 2017). For instance, NMR spectroscopy studies showed that the *Sulfolobus* virus transcription factor (SvtR), a small protein encoded by rudivirus SIRV1, comprises a RHH fold and forms a homodimer in a reminiscent manner to that shown in bacterial RHH transcriptional repressors such as NikR (nickel responsive element) and MetJ (methionine repressor) (Guillière et al. 2009).

Multiple structural and functional characterization approaches have been conducted to unravel the role of the ORF145 of the bicaudavirus ATV. The protein is one of the major virion components in ATV and also turned out to be involved in transcription regulation. Interestingly, ORF145 provided the first example of a viral transcription factor that directly targets the host RNA polymerase and represses its activity independently of the DNA template. This RNA polymerase inhibitory protein (RIP) binds inside the DNA-binding channel of the host RNA polymerase, forming a high-affinity complex, thereby locking the RNA polymerase clamp in one position that prevents the formation of transcription pre-initiation complexes (Sheppard et al. 2016).

6.4. Virus–host interactions

The cycles of reproduction of the hyperthermophilic viruses can be lysogenic, lytic or chronic. The lysogenic cycle of the *Fuselloviridae* on the example of the virus SSV1 is well studied. The circular dsDNA of the virus is integrated as a provirus into the *Sulfolobus* host chromosome, specifically into the host arginine tRNA gene, with the help of the virus-encoded integrase. Unlike the case of temperate bacteriophages, the integrase gene is partitioned in two fragments upon the viral genome integration. The replication of SSV1 can be induced by exposing the lysogenic cells to stress factors, for example, UV irradiation (Martin et al. 1984; Palm et al. 1991, Muskhelishvili et al. 1993). Similarly, proviruses derived from turriviruses and guttaviruses have been detected in *Sulfolobus* and *Aeropyrum* genomes, respectively (Mochizuki et al. 2011; Anderson et al. 2017). A fragment of the circular dsDNA of the bicaudavirus ATV has been also detected in the chromosome of the host *Acidianus hospitalis* (Prangishvili et al. 2006).

Filamentous viruses from the families *Rudiviridae*, *Tristromaviridae*, *Ovaliviridae* and *Turriviridae* are lytic, causing cell death as a result of destruction of cell membranes (Bize et al. 2009; Brumfield et al. 2009; Rensen et al. 2016). Members of other families are believed to continuously release their virions in a state of chronic infection. As a result, a balance between virus replication and cell multiplication is maintained.

6.4.1. *Virus entry*

Observations by EM suggest that the primary interaction of hyperthermophilic viruses with their archaeal hosts is often mediated by cellular appendages. The mode of interaction has been confirmed in the example of the rudivirus SIRV2 and the turrivirus STIV, the infection cycle of which is initiated by binding of the virions to the pilus-like appendages of *Sulfolobus* cells (Quemin et al. 2013; Hartman et al. 2019). In the case of SIRV2, the virion initially attaches to the tips of the cellular appendages with the help of three terminal fibers and subsequently moves along the pili to reach the cell surface within one minute (see Figure 6.3). The mechanisms of virion translocation and DNA injection remain unknown.

The spindle-shaped virions of *Fuselloviridae* and *Bicaudaviridae* do not attach to cellular appendages. Instead, in virus preparations, they are often observed bound to cellular fragments, which implies primary interaction with receptors on the host cell surface (Quemin et al. 2015; Uldahl et al. 2016).

6.4.2. *Virion egress*

The egress of the spindle-shaped virions of the *Fuselloviridae* is coupled with the final steps of virion morphogenesis. The virion of the fusellovirus SSV1 acquires its envelope upon budding of the nucleoprotein inner core through the cytoplasmic membrane of the host cell, which, in a portion of cells, remains ruptured and thus causes cell death (see Figure 6.3). Virions first emerge in the form of elongated particles attached to the cell envelope and only subsequently undergo maturation into characteristic spindle-shaped particles (Quemin et al. 2016). Most likely, the morphogenesis of *Bicaudaviridae* virions also takes place at the host cell surface and is coupled with virus budding. Exceptionally, the virion morphogenesis of the bicaudavirus ATV, as mentioned above, is not completed at the cell surface: after being released, the spindle-shaped virions develop long protrusions from the two pointed ends. High temperature, close to that of the natural environment, is a prerequisite for this morphological transformation, which is accompanied by contraction of the "spindle" (Prangishvili et al. 2006).

By contrast to the spindle-shaped viruses, all stages of virion morphogenesis of the rudiviruses are completed intracellularly, in the host cell cytoplasm. In the example of SIRV2, it was demonstrated that mature virions exit the cells via special gateway structures, the virus-associated pyramids (VAPs) with sevenfold symmetry, which develop on the surface of infected *Sulfolobus* cells, protrude through S-layer and open as "flower petals" at the end of the infection cycle (see Figure 6.3) (Quax et al. 2011). The baseless VAP consists of multiple copies of a 10 kDa virus-encoded protein. The protein lacks a signal sequence and inserts into cell membranes through hydrophobic interactions of its transmembrane domain. The protein shows a remarkable ability to form pyramidal structures in virtually all types of biological membranes from all three domains of life. The same VAP-based mechanism of egress is exploited by the turrivirus STIV (Brumfield et al. 2009) and ovalivirus SEV1 (Wang et al. 2018). Moreover, EM observations of virus-infected archaeal cells suggest that such egress mechanism could be common for hyperthermophilic archaeal viruses (Bize et al. 2008; Rensen et al. 2015).

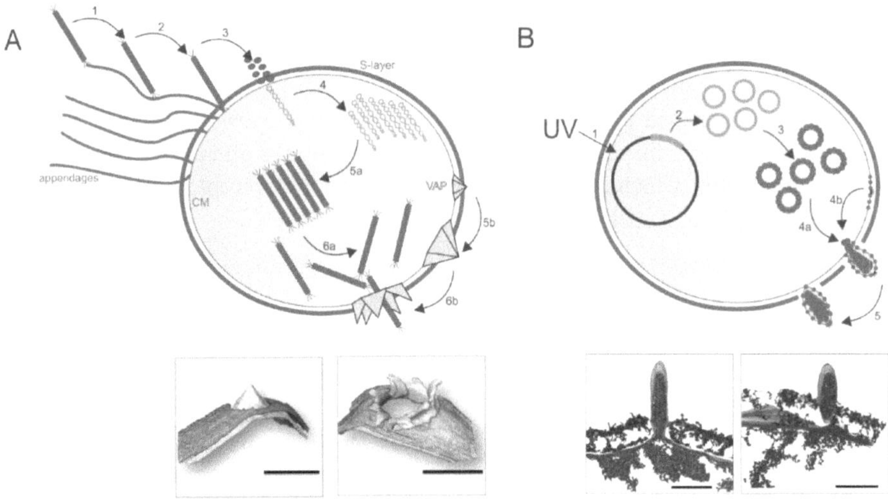

Figure 6.3. *Lifecycles of lytic and temperate archaeal viruses. Image reproduced from Prangishvili et al. (2017). For a color version of this figure, see www.iste.co.uk/clouet/archaea1.zip*

6.5. Conclusions

One main theme, which the hyperthermophilic archaeal viruses share with each other and which distinguishes them from the majority of DNA viruses, is packaging of DNA in a form of nucleocapsid. It appears that the assembly of the virion core of the majority of hyperthermophilic archaeal viruses occurs in concerted fashion with the coating of DNA with copies of MCPs and spontaneous self-assembly of such nucleocapsids into minimum free energy structures of different shapes. Such a pathway of virion morphogenesis differs from the strategy of DNA packaging into the preformed capsid, which separates the processes of capsid assembly and genome packaging and which is common for DNA viruses of bacteria, eukarya and even mesophilic *Euryarchaota*. One more specific feature of the virion morphogenesis of the majority of hyperthemophilic viruses is coating of the nucleoprotein core by lipid-containing envelopes. The coating with lipids for *Fuselloviridae* occurs at the final step of capsid assembly, in the course of budding from the host cell, whereas for filamentous viruses, it takes place apparently intracellularly with the help of unknown molecular mechanism.

The unusual features of hyperthermophilic archaeal viruses extend to the content of their genomes with the vast majority of genes without homologues in viral and cellular genomes. Moreover, the families of hyperthermophilic archaeal viruses, even between themselves, share just a small group of common genes, suggesting that they have evolved independently of one another.

The specific morphological and genomic features of hyperthermophilic archaeal DNA viruses, without counterparts or even strong phylogenetic relationships with other viruses, raises important questions regarding their origins. How and when did these viruses originate? Why do they only infect archaea? These would be the major questions to be answered by more comprehensive studies on the diversity of archaeal viruses, specifically those thriving in extreme geothermal environments, and structural, genomic and functions characterization of these viruses.

6.6. References

Ahn, D.G., Kim, S.I., Rhee, J.K., Kim, K.P., Pan, J.G., Oh, J.W. (2006). TTSV1, a new virus-like particle isolated from the hyperthermophilic crenarchaeote *Thermoproteus tenax*. *Virology*, 351, 280–290.

Anderson, R.E., Kouris, A., Seward, C.H., Campbell, K.M., Whitaker, R.J. (2017). Structured populations of *Sulfolobus acidocaldarius* with susceptibility to mobile genetic elements. *Genome Biology and Evolution*, 9, 1699–1710.

Arnold, H.P., Ziese, U., Zillig, W. (2000a). SNDV, a novel virus of the extremely thermophilic and acidophilic archaeon Sulfolobus. *Virology*, 272, 409–416.

Arnold, H.P., Zillig, W., Ziese, U., Holz, I., Crosby, M., Utterback, T., Weidmann, J.F., Kristjanson, J.K., Klenk, H.P., Nelson, K.E. et al. (2000b). A novel lipothrixvirus, SIFV, of the extremely thermophilic crenarchaeon Sulfolobus. *Virology*, 267, 252–266.

Athukoralage, J.S., McMahon, S.A., Zhang, C., Grüschow, S., Graham, S., Krupovic, M., Whitaker, R.J., Gloster, T.M., White, M.F. (2020). An anti-CRISPR viral ring nuclease subverts type III CRISPR immunity. *Nature*, 577, 572–575.

Baquero, D., Contursi, P., Piochi, M., Bartolucci, S., Liu, Y., Cvirkaite-Krupovic, V., Prangishvili, D., Krupovic, M. (2020). New virus isolates from Italian hydrothermal environments underscore the biogeographic pattern in archaeal virus communities. *The ISME Journal*, 14, 1821–1833.

Bettstetter, M., Peng, X., Garrett, R.A., Prangishvili, D. (2003). AFV1, a novel virus infecting hyperthermophilic archaea of the genus *Acidianus*. *Virology*, 315, 68–79.

Bize, A., Peng, X., Prokofeva, M., Maclellan, K., Lucas, S., Forterre, P., Garrett, R.A., Bonch-Osmolovskaya, E.A., Prangishvili, D. (2008). Viruses in acidic geothermal environments of the Kamchatka Peninsula. *Research in Microbiology*, 159, 358–366.

Bize, A., Karlsson, E.A., Ekefjärd, K., Quax, T.E., Pina, M., Prevost, M.C., Forterre, P., Tenaillon, O., Bernander, R., Prangishvili, D. (2009). A unique virus release mechanism in the Archaea. *Proceedings of the National Academy of Sciences of the United States of America*, 106, 11306–11311.

Bolduc, B., Shaughnessy, D.P., Wolf, Y.I., Koonin, E.V., Roberto, F.F., Young, M. (2012). Identification of novel positive-strand RNA viruses by metagenomic analysis of archaea-dominated Yellowstone hot springs. *Journal of Virology*, 86, 5562–5573.

Bolduc, B., Wirth, J.F., Mazurie, A., Young, M.J. (2015). Viral assemblage composition in Yellowstone acidic hot springs assessed by network analysis. *The ISME Journal*, 9, 2162–2177.

Brumfield, S.K., Ortmann, A.C., Ruigrok, V., Suci, P., Douglas, T., Young, M.J. (2009). Particle assembly and ultrastructural features associated with replication of the lytic archaeal virus Sulfolobus turreted icosahedral virus. *Journal of Virology*, 83, 5964–5970.

DiMaio, F., Yu, X., Rensen, E., Krupovic, M., Prangishvili, D., Egelman, E.H. (2015). A virus that infects a hyperthermophile encapsidates A-form DNA. *Science*, 348, 914–917.

Felisberto-Rodrigues, C., Blangy, S., Goulet, A., Vestergaard, G., Cambillau, C., Garrett, R.A., Ortiz-Lombardía, M. (2012). Crystal structure of ATV(ORF273), a new fold for a thermo- and acido-stable protein from the Acidianus two-tailed virus. *PloS One*, 7, e45847.

Geslin, C., Le Romancer, M., Erauso, G., Gaillard, M., Perrot, G., Prieur, D. (2003). PAV1, the first virus-like particle isolated from a hyperthermophilic euryarchaeote, "*Pyrococcus abyssi*". *Journal of Bacteriology*, 185, 3888–3894.

Geslin, C., Gaillard, M., Flament, D., Rouault, K., Le Romancer, M., Prieur, D., Erauso, G. (2007). Analysis of the first genome of a hyperthermophilic marine virus-like particle, PAV1, isolated from *Pyrococcus abyssi*. *Journal of Bacteriology*, 189, 4510–4519.

Goodman, D.A. and Stedman, K.M. (2018). Comparative genetic and genomic analysis of the novel fusellovirus Sulfolobus spindle-shaped virus 10. *Virus Evolution*, 4, vey022.

Gorlas, A., Koonin, E.V., Bienvenu, N., Prieur, D., Geslin, C. (2012). TPV1, the first virus isolated from the hyperthermophilic genus Thermococcus. *Environmental Microbiology*, 14, 503–516.

Goulet, A., Blangy, S., Redder, P., Prangishvili, D., Felisberto-Rodrigues, C., Forterre, P., Campanacci, V., Cambillau, C. (2009a). Acidianus filamentous virus 1 coat proteins display a helical fold spanning the filamentous archaeal viruses lineage. *Proceedings of the National Academy of Sciences of the United States of America*, 106, 21155–21160.

Goulet, A., Spinelli, S., Blangy, S., Van Tilbeurgh, H., Leulliot, N., Basta, T., Prangishvili, D., Cambillau, C., Campanacci, V. (2009b). The thermo- and acido-stable ORF-99 from the archaeal virus AFV1. *Protein Science: A Publication of the Protein Society*, 18, 1316–1320.

Goulet, A., Pina, M., Redder, P., Prangishvili, D., Vera, L., Lichière, J., Leulliot, N., Van Tilbeurgh, H., Ortiz-Lombardia, M., Campanacci, V. et al. (2010a). ORF157 from the archaeal virus Acidianus filamentous virus 1 defines a new class of nuclease. *Journal of Virology*, 84, 5025–5031.

Goulet, A., Vestergaard, G., Felisberto-Rodrigues, C., Campanacci, V., Garrett, R.A., Cambillau, C., Ortiz-Lombardía, M. (2010b). Getting the best out of long-wavelength X-rays: De novo chlorine/sulfur SAD phasing of a structural protein from ATV. *Acta Crystallographica Section D, Biological Crystallography*, 66, 304–308.

Gudbergsdóttir, S.R., Menzel, P., Krogh, A., Young, M., Peng, X. (2016). Novel viral genomes identified from six metagenomes reveal wide distribution of archaeal viruses and high viral diversity in terrestrial hot springs. *Environmental Microbiology*, 18, 863–874.

Guillière, F., Peixeiro, N., Kessler, A., Raynal, B., Desnoues, N., Keller, J., Delepierre, M., Prangishvili, D., Sezonov, G., Guijarro, J.I. (2009). Structure, function, and targets of the transcriptional regulator SvtR from the hyperthermophilic archaeal virus SIRV1. *The Journal of Biological Chemistry*, 284, 22222–22237.

Happonen, L.J., Redder, P., Peng, X., Reigstad, L.J., Prangishvili, D., Butcher, S.J. (2010). Familial relationships in hyperthermo- and acidophilic archaeal viruses. *Journal of Virology*, 84, 4747–4754.

Häring, M., Peng, X., Brügger, K., Rachel, R., Stetter, K.O., Garrett, R.A., Prangishvili, D. (2004). Morphology and genome organization of the virus PSV of the hyperthermophilic archaeal genera *Pyrobaculum* and *Thermoproteus*: A novel virus family, the Globuloviridae. *Virology*, 323, 233–242.

Häring, M., Rachel, R., Peng, X., Garrett, R.A., Prangishvili, D. (2005a). Viral diversity in hot springs of Pozzuoli, Italy, and characterization of a unique archaeal virus, Acidianus bottle-shaped virus, from a new family, the Ampullaviridae. *Journal of Virology*, 79, 9904–9911.

Häring, M., Vestergaard, G., Brügger, K., Rachel, R., Garrett, R.A., Prangishvili, D. (2005b). Structure and genome organization of AFV2, a novel archaeal lipothrixvirus with unusual terminal and core structures. *Journal of Bacteriology*, 187, 3855–3858.

Hartman, R., Eilers, B.J., Bollschweiler, D., Munson-Mcgee, J.H., Engelhardt, H., Young, M.J., Lawrence, C.M. (2019). The molecular mechanism of cellular attachment for an archaeal virus. *Structure*, 27, 1634–1646, e1633.

Hartman, R., Biewenga, L., Munson-Mcgee, J., Refai, M., Boyd, E.S., Bothner, B., Lawrence, C.M., Young, M. (2020). Discovery and characterization of *Thermoproteus* spherical piliferous virus 1: A spherical archaeal virus decorated with unusual filaments. *Journal of Virology*, 94, e01140–20.

He, F., Bhoobalan-Chitty, Y., Van, L.B., Kjeldsen, A.L., Dedola, M., Makarova, K.S., Koonin, E.V., Brodersen, D.E., Peng, X. (2018). Anti-CRISPR proteins encoded by archaeal lytic viruses inhibit subtype I-D immunity. *Nature Microbiology*, 3, 461–469.

Iranzo, J., Koonin, E.V., Prangishvili, D., Krupovic, M. (2016a). Bipartite network analysis of the archaeal virosphere: Evolutionary connections between viruses and capsidless mobile elements. *Journal of Virology*, 90, 11043–11055.

Iranzo, J., Krupovic, M., Koonin, E.V. (2016b). The double-stranded DNA virosphere as a modular hierarchical network of gene sharing. *MBio*, 7, e00978-16.

Iverson, E.A., Goodman, D.A., Gorchels, M.E., Stedman, K.M. (2017). Extreme mutation tolerance: Nearly half of the archaeal fusellovirus Sulfolobus spindle-shaped virus 1 genes are not required for virus function, including the minor capsid protein gene *vp3*. *Journal of Virology*, 91, e02406-16.

Kasson, P., Dimaio, F., Yu, X., Lucas-Staat, S., Krupovic, M., Schouten, S., Prangishvili, D., Egelman, E.H. (2017). Model for a novel membrane envelope in a filamentous hyperthermophilic virus. *eLife*, 6, e26268.

Krupovic, M., Prangishvili, D., Hendrix, R.W., Bamford, D.H. (2011). Genomics of bacterial and archaeal viruses: Dynamics within the prokaryotic virosphere. *Microbiology and Molecular Biology Reviews*, 75, 610–635.

Krupovic, M., Quemin, E.R., Bamford, D.H., Forterre, P., Prangishvili, D. (2014). Unification of the globally distributed spindle-shaped viruses of the Archaea. *Journal of Virology*, 88, 2354–2358.

Krupovic, M., Cvirkaite-Krupovic, V., Iranzo, J., Prangishvili, D., Koonin, E.V. (2018). Viruses of archaea: Structural, functional, environmental and evolutionary genomics. *Virus Research*, 244, 181–193.

Larson, E.T., Reiter, D., Young, M., Lawrence, C.M. (2006). Structure of A197 from Sulfolobus turreted icosahedral virus: A crenarchaeal viral glycosyltransferase exhibiting the GT-A fold. *Journal of Virology*, 80, 7636–7644.

Larson, E.T., Eilers, B.J., Reiter, D., Ortmann, A.C., Young, M.J., Lawrence, C.M. (2007). A new DNA binding protein highly conserved in diverse crenarchaeal viruses. *Virology*, 363, 387–396.

Liu, Y., Ishino, S., Ishino, Y., Pehau-Arnaudet, G., Krupovic, M., Prangishvili, D. (2017). A novel type of polyhedral viruses infecting hyperthermophilic archaea. *Journal of Virology*, 91, e00589-17.

Martin, A., Yeats, S., Janekovic, D., Reiter, W.D., Aicher, W., Zillig, W. (1984). SAV 1, a temperate u.v.-inducible DNA virus-like particle from the archaebacterium Sulfolobus *acidocaldarius* isolate B12. *The EMBO Journal*, 3, 2165–2168.

Martínez-Alvarez, L., Bell, S.D., Peng, X. (2016). Multiple consecutive initiation of replication producing novel brush-like intermediates at the termini of linear viral dsDNA genomes with hairpin ends. *Nucleic Acids Research*, 44, 8799–8809.

Mochizuki, T., Yoshida, T., Tanaka, R., Forterre, P., Sako, Y., Prangishvili, D. (2010). Diversity of viruses of the hyperthermophilic archaeal genus Aeropyrum, and isolation of the *Aeropyrum pernix* bacilliform virus 1, APBV1, the first representative of the family Clavaviridae. *Virology*, 402, 347–354.

Mochizuki, T., Sako, Y., Prangishvili, D. (2011). Provirus induction in hyperthermophilic archaea: Characterization of *Aeropyrum pernix* spindle-shaped virus 1 and *Aeropyrum pernix* ovoid virus 1. *Journal of Bacteriology*, 193, 5412–5419.

Mochizuki, T., Krupovic, M., Pehau-Arnaudet, G., Sako, Y., Forterre, P., Prangishvili, D. (2012). Archaeal virus with exceptional virion architecture and the largest single-stranded DNA genome. *Proceedings of the National Academy of Sciences of the United States of America*, 109, 13386–13391.

Muskhelishvili, G., Palm, P., Zillig, W. (1993). SSV1-encoded site-specific recombination system in *Sulfolobus shibatae*. *Molecular & General Genetics*, 237, 334–342.

Oke, M., Carter, L.G., Johnson, K.A., Liu, H., Mcmahon, S.A., Yan, X., Kerou, M., Weikart, N.D., Kadi, N., Sheikh, M.A. et al. (2010). The Scottish Structural Proteomics Facility: Targets, methods and outputs. *Journal of Structural and Functional Genomics*, 11, 167–180.

Oke, M., Kerou, M., Liu, H., Peng, X., Garrett, R.A., Prangishvili, D., Naismith, J.H., White, M.F. (2011). A dimeric Rep protein initiates replication of a linear archaeal virus genome: Implications for the Rep mechanism and viral replication. *Journal of Virology*, 85, 925–931.

Palm, P., Schleper, C., Grampp, B., Yeats, S., Mcwilliam, P., Reiter, W.D., Zillig, W. (1991). Complete nucleotide sequence of the virus SSV1 of the archaebacterium *Sulfolobus shibatae*. *Virology*, 185, 242–250.

Peng, X., Basta, T., Häring, M., Garrett, R.A., Prangishvili, D. (2007). Genome of the Acidianus bottle-shaped virus and insights into the replication and packaging mechanisms. *Virology*, 364, 237–243.

Prangishvili, D. and Krupovic, M. (2012). A new proposed taxon for double-stranded DNA viruses, the order "Ligamenvirales". *Archives of Virology*, 157, 791–795.

Prangishvili, D. and Krupovic, M. (2018). ICTV virus taxonomy profile: Ampullaviridae. *The Journal of General Virology*, 99, 288–289.

Prangishvili, D., Arnold, H.P., Götz, D., Ziese, U., Holz, I., Kristjansson, J.K., Zillig, W. (1999). A novel virus family, the Rudiviridae: Structure, virus-host interactions and genome variability of the sulfolobus viruses SIRV1 and SIRV2. *Genetics*, 152, 1387–1396.

Prangishvili, D., Vestergaard, G., Häring, M., Aramayo, R., Basta, T., Rachel, R., Garrett, R.A. (2006). Structural and genomic properties of the hyperthermophilic archaeal virus ATV with an extracellular stage of the reproductive cycle. *Journal of Molecular Biology*, 359, 1203–1216.

Prangishvili, D., Koonin, E.V., Krupovic, M. (2013). Genomics and biology of Rudiviruses, a model for the study of virus-host interactions in archaea. *Biochemical Society Transactions*, 41, 443–450.

Prangishvili, D., Bamford, D.H., Forterre, P., Iranzo, J., Koonin, E.V., Krupovic, M. (2017). The enigmatic archaeal virosphere. *Nature Reviews Microbiology*, 15, 724–739.

Prangishvili, D., Mochizuki, T., Krupovic, M. (2018). ICTV virus taxonomy profile: Guttaviridae. *The Journal of General Virology*, 99, 290–291.

Prangishvili, D., Mochizuki, T., Krupovic, M. (2020). ICTV virus taxonomy profile: Spiraviridae. *The Journal of General Virology*, 101, 240–241.

Ptchelkine, D., Gillum, A., Mochizuki, T., Lucas-Staat, S., Liu, Y., Krupovic, M., Phillips, S.E.V., Prangishvili, D., Huiskonen, J.T. (2017). Unique architecture of thermophilic archaeal virus APBV1 and its genome packaging. *Nature Communications*, 8, 1436.

Quax, T.E., Lucas, S., Reimann, J., Pehau-Arnaudet, G., Prevost, M.C., Forterre, P., Albers, S.V., Prangishvili, D. (2011). Simple and elegant design of a virion egress structure in archaea. *Proceedings of the National Academy of Sciences of the United States of America*, 108, 3354–3359.

Quemin, E.R., Lucas, S., Daum, B., Quax, T.E., Kühlbrandt, W., Forterre, P., Albers, S.V., Prangishvili, D., Krupovic, M. (2013). First insights into the entry process of hyperthermophilic archaeal viruses. *Journal of Virology*, 87, 13379–13385.

Quemin, E.R., Pietilä, M.K., Oksanen, H.M., Forterre, P., Rijpstra, W.I., Schouten, S., Bamford, D.H., Prangishvili, D., Krupovic, M. (2015). Sulfolobus spindle-shaped virus 1 contains glycosylated capsid proteins, a cellular chromatin protein, and host-derived lipids. *Journal of Virology*, 89, 11681–11691.

Quemin, E.R., Chlanda, P., Sachse, M., Forterre, P., Prangishvili, D., Krupovic, M. (2016). Eukaryotic-like virus budding in Archaea. *MBio*, 7, e01439-16.

Redder, P., Peng, X., Brügger, K., Shah, S.A., Roesch, F., Greve, B., She, Q., Schleper, C., Forterre, P., Garrett, R.A. et al. (2009). Four newly isolated fuselloviruses from extreme geothermal environments reveal unusual morphologies and a possible interviral recombination mechanism. *Environmental Microbiology*, 11, 2849–2862.

Rensen, E.I., Krupovic, M., Prangishvili, D. (2015). Mysterious hexagonal pyramids on the surface of Pyrobaculum cells. *Biochimie*, 118, 365–367.

Rensen, E.I., Mochizuki, T., Quemin, E., Schouten, S., Krupovic, M., Prangishvili, D. (2016). A virus of hyperthermophilic archaea with a unique architecture among DNA viruses. *Proceedings of the National Academy of Sciences of the United States of America*, 113, 2478–2483.

Rice, G., Tang, L., Stedman, K., Roberto, F., Spuhler, J., Gillitzer, E., Johnson, J.E., Douglas, T., Young, M. (2004). The structure of a thermophilic archaeal virus shows a double-stranded DNA viral capsid type that spans all domains of life. *Proceedings of the National Academy of Sciences of the United States of America*, 101, 7716–7720.

Schleper, C., Kubo, K., Zillig, W. (1992). The particle SSV1 from the extremely thermophilic archaeon Sulfolobus is a virus: Demonstration of infectivity and of transfection with viral DNA. *Proceedings of the National Academy of Sciences of the United States of America*, 89, 7645–7649.

Sheppard, C. and Werner, F. (2017). Structure and mechanisms of viral transcription factors in archaea. *Extremophiles: Life Under Extreme Conditions*, 21(5), 829–838.

Sheppard, C., Blombach, F., Belsom, A., Schulz, S., Daviter, T., Smollett, K., Mahieu, E., Erdmann, S., Tinnefeld, P., Garrett, R. et al. (2016). Repression of RNA polymerase by the archaeoviral regulator ORF145/RIP. *Nature Communications*, 7, 13595.

Stedman, K.M., She, Q., Phan, H., Arnold, H.P., Holz, I., Garrett, R.A., Zillig, W. (2003). Relationships between fuselloviruses infecting the extremely thermophilic archaeon Sulfolobus: SSV1 and SSV2. *Research in Microbiology*, 154, 295–302.

Uldahl, K.B., Jensen, S.B., Bhoobalan-Chitty, Y., Martínez-Álvarez, L., Papathanasiou, P., Peng, X. (2016). Life cycle characterization of Sulfolobus monocaudavirus 1, an extremophilic spindle-shaped virus with extracellular tail development. *Journal of Virology*, 90, 5693–5699.

Veesler, D., Ng, T.S., Sendamarai, A.K., Eilers, B.J., Lawrence, C.M., Lok, S.M., Young, M.J., Johnson, J.E., Fu, C.Y. (2013). Atomic structure of the 75 MDa extremophile Sulfolobus turreted icosahedral virus determined by CryoEM and X-ray crystallography. *Proceedings of the National Academy of Sciences of the United States of America*, 110, 5504–5509.

Vestergaard, G., Aramayo, R., Basta, T., Häring, M., Peng, X., Brügger, K., Chen, L., Rachel, R., Boisset, N., Garrett, R.A. et al. (2008). Structure of the acidianus filamentous virus 3 and comparative genomics of related archaeal lipothrixviruses. *Journal of Virology*, 82, 371–381.

Wang, H., Guo, Z., Feng, H., Chen, Y., Chen, X., Li, Z., Hernández-Ascencio, W., Dai, X., Zhang, Z., Zheng, X. et al. (2018). Novel Sulfolobus virus with an exceptional capsid architecture. *Journal of Virology*, 92, e01727-17.

Wang, F., Liu, Y., Su, Z., Osinski, T., De Oliveira, G.A.P., Conway, J.F., Schouten, S., Krupovic, M., Prangishvili, D., Egelman, E.H. (2019). A packing for A-form DNA in an icosahedral virus. *Proceedings of the National Academy of Sciences of the United States of America*, 116, 22591–22597.

Wang, F., Baquero, D., Su, Z., Osinski, T., Prangishvili, D., Egelman, E.H., Krupovic, M. (2020). Structure of a filamentous virus uncovers familial ties within the archaeal virosphere. *Virus Evolution*, 6, veaa023.

List of Authors

Karine ALAIN
BEEP-BIOMEX, CNRS
Ifremer
Université de Bretagne Occidentale
Plouzané
France

Maxime ALLIOUX
BEEP-BIOMEX, CNRS
Ifremer
Université de Bretagne Occidentale
Plouzané
France

Diana BAQUERO
Institut Pasteur
Université Paris Cité
France

Guillaume BORREL
Institut Pasteur
Université Paris Cité
UMR CNRS 6047
Evolutionary Biology of the Microbial Cell
Paris
France

Béatrice CLOUET-D'ORVAL
MCD-CBI, CNRS
Université de Toulouse
France

Marc COZANNET
BEEP-BIOMEX, CNRS
Ifremer
Université de Bretagne Occidentale
Plouzané
France

Patrick FORTERRE
Institut Pasteur
Université Paris Cité
and
Université Paris-Saclay
Gif-sur-Yvette
France

Bruno FRANZETTI
IBS, CNRS
Université Grenoble Alpes
France

Claire GESLIN
Univ Brest
Ifremer
BEEP
F-29280
Plouzané
France

Jordan HARTUNIANS
BEEP-BIOMEX, CNRS
Ifremer
Université de Bretagne Occidentale
Plouzané
France

Mart KRUPOVIC
Institut Pasteur
Université Paris Cité
France

Philippe OGER
MAP, CNRS
INSA Lyon
Villeurbanne
France

David PRANGISHVILI
Institut Pasteur
Université Paris Cité
France
and
Ivane Javakhishvili TSU
Tbilisi
Georgia

Sarah THIROUX
BEEP-BIOMEX, CNRS
Ifremer
Université de Bretagne Occidentale
Plouzané
France

Laurent TOFFIN
Univ Brest
Ifremer
CNRS
Unité Biologie et Ecologie des
Ecosystèmes Marins Profonds
F-29280
Plouzané
France

Tristan WAGNER
Max Planck Institute
for Marine Microbiology
Bremen
Germany

Index

T, V